LINEAR PROGRAMMING IN SINGLE- & MULTIPLE-OBJECTIVE SYSTEMS

PRENTICE-HALL INTERNATIONAL SERIES
IN INDUSTRIAL AND SYSTEMS ENGINEERING

W. J. Fabrycky and J. H. Mize, Editors

BEIGHTLER, PHILLIPS AND WILDE *Fundamentals of Optimization, 2/E*
BLANCHARD *Logistics Engineering and Management 2/E*
BLANCHARD AND FABRYCKY *Systems Engineering and Analysis*
BROWN *Systems Analysis and Design for Safety*
FABRYCKY, GHARE, AND TORGERSEN *Industrial Operations Research*
FRANCES AND WHITE *Facility Layout and Location: An Analytical Approach*
GOTTFRIED AND WEISMAN *Introduction to Optimization Theory*
HAMMER *Product Safety Management and Engineering*
IGNIZIO *Linear Programming in Single- and Multiple-Objective Systems*
KIRKPATRICK *Introductory Statistics and Probability for Engineering, Science, and Technology*
MIZE, WHITE, AND BROOKS *Operation Planning and Control*
MUNDEL *Motion and Time Study: Improving Productivity, 5/E*
OSTWALD *Cost Estimating for Engineering and Management*
PHILLIPS AND GARCIA-DIAZ *Fundamentals of Network Analysis*
SIVAZLIAN AND STANFEL *Analysis of Systems in Operations Research*
SIVAZLIAN AND STANFEL *Optimization Techniques in Operations Research*
THUESEN, FABRYCKY, AND THEUSEN *Engineering Economy, 5/E*
TURNER, MIZE, AND CASE *Introduction to Industrial and Systems Engineering*
WHITEHOUSE *Systems Analysis and Design Using Network Techniques*

LINEAR PROGRAMMING IN SINGLE- & MULTIPLE- OBJECTIVE SYSTEMS

JAMES P. IGNIZIO
The Pennsylvania State University

PRENTICE-HALL, INC., *Englewood Cliffs, N.J. 07632*

Library of Congress Cataloging in Publication Data

IGNIZIO, JAMES P.
Linear programming in single- & multiple-objective systems.

Includes bibliographical references and index.
1. Linear programming. I. Title. II. Title: Linear programming in single- and multiple-objective systems.
T57.74.I36 519.7′2 81-7314
ISBN 0-13-537027-2 AACR2

Editorial/production supervision: Nancy Moskowitz
Manufacturing buyer: Joyce Levatino

Printed in the United States of America

10 9 8 7 6 5 4

PRENTICE-HALL INTERNATIONAL, INC., *London*
PRENTICE-HALL OF AUSTRALIA PTY. LIMITED, *Sydney*
PRENTICE-HALL OF CANADA, LTD., *Toronto*
PRENTICE-HALL OF INDIA PRIVATE LIMITED, *New Delhi*
PRENTICE-HALL OF JAPAN, INC., *Tokyo*
PRENTICE-HALL OF SOUTHEAST ASIA PTE. LTD., *Singapore*
WHITEHALL BOOKS LIMITED, *Wellington, New Zealand*

To My Grandparents:
W. C. and A. Blanche Roberson

CONTENTS

10 Sensitivity Analysis in Linear Programming 216

11 Selected Applications and Some Computational Considerations 246

Part Three: Special-Models in Single-Objective Linear Programming 271

12 Characteristics of the "Special Model" 272

13 The Transportation Problem 278

14 The Assignment Problem 318

15 Network Analysis 344

Part Four: Multiple-Objective Linear Programming 371

16 Formulation of the Multiple-Objective Model 372

17 Methods of Solution 392

18 Duality in Linear Goal Programming 430

19 Sensitivity Analysis in Linear Goal Programming 452

PREFACE

Within the very broad subject area known as "optimization," the most widely known and implemented technique for modeling and solution is, by far, the methodology denoted as "linear programming." Linear programming, in turn, deals with the optimization (in terms of a measure or measures of performance) of a system that may be modeled as a set of linear, mathematical functions.

At the time of the writing of this text, the field of linear programming was already midway into its fourth decade and, as such, it should not be surprising that a substantial number of linear programming textbooks have already been published. These other books differ from one another primarily in regard to style, length, depth of coverage, and level of mathematical sophistication. They differ little, however, in the basic subject matter: the development of the linear programming model and the method of solution. This is neither surprising nor intended as a criticism, since the traditional presentations of the modeling and solution process should not be radically changed simply for the sake of change. Consequently, the reader may well ask why the need for another textbook on linear programming, and how this book differs from the others in the field.

This textbook was written with two primary purposes in mind. First, it has been written as an introductory, yet relatively complete exposition of linear programming for the practitioner. That is, the text is intended for those individuals who wish to both understand and implement the methodology. The second purpose of the text is to incorporate, in the coverage, a *unified* presentation of both traditional (i.e., single-objective) linear programming and multiple-objective linear programming. These two purposes, and in particular the latter, are what serve to differentiate this text from most other linear programming textbooks presently in print.

To accomplish the first purpose, the author has attempted to present the mathematical background, and the modeling and solution process, in a clear and concise fashion and accompanied by illustrative examples that serve to demonstrate and reinforce the discussions. The accomplishment of the second purpose rests upon the initial presentation of a unified mathematical model, known as the baseline model. The reader is then provided with the set of assumptions necessary to either convert this baseline model to the traditional linear programming model or to the multiple-objective version and, in either case, to then solve, interpret, and evaluate the results obtained as well as the impact of assumptions used in the conversion. The inclusion of the multiple-objective model and the use of a unified approach are factors that contribute to making this linear programming textbook unique and, hopefully, more flexible then texts dealing primarily with the traditional, single-objective model.

The textbook has been written so as to support either a sequence of courses or as a text for a course either in strictly traditional linear programming or multiple-objective linear programming. For the latter course, it is assumed that the student has had a background in single-objective linear programming. The following diagram serves to indicate these options as well as prerequisite material.

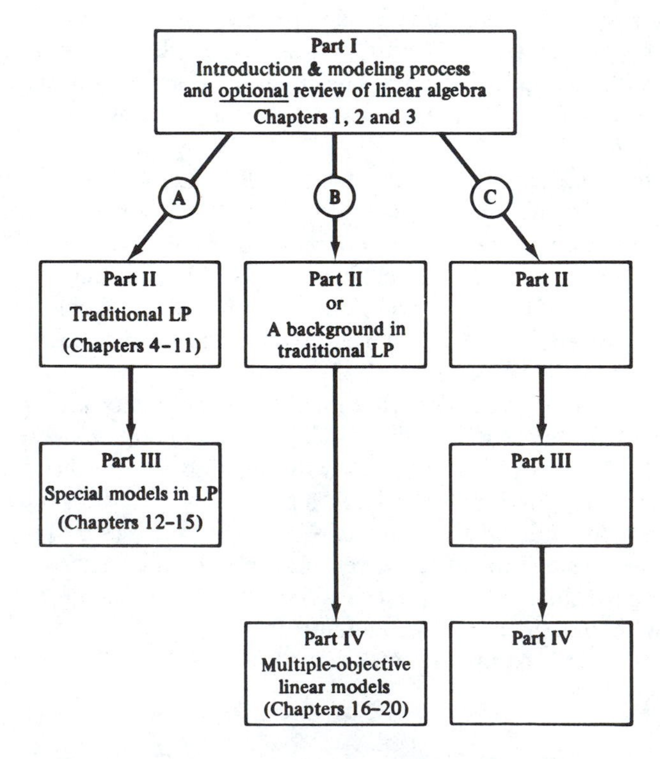

In the diagram, three alternative plans for course content are given. Plan A consists of a one-term course in, primarily, traditional linear programming. Since there are a total of 15 chapters in this plan, there is more than enough material for the term, and the instructor may choose to delete some topics or give them less emphasis. Plan B is also a one-term course, in which the concentration of the course is placed on Parts One and Four, thus leading to a plan of study that emphasizes the multiple-objective models and methods in linear programming. Under such a plan, the material in Part Two (i.e., the traditional linear programming background) need not be covered if the student has a previous background in this area. The final plan of study, C, is intended as a two-term sequence of linear programming courses and covers the entire text.

Regardless of the study plan selected, the alternatives are appropriate for either senior-level or first-year graduate students having some familiarity with linear algebra. When directed specifically at the graduate student, the course may be enhanced by individual or group projects.

JAMES P. IGNIZIO
University Park, Pennsylvania

ACKNOWLEDGMENTS

The author wishes to thank Catherine M. Murphy for reading and commenting on the final draft of the manuscript, and those students who used the draft manuscript in their linear programming classes at Penn State. Special thanks are also given to the three people who shared the typing of the several drafts and final manuscript: Kim Williams, Cathi Moyer, and Cynthia Ignizio.

LINEAR PROGRAMMING IN SINGLE- & MULTIPLE-OBJECTIVE SYSTEMS

part one

THE UNIFIED LINEAR MODEL

INTRODUCTION

The Linear Decision Model

Despite the claims of "seat-of-the-pants" decision making, divine revelations, and woman's intuition, the human mind is simply not equipped to perform a thorough, systematic, and objective analysis of most of the large and complex decisions that we often face. Consequently, the majority of credible approaches to decision making must employ an aid: a *model* of the problem under investigation. Such models do not, as some managers fear, *make* decisions. Rather, they are, or should be, used to *complement* the judgment process, to help clarify the situation, and thus to provide for an improvement in the decisions and policies ultimately set forth. It would then seem obvious that, the better the model, the better should be the resulting decision.

In this text we restrict our attention to a single, yet exceptionally useful and important type of quantitative model. This is the *linear* mathematical model: a mathematical model consisting solely of linear functions. A linear function, in turn, is one in which all terms consist of a single variable and where each variable is raised to the power 1. Thus, functions (1.1) and (1.2) are linear functions:

$$x_1 + x_2 = 10 \tag{1.1}$$

$$3x_1 + 7x_2 - 8x_3 = 7 \tag{1.2}$$

$$2x_1^2 + 3x_2^{3.7} = 82 \tag{1.3}$$

$$4x_1 - 3x_1x_2 + 2x_2^2 = 30 \tag{1.4}$$

1

However, function (1.3) is not linear because both x_1 and x_2 are raised to powers other than 1. Function (1.4) is also not linear because it contains a term $(-3x_1x_2)$ that consists of more than a single variable. (A more detailed discussion of linear functions and their components appears later in this chapter and in the chapters that follow.)

Not only do we learn how to construct linear models of various decision problems, but we also discover how to obtain the solutions to such problems—and then to evaluate the implications of these solutions. Generally, the solutions obtained represent optimal (with respect, at least, to the model, if not always to the actual problem) courses of action to be taken.

The general approach to the modeling and solution of such linear mathematical models is denoted as *linear programming* or, less often (although possibly more appropriately), as linear optimization. Traditionally, the linear mathematical models of interest have had only a *single* objective (to be optimized). However, in this text we also use the term "linear programming" in a broader sense, to describe, in addition, linear mathematical models in which there exist *multiple* objectives. The word *linear* refers to the fact that we are addressing a linear model, whereas the word *program* is used to denote the values of the variables for the linear model. We seek then, to find the program that serves to either optimize (i.e., maximize or minimize) a single objective *or* which provides the optimal compromise to a set of objectives. Consequently, "linear programming" is an appropriate descriptor for either single- or multiple-objective analysis. Note particularly that, although the solution process may be (and often is) carried out on a digital computer, the term "linear programming," itself, has nothing to do with computer programming.

It should be recognized that not all decisions may be (reasonably) represented by a linear model. However, a sufficient number of very important classes of problems are actually linear (or, at least, capable of being reasonably approximated by linear functions) so that the study of the linear model and linear programming is certainly worthwhile. In addition, the modeling and solution process involved in linear programming is similar in many respects to that used for the remaining class of mathematical models: the nonlinear model. However, the work in linear programming is generally both more refined and complete than that available for the usually more difficult area of nonlinear programming. Finally, extensions of the linear programming process have found use in solving (even if perhaps only approximating the solution) the nonlinear models. Consequently, the reader wishing, at a later date, to investigate nonlinear programming will find that the material in this text will provide a foundation helpful for an understanding of this related area.

Purpose of the Text

The obvious purpose of this text is to teach the reader how to:

1. Develop models of linear decision problems.
2. Solve these models and relate their solution to the actual problem.

However, in addition, it is the author's desire to provide a text, useful to a wide audience, that is both readable and understandable. Most important, this is a text intended for the *practitioner*. As such, the mathematician may find that the text lacks the rigor and sophistication with which he or she feels comfortable. Those wishing to delve into the theory, the rigorous details, and the more esoteric aspects of linear programming are thus directed to the references. The author has attempted, in this text, to limit the mathematics to only that really required for a serious practitioner's understanding of the methods and procedures involved. In addition, rather than concentrating solely on problem solution, an attempt has been made to clarify and stress the *interpretation* and *implementation* of the results obtained. Fortunately, linear programming is not a particularly complex topic and thus this straightforward, rather than elegant, approach will not affect the reader's ability to understand or actually apply the methodology.

In addition to the objectives listed above, the author has also endeavored to design this text so that the reader may:

1. Recognize both the scope and limitations of linear programming.
2. Establish a single, unified, baseline model of both single- and multiple-objective problems.

3. Recognize the simplifying assumptions necessary to convert the baseline model into either a traditional single-objective or the more recent multiple-objective problem, and then to solve the problem—recognizing the impact of these assumptions on the solution obtained.

Intended Audience and Prerequisites

The intended audience for this text is virtually anyone who has a serious desire to learn about linear programming and the solution of a diverse class of linear decision models. The techniques of linear programming have, for the most part, been developed over the past three decades (although linear algebra, which forms the foundation for linear programming, is, of course, considerably older). Initially, the methodology was of interest only to those in applied mathematics, operations research, and economics. However, as word of the power and success of the method spread and as the number of actual implementations increased, interest grew in many other disciplines. As a consequence, the drafts of this particular text have already been used by graduate and undergraduate students in the fields of:

—Business administration
—Marketing
—Logistics
—Economics
—Virtually all fields of engineering
—Operations research/management science
—Geography
—Forestry
—Computer science
—Statistics
—Acoustics
—Accounting
—Food services and hotel management
—Agriculture
—Earth and mineral science
—Meteorology
—Anthropology
—Biology

The prerequisites for the text are, primarily, a knowledge of elementary linear algebra and, perhaps most important, a serious desire to learn. The actual mathematical operations involved in linear programming are surprisingly few and simple, being limited primarily to addition, subtraction, multiplication, and division. The text is designed for upper-level (juniors or seniors) undergraduates or first-year graduate students and contains sufficient material for either a one-quarter or one-semester introductory course or a two-quarter or two-semester sequence of courses.

Differences

All authors believe that their book is unique—and significantly so. It would be hard to justify the labor involved in the preparation of a textbook if one did not feel this way. This author is no exception. Consequently, those aspects that are believed to set this book at least somewhat apart from others are described below.

There are quite a few published texts on linear programming. However, the vast majority of them approach the topic from a relatively traditional point of view that has not changed substantially since the origin of the field. In addition, it is the author's opinion that many of these texts are at an unnecessarily high level of sophistication (or, depending on one's viewpoint, complexity). As such, they are of more interest to the theoretician than to the practitioner. Some of the remaining set of linear programming texts lie at the other extreme. In fact, in some cases the reader is given the impression of being "talked down to." Such texts are of little, if any interest to the mathematically inclined, nor are they really of much help to the practitioner. Consequently, in writing this book, the author has attempted to fill the relative void between these two extremes. As a result, this text has been designed to satisfy the needs of the serious practitioner who wishes to not only use but to understand the tool that he or she is employing.

An equally important difference (and probably the least subjective) is that, while virtually all other linear programming texts concentrate on the traditional, single-objective linear decision model, this text both prepares the reader for and presents him or her with *both* the single-objective model and the newer and often more flexible multiple-objective model. Further, this is accomplished through the introduction of a *single*, *unified*, *baseline model* from which either the traditional or multiple-objective model may be conveniently developed. The assumptions required to convert the baseline model into either the traditional or multiple-objective model are also explained in detail. Since many real-world problems involve several, conflicting objectives, the importance of such an approach cannot be overstated. The analyst with an exposure to only traditional linear programming has limited the scope of his or her knowledge as well as limited his or her problem-solving abilities. This observation should become increasingly more evident in the very near future as awareness of the value of the multiple-objective model increases.

A Historical Perspective

Although it will be shown that traditional, single-objective linear programming and multiple-objective linear programming both stem from the same baseline model, and, although both models are highly interrelated, we divide our brief historical sketches of these models into two subsections. This is done simply to clarify the discussion. We first introduce the traditional model and the contributions leading both to its development and to later refinements.

Traditional Linear Programming

The human race has been building models of decision problems for thousands of years. However, until this century, most of these models have, of necessity, been restricted to relatively small problems involving but a few variables. These earlier mathematical methods (such as calculus, Lagrange multipliers, etc.) relied on a *sophistication of technique* to solve the problem classes for which they were suited. However, linear programming, and newer methods of mathematical optimization, rely far less on mathematical sophistication than they do on an unusual *adaptability to the mode of solution inherent in the modern digital computer*. As such, those familiar with classical mathematics will observe some striking differences, particularly in the simplicity of these new methods coupled with their iterative processes (i.e., the repeated performance of a series of relatively simple operations).

The development of the techniques that provide the foundations of linear programming, other than linear algebra, occurred primarily during this century. In 1928, John von Neumann published the central theorem of *game theory*. Although the original intent of game theory was to explain rather specific problems encountered in physics, it was later recognized that such games and linear programming were interrelated and, in fact, some of the properties of linear programming (particularly duality) may be explained through game theory. In 1944, von Neumann and Morgenstern published the *Theory of Games and Economic Behavior* [23]. The interrelationship between games and economics is what really motivated interest in this area (i.e., game theory) and undoubtedly helped to encourage later interest in linear programming.

Another related development was the concept of input–output analysis as proposed by W. W. Leontief in his 1936 paper: "Quantitative Input and Output Relations in the Economic Systems of the United States" [19]. The input–output concept related a linear model to the interindustry system. However, this model, unlike linear programming models, had no objective function to be optimized.

Recognition of the linear programming problem (i.e., the optimization of a linear function subject to a set of linear restrictions or constraints) seemed to first receive serious attention in this century, particularly in the late 1930s and the 1940s. A number of interesting problems were formulated in conjunction with the growth of a serious interest in obtaining a general method for their solution.

The Soviet mathematician and economist L. V. Kantorovich [13] actually formulated and solved a linear programming problem in 1939. This problem dealt with the specific problem of organization and planning for production. In 1941, Hitchcock [7] posed a special class of linear programming problem known as the transportation problem (which was also posed independently some six years later by Koopmans [15]). Stigler [22], in 1945, presented yet another special linear programming formulation, the so-called diet problem. During the conduct of World War II, British and, later, American research teams encountered a wide variety of military-related linear programming problems (which, however, were often not recognized as such). Shortly after World War II, a group of scientists were called upon by the U.S. Air Force to investigate the feasibility of applying mathematical techniques to the problems of military programming budgeting and planning. The wide-ranging problem of military logistics was of particular concern. George Dantzig was one member of this research team. Dantzig had earlier proposed that the interrelations between activities of a large organization be viewed as a linear programming model and that the optimal program (or solution) be determined by minimizing a (single) linear objective function.[a] Such ideas led the Air Force to set up a team under the project name SCOOP (for Scientific Computation of Optimum Programs).

Project SCOOP began in June 1947. By the end of that *same* summer, Dantzig and his associates had developed not only an initial mathematical model of the *general* linear programming problem, but, in addition, developed a general method of solution which was designated the *simplex method* [4].

Interest in linear programming quickly spread among, primarily, mathematicians, economists, operations researchers, and individuals in a number of government organizations. In the summer of 1949, only two years after the development of the simplex method, a conference on linear programming was held under the sponsorship of the Cowles Commission for Research in Economics. The papers presented at this conference were published in the text *Activity Analysis of Production and Allocation* [15].

The earliest applications of linear programming fell into three main classes:

1. Military applications (as generated by Project SCOOP).
2. Interindustry economics as based on Leontief's input–output model.
3. Relationships between linear programming and two-person, zero-sum games.

More recently, emphasis has shifted to such problems as those encountered in the environment, transportation, energy, social problems, and the general industrial area.

The development of linear programming, specifically the simplex method, was a "giant step for mankind" in its model representation and problem-solving abilities. However, it might well have received little attention had it not come about at certainly the most opportune time conceivable. The large-scale electronic (digital) computer first became a practical reality in 1946 at the University of Pennsylvania. This was just *one year* prior to the development of simplex. Without the digital computer, the method of simplex would most probably hold only academic interest.

---

[a]This proposal, stressing a single-objective function, led naturally enough to what we have termed traditional linear programming.

The simplex method consists of only a few steps and these steps, themselves, require only the most basic of mathematical operations—operations that a digital computer is best able to handle. The only difficulty inherent in the simplex method is in a practical sense for, although its steps are simple and few, they must be repeated over and over before one finally obtains an answer. This repetition of the steps is what removes the possibility of solving problems (i.e., other than those that are quite small) by hand. However, the digital computer can easily and efficiently perform this repetitive process and, as a consequence, the simplex method in conjunction with the digital computer is a practical means for solving extremely large, real-life problems. The first successful solution of a linear programming problem took place in January 1952 on the National Bureau of Standards SEAC computer. In 1953, computers became available on a production-line basis. Also, at about this same time, the popular journals and magazines began to expound on the power and potential of the new tool of linear programming. Today, almost all computers come with a linear programming routine and it is one of the most highly utilized of their abilities.

The simplex method of Dantzig, although general, still has several limitations. One of these, of course, was its inability to deal with problems having more than a single objective. Another is its inability to handle problems in which the variables must take on *integer* values. The latter deficiency has probably received the most interest. In 1958, R. E. Gomory [5] developed the so-called "cutting-plane algorithms" for solving linear models with integer variables. In 1960, A. H. Land and A. G. Doig [17] published a paper on the solution of such problems by a quasi-enumerative technique known as "branch-and-bound" (the algorithm itself, according to Land and Doig, was actually developed in 1958). In 1965, Egon Balas [1] introduced an algorithm, of the quasi-enumerative nature, for solving linear problems in which the variables take on only binary (zero or one) values. Since the early 1960s, an extremely wide variety of such integer and zero–one algorithms have appeared, based primarily on refinements to or extensions of either the Gomory, Land and Doig, or Balas algorithms. Unfortunately, although this effort has attracted many investigators and resulted in an abundance (or even an overabundance) of papers, solution methods to linear models with integer or zero–one variables are still far less efficient than those obtained by the simplex method on linear models without such restricted variables. The rapid increase in the complexity and mathematical elegance of such special-purpose algorithms has, unfortunately, not always been reflected in any truly significant increase in practical problem-solving ability [11]. Hopefully, this situation will change, as there are a large number of real problems that involve integer variables.

After the development of the simplex algorithm by Dantzig, many investigators concluded that simplex was likely the last word in an algorithm for the general linear programming problem. However, other analysts still pursued a more computationally efficient procedure. This was likely motivated by the work of Klee and Minty [14], who illustrated, through a pathological example, that the simplex algorithm *can* exhibit exponential growth in computation time. Consequently, renewed interest in the development of a provable polynomial time algorithm was generated. It should be stressed that the interest in such polynomial algorithms is primarily academic, because *in actual practice*, the simplex algorithm does exhibit (low-order) polynomial computation time.

In 1979, the mathematical world was told of the development of a true polynomial time algorithm for linear programming as developed by a Soviet doctoral candidate, L. G. Khachian (or Hacijan, or Khatchian, or Khachiyan, depending on the reference). The result was considered so significant, in fact, that it made the front page of *The New York Times* [2] as well as being featured in several journals [6]. Not only was the Khachian algorithm hailed as the potential replacement for the simplex algorithm, but there were also claims that it would soon solve computationally difficult problems involving nonlinear functions as well as integer variables.

The actual facts concerning this new algorithm are, however, not quite as dramatic. Although it does, indeed, guarantee the solution of a linear programming problem in polynomial time, for practical, real-life problems, the simplex algorithm is still considerably more efficient and, in all likelihood, will remain the dominant linear programming algorithm for at least the near future [21]. For this reason, our emphasis is directed toward the simplex algorithm, although a brief summary of the Khachian algorithm is provided in Chapter 11.

Multiple-Objective Linear Programming

As mentioned earlier, Dantzig's initial concept was centered about the development of a linear programming model with but a single objective. Evidently, this so set the tone for the development of traditional linear programming that many (if not most) linear programming texts completely ignore even the possiblity of more than one objective. The single-objective approach is so common (and has become such a habit) that many students, and even their professors, may often be unaware that it is not necessarily the one, best way to model a problem, particularly a real-world problem. Quite often, regardless of the number of objectives that may actually exist, we see the investigator proceed directly to the single-objective model with little, if any, thought as to the consequences of his or her decision or of the deficiencies of such a model.

Some of the early investigators who did recognize this deficiency in the single-objective approach suggested a rather brute-force approach to circumvent (but not to actually resolve) the problem. Their proposal was to:

1. List all the problem objectives.
2. Pick *one* of these objectives as the "single objective" in the traditional model and consider all the other objectives to be rigid constraints (we discuss later how an objective may be converted into a "constraint").
3. Solve the resultant model.
4. Repeat steps 2 and 3 with another objective as the single objective in the model.

5. Pick the solution, of those found in step 3, that appears to "best" satisfy all the objectives.

There are at least two drawbacks to this approach. First, we must solve a possibly large number of linear programming problems (i.e., one for each objective). Second, and most critical, the resultant solution as achieved in step 5 may very well not represent that solution that *best* satisfies all the objectives (the larger the problem, the more likely that this is so). It is rather obvious that something far more systematic is required.

In the early 1950s, Kuhn and Tucker [16] addressed the multiple-objective problem via their formulation of the so-called vector-maximum model. However, most early interest in their paper centered not about the vector maximum but, instead, about their results with regard to nonlinear models.

At about the same time as the development of the vector-maximum model, A. Charnes and W. W. Cooper [3] proposed a nonparametric curve and response surface fitting tool which they called the method of "constrained regressions." The results of their work, reported in a conference held in 1952, also, as the case with the vector-maximum model, evidently attracted relatively meager attention. However, the method of constrained regressions is, in actuality, one version of the multiple-objective optimization technique known as goal programming.

Also, in the 1950s there occurred a philosophical development that forms the basis for much of our present-day work in multiple-objective models. This philosophy is stated in the hypothesis, as advanced by March and Simon [20], that: "Most human decision making, whether organizational or individual, is concerned with the discovery and selection of satisfactory alternatives; only in exceptional cases is it concerned with the discovery and selection of optimal alternatives." This author would go a bit further in adding that, even if one really wished to determine the set of optimal alternatives, such a set is unlikely to be obtained, from a practical sense, for other than unusually simple real-world problems. The foregoing hypothesis is commonly known as the "satisficing" concept and, as we shall see in later chapters, is vital to the foundations of many of our present-day multiple-objective models and methods of solution.

In 1961, Charnes and Cooper published their two-volume text *Management Models and Industrial Applications of Linear Programming* [3]. A brief section in one of these volumes discussed an approach to the solution of managerial-level problems involving multiple, conflicting objectives (or goals). Its originators dubbed this technique *goal programming*. However, goal programming is, in essence, simply a new name for their earlier tool (as discussed above) known as constrained regressions. Rather than addressing the curve or response surface fitting problem,[b] the "goal programming" method addressed a multiple-objective linear management-level problem. Despite the new name and broader area of implementation, goal programming still did not receive the attention that one might expect.

---

[b]It is interesting to note that, even quite recently, some investigators have reinvented the original use of goal programming (i.e., in curve fitting).

The author's first exposure to the field of multiple-objective methodology came as a result of exposure to the Charnes and Cooper work as well as the immediate pressure dictated by a very real multiple-objective problem.

While working as an engineer on the Saturn/Apollo program (the U.S. moon-landing mission), the author was faced with the problem of designing an antenna and communications system that had to satisfy a number of conflicting objectives. However, rather than fitting directly into the Charnes and Cooper model, which was intended for linear functions, this problem was represented by an extremely large (in terms of the number of variables and functions) *nonlinear* model. The solution was ultimately obtained by extending the original goal programming concept to a nonlinear model. This occurred in 1962 and was, at least to the author's knowledge, the first application of goal programming to an engineering design problem [8, 10].

The results of the antenna design effort were quite successful and led the author, first, to other applications and then to the motivation of interest in developing further extensions of the goal programming approach. Included among these were extensions to models involving discrete variables and an alternative approach to the measurement of the achievement of a multiple-objective model that involved the lexicographic minimum of an ordered vector representation of goal "deviations" [9, 10].

In 1965, Y. Ijiri [12] proposed the inclusion of the so-called concept of "preemptive priorities." In essence, he suggested that a priority be given to each objective, or set of commensurable objectives, in the problem. If P_1 represents the priority associated with the top-priority set of objectives and P_2 with the next set of objectives, and so on, it is understood that the satisfaction of the objectives associated with P_k preempts the satisfaction of those at a lower priority (i.e., $P_{k+1}, P_{k+2}, \ldots$). That is, the satisfaction of a higher-priority goal set is always preferred to the satisfaction of all lower-priority sets. (Actually, although common use is to define this as a preemptive-priority structure, the *structure* itself is nonpreemptive.) In practice, the preemptive-priority concept is achieved through finding the lexicographic minimum of an ordered vector (as mentioned above), and thus the terms "lexicographic minimum" and "preemptive priority" structure may be used interchangeably.

The goal programming approach [3, 9, 18], although one of the earliest and most widely used (particularly in actual practice) of the multiple-objective methods, is now just one among many approaches to the problem. One alternative approach that has generated considerable interest is based on the concept of combining objectives by means of utility theory. Another technique attempts, at least in theory, to generate the so-called set of efficient (or nondominated, Pareto optimal or unimprovable) solutions [24]. Yet another, more recent approach utilizes the concept of the fuzzy membership function and results in a technique known as fuzzy programming [25]. In this text we address all of the approaches described above as well as several other concepts for the multiple-objective model.

Some Examples

The number (and surprising variety) of applications of either traditional or multiple-objective linear programming is now so large as to prohibit any attempt at a complete, exhaustive listing.[c] However, in order that the reader be made aware of both the importance and potential of these tools, we present a list of just a few actual implementations and then discuss in more detail some of the better known and common applications.

Included among the applications encountered in traditional and multiple-objective linear programming have been such problems as:

—Establishment of a proper diet
—Transportation of goods
—Assignment of personnel, material, and jobs
—Blending of gasoline
—Media planning
—Personnel planning
—Production planning
—Academic resource allocation
—Financial planning
—Municipal planning
—Program selection and capital budgeting
—Portfolio selection
—Transportation system planning and routing
—Curve and response surface fitting
—Energy models and planning
—Deployment of antennas on space vehicles
—Deployment of airborne and ground-based radar systems for defense
—Design of command and control centers for the military (including the design and layout of cathode-ray-tube (CRT) displays)
—Social health care delivery systems
—Location and staffing of regional health care centers
—Location of civil defense systems
—Establishment of the grading and cutability of meat
—Planning and planting of crops
—Contract awards
—Design of electrical filters
—Design of sonar systems
—Airline crew scheduling
—Coal mining operations

---

[c]A portion of such applications have been listed in a "bibliography of linear programming applications" by Saul Gass in his text *Linear Programming: Methods and Applications* (New York: McGraw-Hill, 1975).

—Cupola design
—Defensive missile allocation and engagement planning
—Aircraft overhaul planning
—Inventory planning and control
—Structural design
—Traffic signal planning
—Establishment of racial mixture in schools
—Criminal justice system
—Solid waste collection and routing
—Firing order and duration of rockets on spacecraft
—Water quality control
—Police patrols
—Political redistricting
—Electrical network design
—Radar ambiguity function design
—Air pollution control
—Reliability in multicomponent systems
—Forest management and harvest policies

Among the list of applications given above are a number of examples that are so common and well known that they are often presented to characterize the types of problems that might be attacked by linear programming. We shall consider a few of these: the diet problem, the production scheduling and inventory control problem, the blending problem, and the routing and assignment problem.

The Diet Problem

In the diet problem we are given the nutrient content of a number of different foods together with the minimum daily requirement of each nutrient. Knowing also the cost per unit weight of each type of food, the problem is to determine the diet that both satisfies the minimum daily requirements and minimizes total cost. The multiple-objective diet problem might consider other objectives, such as the minimization of cholesterol and the maximization of the "desirability" of the foods selected (i.e., their taste and/or appearance).

The Production Scheduling and Inventory Control Problem

Many firms—for example, air-conditioning manufacturers—produce products that are highly subject to seasonal sales fluctuations. When they try to follow such fluctuations with fluctuating production rates, their production costs tend to increase. However, with a uniform production rate they usually discover a buildup in inventory, resulting in large storage costs. Consequently, in the production scheduling and inventory control problem, we are given the predicted demand (perhaps for each month or quarter), the predicted production costs per unit for each period, and the production capacity for each period. The problem is to minimize the combined production and inventory costs while not exceeding production limits (and, if possible, satisfying customer demand). Other objectives that might be included are the minimization of labor turnover and the minimization of overtime.

The Blending Problem

There are a wide variety of problems in which certain basic components or raw materials are combined, or blended, to produce a product that satisfies certain specifications. Typical of such problems are the blending of gasolines, the mixture of nuts in a canned assortment, the blending of feeds for animals, and the mixture of meats to produce sausage or lunch meats. We are usually given the specifications on the blend (i.e., the recipe) together with restrictions (such as government requirements) and our task is to produce a blend that minimizes total cost (or maximizes total profit) while satisfying these restrictions. The multiple-objective blend problem might consider additional objectives, such as the capture of the maximum number of shares of the market and the minimization of blending operations.

The Routing and Assignment Problem

The routing and assignment problems are closely related, with the assignment problem generally being a special class of routing problem. Typical examples include the routing of goods from warehouses to distributors, the assignment of jobs to machines or personnel to jobs or the routing of a calibration van. Generally, we seek to find the minimal cost (or distance) route or assignment that still satisfies such constraints as customer demand and warehouse capacity.

Hopefully, the reader has noticed the one thing that all these problems have in common. That is, we usually wish to allocate scarce resources so as to optimize one or more objectives wherein the mathematical model representing the problem consists solely of linear functions. Most linear programming problems may be so characterized.

The Material to Follow

This text is divided into four parts. The first, which includes Chapters 1 through 3, presents an introduction to the background of linear programming in both single- and multiple-objective problems. Particular emphasis is given, in Chapter 2, to the development of a unified, *baseline* model for *either* the traditional single-objective linear programming problem *or* the newer multiple-objective concept. The approach proposed is a definite deviation from the normal procedure employed in other linear programming texts. The concluding chapter in Part One provides a review of linear algebra, the basic foundation of the methodology employed in linear programming.

Part Two encompasses Chapters 4 through 11. The emphasis of these chapters centers about the traditional, single-objective linear model and its methods for solution and analysis. Initially (in Chapter 4), it is shown how one transforms, via a number of assumptions, the baseline model into the traditional linear programming model. The remaining chapters in Part Two concentrate on the modeling and solution processes used in traditional linear programming. Topics include the simplex method and its foundation, various tableau representations, duality, economic analysis, dual simplex, the primal–dual algorithm, sensitivity analysis, and (in Chapter 11) a brief discussion of several applications, some computational considerations, and an overview of the new Khachian algorithm.

In Part Three we address a selection of special subclasses of the traditional linear programming problem. Included are the transportation problem, assignment and matching problems, the transshipment problem, and a variety of other problems that are best represented by network models.

Finally, in Part Four, the multiple-objective problem is introduced. However, it should be noted that, even if this material is not covered, the reader still has received the necessary prerequisites to investigate the multiple-objective linear model through his or her introduction to the baseline model of Chapter 2. Topics covered in Part Four include an introduction to various approaches to multiple-objective linear programming as well as duality (i.e., the multidimensional dual) and sensitivity analysis for the multiple-objective problem. Included among the techniques discussed are weighted linear goal programming, lexicographic linear goal programming, interval goal programming, augmented goal programming, the method of efficient solutions, and fuzzy programming. Much of the material in Part Four is based on original contributions of the author.

REFERENCES

1. Balas, E. "An Additive Algorithm for Solving Linear Programs with Zero–One Variables," *Operations Research*, Vol. 13, 1965, pp. 517–546.
2. Browne, M. W. "Mathematic Problem-Solving Discovery Reported," *The New York Times*, November 7, 1979, p. 1.
3. Charnes, A., and Cooper, W. W. *Management Models and Industrial Applications of Linear Programming*, Vols. 1 and 2. New York: Wiley, 1961.
4. Dantzig, G. B. "Programming in a Linear Structure," Comptroller, U.S. Air Force, Washington, D.C., February 1948.
5. Gomory, R. E. "Outline of an Algorithm for Integer Solutions to Linear Programs," *Bulletin of the American Mathematical Society*, Vol. 64, 1958, pp. 275–278.

6. HACIJAN, L. G. "A Polynomial Algorithm in Linear Programming," *Doklady Akademii Nauk SSSR*, Vol. 244, No. 5, 1979, pp. 192–194.
7. HITCHCOCK, F. L. "The Distribution of a Product from Several Sources to Numerous Localities," *Journal of Mathematical Physics*, Vol. 20, 1941, pp. 224–230.
8. IGNIZIO, J. P. "S-II Trajectory Study and Optimum Antenna Placement," North American Aviation Report SID-63, Downey, Calif., 1963.
9. IGNIZIO, J. P. *Goal Programming and Extensions*. Lexington, Mass.: Heath (Lexington Books), 1976.
10. IGNIZIO, J. P. "Goal Programming: A Tool for Multiobjective Analysis," *Journal of the Operational Research Society*, Vol. 29, II, 1978, pp. 1109–1119.
11. IGNIZIO, J. P. "Solving Large-Scale Problems: A Venture into a New Dimension," *Journal of the Operational Research Society*, Vol. 31, 1980, pp. 217–225.
12. IJIRI, Y. *Management Goals and Accounting for Control*. Chicago: Rand McNally, 1965.
13. KANTOROVICH, L. V. "Mathematical Methods of Organization and Planning Production," Publication House of the Leningrad State University, 1939, translated in *Management Science*, Vol. 6, No. 4, 1960, pp. 366–422.
14. KLEE, V., AND MINTY, G. "How Good Is the Simplex Algorithm?" in O. Shisha (ed.), *Inequalities III*. New York: Academic Press, 1972, pp. 159–175.
15. KOOPMANS, T. C. (ed.). *Activity Analysis of Production and Allocation*. New York: Wiley, 1951.
16. KUHN, H. W., AND TUCKER, A. W. "Nonlinear Programming," in J. Neyman (ed.), *Proceedings of Second Berkeley Symposium on Mathematical Statistics and Probability*. Berkeley, Calif.: University of California Press, 1951, pp. 481–491.
17. LAND, A. H., AND DOIG, A. G. "An Automatic Method of Solving Discrete Programming Problems," *Econometrica*, Vol. 28, No. 3, 1960, pp. 497–520.
18. LEE, S. M. *Goal Programming for Decision Analysis*. Philadelphia: Auerbach, 1972.
19. LEONTIEF, W. W. "Quantitative Input and Output Relations in the Economic Systems of the United States," *Review of Economics and Statistics*, August 1936.
20. MARCH, J. G., AND SIMON, H. A. *Organizations*. New York: Wiley, 1958.
21. PEGDEN, C. D., AND IGNIZIO, J. P. "Khachian's Polynomial Algorithm for Linear Programming," *AIIE News: Operations Research*, Vol. 14, No. 4, Spring 1980, pp. 1–4.
22. STIGLER, G. J. "The Cost of Subsistence," *Journal of Farm Economics*, Vol. 27, No. 2, May 1945, pp. 303–314.
23. VON NEUMANN, J., AND MORGENSTERN, O. *Theory of Games and Economic Behavior*. Princeton, N.J.: Princeton University Press, 1944.
24. YU, P. L. "Introduction to Domination Structures in Multicriteria Problems," *Proceedings of Seminar on Multiple Criteria Decision Making*, University of South Carolina, 1972.
25. ZIMMERMANN, H.-J. "Fuzzy Programming and Linear Programming with Several Objectives," *Fuzzy Sets and Systems*, Vol. 1, 1978, pp. 45–55.

THE BASELINE MODEL

The Need for a Baseline Model

In this text the *baseline* model is the *initial*, unified mathematical model of a problem. Such a baseline model encompasses all classes of mathematical models but, because of the thrust of our text, we concentrate primarily on linear models. The baseline model is the springboard by which we move on to either the traditional, single-objective linear programming problem or to the multiple-objective linear programming problem. As such, it should be considered a vital phase of model development and consequent problem solution (see Figure 2-1).

It should be noted, however, that the baseline model (or its equivalent) is either not discussed or is simply mentioned in passing in most linear programming texts. In such texts, the reader is, rather, introduced immediately to the traditional single-objective mathematical form of the linear programming problem. The main reason for this neglect is probably because, until fairly recently, there simply were no effective, practical approaches to multiple-objective models. However, the omission of the baseline model can serve to both limit and distort the reader's concept of actual model development. The result is that the reader is generally unprepared for the multiple-objective models and may even suffer from a limited awareness of the steps required to build mathematical models in actual practice.

For these and other reasons, the author believes that this chapter may well be one of the most important of this text. It, at the least, is indicative of some

2

of the differences between this text and others. Consequently, for those who might be tempted to move directly to Part Two (and this seems particularly true for those with a previous exposure to linear programming), it is advised that they reconsider and study this chapter carefully.

Models and Model Types

Most decision problems of interest, and virtually all of those of importance in the real world, can be thought of as occurring in large and complex systems. One could hardly hope to evaluate alternative courses of action by actually trying these in the system. Instead, a model that, hopefully, represents the system is normally used for such experimentation. Consequently, models perform a valuable, if not essential role in the decision-making process.

We issue one caution, however. Quite often a reader becomes so involved with the model and *its* solution that he or she forgets that it is indeed the model, not the actual system, that is being analyzed. If the model is a "good" model, its solution should (hopefully at least) represent a good approximation to the solution of the actual system. But it is rarely *the* solution to the actual system. This does not negate the importance and usefulness of the model; but it is a factor that must always be kept in mind.

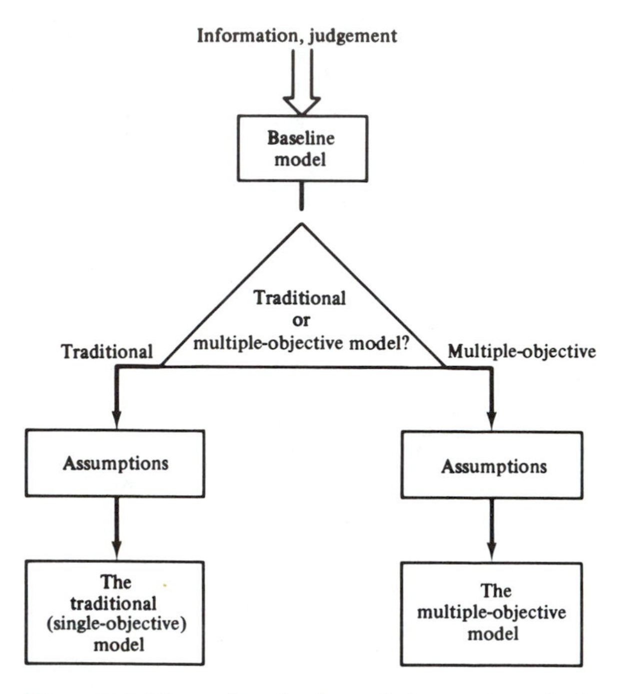

Figure 2-1. Phases in selection of the mathematical model.

Models may take on a wide variety of forms, including among the best known:

—The scale model
—The pictorial model
—The flow chart (or network)
—The matrix
—The mathematical model

A typical scale model could be characterized by the aerodynamic model of an airplane. Such a model "looks" almost exactly like the actual aircraft except for its smaller size. By employing the model in a wind tunnel, the aerodynamic stability of the model, and thus of the actual aircraft, may be observed. As such, the scale model allows the investigator to observe, economically, the performance of the model under certain conditions. Such a model, however, does not generally lend itself to a systematic *optimization* of the system being modeled. That is, it serves as an *evaluative* tool rather than a means for system optimization.

The pictorial model is usually a two-dimensional photograph or sketch of a system. An example might be an aerial photograph of a region. Such a photo could then be used by an electrical power company to decide where to construct their power lines. Again, although such a model aids the decision process, it does not actually lend itself to an optimizing procedure.

The flow chart is a special type of pictorial model. However, rather than simply depicting spatial relationships, the flow chart illustrates the *interrelationships* among the components. Such charts are useful in a wide variety of areas, including:

—Depiction of the inputs and outputs of system components and the flows between these components
—Timing of activities, and their sequence, in the scheduling of a project
—Steps and logic flow of a procedure that is to be simulated

In some instances the flow chart is used only as a decision aid but, with the aid of the methodology of network analysis, one can often actually perform an optimization of the system (actually, of the model of the system).

The matrix is also a common model for decision analysis. Like the flow chart, one may also sometimes employ optimizing procedures on the matrix model. In many instances, the flow-chart model and the matrix model may be interchangeable. Consider, for example, the flow chart (or network) of Figure 2-2. The nodes (circles) on the left represent workers (designated as *A*, *B*, *C*, and *D*) and the nodes on the right represent jobs (designated as I, II, III, and IV). If a worker can perform a given job, a link exists between the worker node and the job node. Otherwise, there is no link. For example, in Figure 2-2, worker *A* can perform jobs I, III, and IV but not job II.

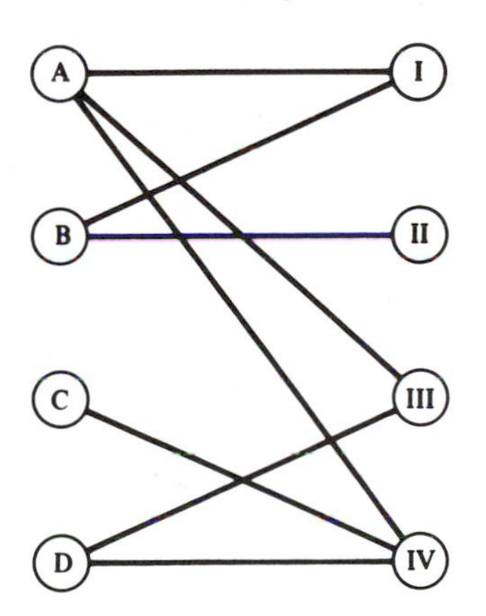

Figure 2-2. Flow graph or network model.

The same network could also be represented by the matrix of Table 2-1. Here the workers head the rows of the matrix and the jobs head the columns. If a worker can perform a given job, there is a 1 in the cell at the intersection of the worker row and job column. Otherwise, a 0 is placed in the cell.

Table 2-1. EQUIVALENT MATRIX MODEL

	Job			
Worker	*I*	*II*	*III*	*IV*
A	1	0	1	1
B	1	1	0	0
C	0	0	0	1
D	0	0	1	1

The primary model of interest in this text is the mathematical model (i.e., a set of mathematical functions that represent the problem under consideration). For example, the statement that profit equals sales minus costs may, in some cases, be written mathematically as

$$P = sx_1 - c(x_1 + x_2)$$

or (2.1)

$$P = sx_1 - cx_1 - cx_2$$

where P = profit, dollars

s = sales price of each unit of product

c = total cost of each unit of product

x_1 = total amount of products sold

x_2 = total amount of products not sold

Thus, $x_1 + x_2$ = the total amount of products produced. Equation (2.1) is a linear function[a] because each term contains only a single variable and, in turn, each variable is to the power 1. A number of additional assumptions must also hold in such a case (we discuss these in Chapter 3).

The mathematical model is the most abstract of all the models we have discussed. That is, whereas the scale model of a building or airplane is immediately recognizable, one cannot so readily identify the problem that is being mathematically modeled. However, the mathematical model more than compensates for this shortcoming by being (generally speaking) the easiest model to manipulate, to analyze, and to optimize. Further, it can be used to represent an extremely wide variety of actual problems.

Guidelines in Model Building

Before discussing the construction of our baseline linear model, it is important to take note of several rules and/or guidelines in model development. These rules form a philosophy of model construction that, if followed, should

[a]Strictly speaking, (2.1) is linear only if the variables x_1 and x_2 are continuous. This is discussed in detail later.

result in a model that is more credible, more useful, and less costly than if a more-or-less brute-force approach were taken.

We should first consider the primary *purpose* of the model. Is it to be used to simulate a system? to evaluate a system? to optimize a system? or to simply describe a system? What are the results going to be used for? How much accuracy is required? Over what time scale is the model to be used? Over what range of inputs must it respond? What are the budget and time restrictions on model development? Who will use the final model and/or its results?

Usually, in answering such questions as those listed above, a general class of model, to best suit all requirements, may be identified. One must next determine, more specifically, the actual *type* of model to use. That is, if we wish to optimize a system, should a network model, or matrix model, or mathematical model be employed? Although the mathematical model is usually the most powerful and flexible, this power and flexibility may not be needed for the specific problem under consideration. When isolating the specific type of model, we should try to avoid one of the most common and costly of all mistakes encountered in model construction. This is to *force* the problem to fit a particular model type. All too often one becomes so enthralled with one type of model (or perhaps it is the only model the analyst is familiar with) that he or she tries to place every problem into that particular format. Rather than fitting the problem to a model, we should fit the most appropriate model to the problem. That is, the model that seems to most *naturally* fit the problem should usually be employed.

The *level* of the model is an extremely important, often abused, factor in model building. By "level" we refer to the amount of "detail" presented by the model, the number of variables considered, the number of relationships and interrelationships presented, and so forth. It is an unfortunate misconception that the more "detailed" (or, in other words, the larger, more complex, and "sophisticated") the model, the "better" the model. Often, just the reverse is true. These highly "detailed" models may be but a cluttered mass of assumptions and inaccuracies whose results are meaningless.

The proper way to determine the level of a model is to begin with a model that is as simplified as reasonable. That is, try to absolutely minimize the number of variables (and other factors) considered. In general, even in very large and complex systems, only relatively few variables really have a significant impact on the system's output. Once this preliminary model is constructed, we evaluate it. We then add to the model *only* that detail believed absolutely necessary to refine the model (and its results) to the level necessary for its actual use.

Those who attempt to build highly detailed models usually find (after considerable time and expense) that the model is too cumbersome to use or too computationally burdensome to employ even on the largest computer. They then attempt to "strip" the level of detail until they obtain a "workable" model. On the other hand, when following our guidelines, a decrease in time and cost is achieved and one should also obtain a more credible and efficient model.

Another consideration in model development, closely related to model level, is the definition of the *system*, its limits, its inputs and outputs, and its components or subsystems. A particularly useful aid here is the use of a flow-chart or network model—regardless of the final model to actually be used to represent the system. To illustrate this concept, consider the following example.

Example 2-1: System Definition Illustration

We shall assume that we wish to determine an (optimal) production schedule (or sequence) for a small flow-shop[b] operation. To define the extreme boundary or limits of our system to be modeled, we first determine the *extent of our actual control*. For clarity of illustration let us assume that our only control (at least from a practical sense) is over the *sequence* of the jobs to be accomplished. Consequently, we might define the limit of our system as shown in Figure 2-3. The box is our system; the arrows leading into and out of the box are the system inputs and outputs.

Our next step would probably be to decompose the system (our box in Figure 2-3) into its *pertinent* (to our sequencing decision) components. An attempt to do this is reflected in Figure 2-4, wherein each component is a machine or operation that is used in the processing of at least one of the jobs. Notice also that the flow between operations is shown.

Although an actual system would generally be much more involved, this example does indicate the typical approach to be taken.

Definitions

Prior to describing our approach to the development of the baseline linear model, let us establish some working definitions of the terms commonly employed in model construction.

VARIABLES A variable, usually denoted as x_j ($j = 1, 2, \ldots, n$), is a factor subject to change within problem. That is, its value may change, or at least change within certain limits.

DECISION (or CONTROL) VARIABLE A variable that is both under the control of the decision maker and could have an impact on the solution to the problem of interest is termed a decision or control variable.

CONTINUOUS VARIABLES Variables that may take on *any* values between an upper and lower limit are said to be continuous.

DISCRETE VARIABLES Variables that may take on only certain prescribed values are said to be discrete. For example, if x_1 can take on only the values 0, 1, $\frac{5}{2}$, and 10.32, then x_1 is discrete. A special subclass of discrete variables are those that can take on only integer values (such as $-5, 0, 2, 3, 8$), and these are often called *integer* variables.

---

[b]Reference [2] provides a more rigorous definition of a "flow shop." In simple terms, it is a shop in which a limited variety of jobs are undertaken and, as a consequence, their flow between operations does not vary from job to job.

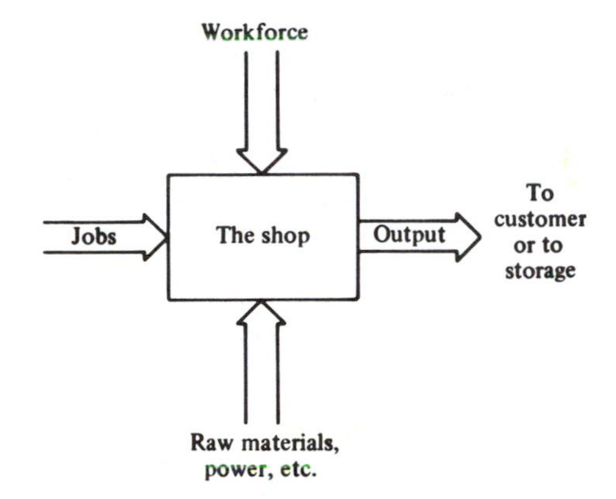

Figure 2-3. System definition—system boundaries.

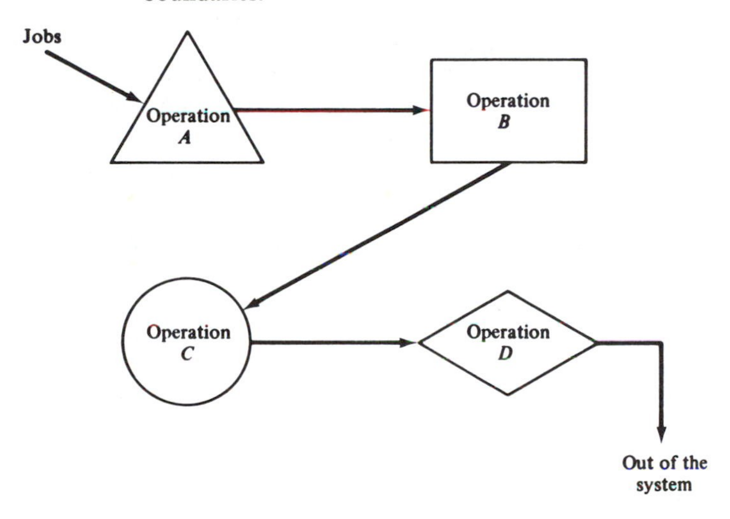

Figure 2-4. System definition—systems components and flow.

LINEAR FUNCTIONS A linear function contains terms each of which is composed of only a single, continuous variable raised to (and only to) the power of 1. No functions such as cos x, log x, or exp x may be involved.

NONLINEAR FUNCTION A nonlinear function is basically the complement of a linear function. That is, more than a single variable may appear in a single term and the variables may be raised to any power. *Strictly* speaking, even if a function satisfies the conditions listed in the previous definition, it is not considered linear if any of the variables involved are discrete. However, it is common practice to term such a function a *linear discrete* (*or integer*) *function.*

MATHEMATICAL MODEL A mathematical model consists of a set of related mathematical functions whose purpose is to simulate the response of the system being modeled. A *linear mathematical model* consists of solely linear functions; a *nonlinear mathematical model* involves one or more nonlinear functions.

EQUATION A mathematical equation, represented in general as[c] $f(\mathbf{x}) = b$ (i.e., some function of the variables $\mathbf{x} = x_1, x_2, \ldots, x_n$ is *equal* to a constant right-hand side value, b) expresses the equivalence between the function on the left and the function on the right (which, in this text, is usually a constant).

INEQUALITIES Consider the previous definition. If the function on the left can *equal or exceed* the function or constant on the right, this is called an inequality and, more specially, a type II inequality. On the other hand, if the function on the left can be *less than or equal* to the function or constant on the right, this is a type I inequality. Mathematically, they are represented as:

$$f(\mathbf{x}) \leq b \qquad \text{(type I)}$$
$$f(\mathbf{x}) \geq b \qquad \text{(type II)}$$

OBJECTIVES Objectives are represented by mathematical functions of the decision variables. Such functions usually represent the desires of the decision maker—such as to maximize profit or minimize cost. Objective functions may also be linear or nonlinear in form, although herein we concentrate primarily on those of a linear form. It is important to note that the right-hand side of an objective function (i.e., its value) is left unspecified. That is, the two most typical forms of objective functions are:

$$\text{maximize } f(\mathbf{x}) \qquad \text{or} \qquad \text{minimize } f(\mathbf{x})$$

GOALS A goal is a mathematical function of the decision variables which represents the *combination of an objective with a target* (i.e., right-hand side) *value*. For example, if we say we wish to maximize profit, that is an objective. However, if we wish to achieve a profit level of at least, say, \$1,000, we have established a goal. Notice in particular that, given a target value, any objective may be converted into a goal. The mathematical form of a goal is either

$$f(\mathbf{x}) \leq b$$

or

$$f(\mathbf{x}) \geq b$$

or

$$f(\mathbf{x}) = b$$

depending upon the situation.

CONSTRAINTS A constraint has the same mathematical appearance as a goal (i.e., it is either a type I or type II inequality or an equality). However, the difference between a goal and constraint is that a goal implies some flexibility, whereas a constraint, at least in the mathematical sense, is absolute or inflexible. Consequently, if a goal *must* be satisfied, we term it as a *rigid constraint* (or, alternatively, as an absolute goal). We discuss this in more detail below.

---

[c]Boldface roman type is used to denote a vector. Thus, $\mathbf{x} = (x_1, x_2, \ldots, x_n)$, $\mathbf{0} = (0, 0, \ldots, 0)$, and so forth.

Distinctions Among Objectives, Goals, and Constraints

It is vital that the reader establish in his or her mind the differences among objectives, goals, and constraints because in this text such distinctions play a key role in model development and appreciation. An objective differs from a goal in that, for a goal, we ascribe a minimum acceptable or target value for its level of performance, whereas in the case of an objective, we simply state that we wish to "optimize" its measure of performance (e.g., maximize, minimize, maximize the minimum level, etc.). As a result, the right-hand side of an objective is left unspecified.

The difference between a goal and a constraint can be more subtle. A goal looks, in terms of its mathematical formulation, exactly like a constraint. However, the concept of a "goal" function implies more flexibility and less rigidity than that of a "constraint" function. For a goal, the right-hand side value is simply a target level to which we aspire. However, under the strict, mathematical notion of a constraint, we must always achieve the right-hand side value; otherwise, the constraint is considered "violated," with the further implication of mathematical infeasibility.

Conventional single-objective mathematical programming draws no such distinction, as it ignores the concept of a goal. Rather, we optimize one objective subject to the absolute satisfaction of a set of rigid constraints. Multiple-objective models, on the other hand, relax this rigidity and are based on the belief that, in the real world, not all "constraints" are inflexible or as binding as is implied by their strictly mathematical interpretation.

Steps in Baseline Model Formulation

As we mentioned earlier, the establishment of the baseline mathematical model *should be*, in our opinion, a key part of the very first phase of model formulation. In this phase we attempt to isolate and mathematically define the decision or control variables, the objectives, the goals, and the rigid constraints (or absolute goals) that best represent the problem at hand. In this section we present an approach to the formulation of the baseline model. Following this, a number of simplified but illustrative examples are provided.

The steps in baseline model formulation are:

1. Determine the decision (or control) variables.
2. Formulate all pertinent objectives and/or goals.
3. From the list of *goals*, as identified in step 2, isolate those that are absolute (i.e., the rigid constraints).

These three basic steps lead to what we term the baseline mathematical model. In Part Two we describe how to place this model into the traditional (single-objective) linear programming model format, and in Part Four we show how to establish the multiple-objective or goal programming model.

Determination of the Decision Variables

The decision variables within a problem are those over which we actually have *control*. Consequently, they are often also referred to as control variables. Any set of decision variables is denoted as **x**, or the solution vector, or the program. The optimal set or program is termed **x***. The main thrust (at least initially) in linear programming is to determine the values of **x***.

Consider, for example, the problem involved in insulating your house so as to reduce utility costs. Numerous variables exist in such a problem, including:

1. The amount of attic insulation installed.
2. The amount of side-wall insulation to be installed.
3. The amount of caulking to be done.
4. The number (and perhaps type) of storm windows.
5. The number (and perhaps type) of insulating drapes or curtains used.
6. The amount of insulation around your hot-water tank.
7. The temperatures experienced.
8. The wind velocity and direction.
9. The amount of sunshine incident on the house.
10. The number of individuals within the house.
11. The number of times per day that a door or garage door is opened.
12. The cost of utilities furnished.

Now, of these variables, only the first six are directly controllable. Thus, although variables 7 through 12 will certainly combine to determine the resultant cost of your utilities, they are not within your control. Consequently, *only* the first six variables should appear, in your model, as *decision* variables.

It is important that you both identify and *define* your decision variables. The definition of the decision variables should appear directly prior to the actual mathematical formulation. For example, in our home insulation example we would define our variables as follows:

x_1 = amount, feet, of 6-inch attic insulation employed

x_2 = amount, pounds, of side-wall insulation installed

x_3 = amount, number of tubes, of caulking employed

x_4 = total number of storm windows (this may have to be broken down further into types and sizes)

x_5 = total number of yards of drapery material employed (again, this may very well have to be broken down further)

x_6 = amount, feet, of hot-water-tank insulation used

FORMULATION OF THE OBJECTIVES AND GOALS

Keeping in mind the distinction drawn between an objective and a goal, we next proceed to list all pertinent objectives and goals within the problem. Such objectives and goals are generally the results of one of the following:

1. The desires (or aspirations) of the decision maker.
2. Limited resources.
3. Any other restrictions either explicitly or implicitly placed on the choice of decision variables.

The recommended approach in the formulation of the objectives and goals is to first simply list those objectives and goals associated with each class. Those typically associated with the first class include:

—Maximization of profit
—Minimization of costs
—Minimization of overtime
—Maximization of resource utilization (personnel, machinery, or processes)
—Minimization of labor turnover
—Minimization of machine downtime
—Minimization of risk (to the firm, to the environment, etc.)
—Maximization of the probability that a given process remains within certain control limits
—Minimization of the deviation from a standard

With the implication of "maximization" or "minimization," such functions are typically objectives (i.e., an unspecified right-hand side) rather than goals.

The second class is generally associated with goals wherein such goals reflect our desire not to violate or, at least, minimize the violation or non-achievement of a goal as defined by various resource limitations, such as:

—Limited raw material
—Limited budget
—Limited time
—Limited personnel
—Limited ability or skills

The third, and final, class of functions are typically goals associated with an attempt to satisfy various "legal" or physical restrictions, such as:

1. A physical requirement that a variable or variables be nonnegative.
2. A contractual requirement which specifies that a function must equal or exceed (or perhaps equal or be less than) a certain minimum (maximum) value.
3. A physical limit on the upper and/or lower value that a variable or variables may take (e.g., the temperature in a system component might be restricted to such limitations so as to avoid component failure).

Once these functions have been listed, one should next examine the list for such things as obvious redundancy (i.e., are some objectives or goals dominated by others?), for possible combinations (e.g., can some objectives or goals be *meaningfully* combined into a single valid objective or goal?), or for elimination due to relative unimportance. Our overall guideline here should always be to try to reduce the final set of functions to the minimum number necessary to adequately represent the problem.

The result of this phase of the baseline modeling process will be a set of objectives (to be maximized or minimized) and a set of goals (to be satisfied). However, as discussed earlier, some goals are less flexible than others, so we now move to the next phase of the process.

Isolation of Rigid Constraints (Absolute Goals)

Our final step before the mathematical formulation is to determine, from the set of goals, those goals that *must* be satisfied (i.e., the set of rigid constraints). One must take care to identify *only* those goals that *must absolutely* be satisfied (i.e., achieved, or not violated according to their form). All too often the individual (particularly one exposed only to conventional, single-objective modeling) will designate a goal as absolute or rigid when, in fact, it would be possible to live with some degree of nonachievement. Consequently, *a goal should be designated as absolute, or as a rigid constraint, if and only if its nonachievement would render the resulting solution unimplementable in actual practice*. As a general rule of thumb, it is suggested that, if there is any doubt in the analyst's mind as to the rigidity of a goal, it not be designated as a rigid constraint.

The General Form of the Baseline Model

We are now prepared to state, mathematically, the general form of the baseline model.

Find $\mathbf{x} = (x_1, x_2, \ldots, x_n)$ so as either to maximize or minimize the model objectives subject to the satisfaction of the model goals (and rigid constraints) as indicated:

Objectives:

$$\text{maximize: } a_{r,1}x_1 + a_{r,2}x_2 + \ldots + a_{r,n}x_n \quad \text{for all } r \tag{2.2}$$

$$\text{minimize: } a_{s,1}x_1 + a_{s,2}x_2 + \ldots + a_{s,n}x_n \quad \text{for all } s \tag{2.3}$$

Goals and Rigid Constraints:

$$\text{satisfy: } a_{t,1}x_1 + a_{t,2}x_2 + \ldots + a_{t,n}x_n \ (*)\ b \quad \text{for all } t \tag{2.4}$$

$$\text{where: } x_1 x_2, \ldots, x_n \geq 0 \tag{2.5}$$

and in (2.4) the expression (*) is either a type I inequality ($\leq$), or a type II inequality ($\geq$), or an equality ($=$) for each t.

It is stressed that an actual baseline model may then consist of:

1. 0, 1, 2, or more maximizing objectives
2. 0, 1, 2, or more minimizing objectives
3. 0, 1, 2, or more goals and rigid constraints

and, in some instances, the nonnegativity restrictions (2.5) on the decision variables may not be appropriate. That is, there may be cases in which the decision variables may take on negative values.

Finally, realize that the purpose of the baseline model is to reflect, as closely as possible, the problem as perceived by the decision maker(s). The implication of this factor will become more apparent in the development of the models for the examples that follow.

Examples of Baseline Model Formulation

In lieu of real-world experience, the best way to learn how to establish mathematical formulations is through illustrative examples. In this section we present a number of such illustrations wherein it is assumed that all prior steps of model construction and refinement (i.e., determining model form, level, type, system boundaries, and components) have been performed, and we have determined that a baseline mathematical model is to be developed.

Example 2-2: A Baseline Linear Model

An investor has decided to invest a total of $50,000 among three investment opportunities: savings certificates, municipal bonds, and stocks. The annual return on each investment is estimated to be 7%, 9%, and 14%, respectively. The investor does not intend to reinvest his annual interest returns (i.e., he plans to use the interest to finance his desire to travel). His desires, then, and not necessarily in order of priority, include:

1. Obtain a yearly return of at least $5,000.
2. Invest a minimum of $10,000 in bonds.
3. The investment in stocks should not exceed more than the combined total investment in bonds and savings certificates.
4. Invest between $5,000 and $15,000 in savings certificates.

Determination of the decision variables. The problem is obviously to determine the proper allocation of our resources ($50,000) among three investment opportunities. The variables directly under our control are:

$$x_1 = \text{dollars invested in savings certificates}$$
$$x_2 = \text{dollars invested in municipal bonds}$$
$$x_3 = \text{dollars invested in stocks}$$

Formulation of the objectives and goals. Before we rush headlong into formulating the objectives and goals, let us first pause a moment to reflect on two aspects of this problem. First, note that our decision maker did not mention anything in regard to a desire to either maximize or minimize a performance measure. Rather, he placed aspiration values on each of his desires and thus provided us with a set of goals.

Second, note that our decision maker has neglected to mention a fundamental restriction on investment alternatives. That is, the total amount invested is limited by the total funds available ($50,000). Further, obvious physical restrictions are that each variable or amount invested be nonnegative. The omission of such factors is typical, and thus the analyst must always review the preliminary listing with care. Consequently, we may rewrite our list of objectives and goals (where, again, it is noted that in this problem we simply have a list of goals) as follows:

1. Obtain a yearly return of at least $5,000.
2. Invest a minimum of $10,000 in bonds.
3. The investment in stocks should not exceed more than the combined total investment in bonds and savings certificates.
4. Invest between $5,000 and $15,000 in savings certificates.
5. The total amount invested in all three opportunities must not exceed $50,000.
6. All investment amounts must be nonnegative.

We are now ready to proceed to the mathematical formulation of each function.

The first stated goal was to achieve a yearly return of at least $5,000. This return is determined by the amount in each investment times the yearly interest:

$$\text{Yearly return:}\quad 0.07x_1 + 0.09x_2 + 0.14x_3 \geq 5{,}000$$

Next, the goal to invest a minimum of $10,000 in bonds is written as:

$$\text{Investment in bonds:}\quad x_2 \geq 10{,}000$$

The investment in stocks (x_3) should not exceed more than the combined total investment in bonds and savings certificates ($x_1 + x_2$):

$$\text{Stock restriction:}\quad x_3 \leq x_1 + x_2$$

$$\text{or}\quad x_3 - x_1 - x_2 \leq 0$$

Finally, between $5,000 and $15,000 is to placed in the savings certificates:

$$\text{Savings certificates:}\quad 5{,}000 \leq x_1 \leq 15{,}000$$

This last function is best formulated as two functions, as follows:

$$x_1 \geq 5{,}000$$

$$x_1 \leq 15{,}000$$

The function restricting the total investment to $50,000 is given simply as

$$x_1 + x_2 + x_3 \leq 50{,}000$$

and the nonnegative restrictions on each variable are written as

$$x_1 \geq 0$$
$$x_2 \geq 0$$
$$x_3 \geq 0$$

As a result, the baseline mathematical model for this problem may be summarized as shown below:

Find $\mathbf{x}$ so as to satisfy

$$
\begin{aligned}
0.07x_1 + 0.09x_2 + 0.14x_3 &\geq 5{,}000 && (2.6)\\
x_2 &\geq 10{,}000 && (2.7)\\
-x_1 - x_2 + x_3 &\leq 0 && (2.8)\\
x_1 &\geq 5{,}000 && (2.9)\\
x_1 &\leq 15{,}000 && (2.10)\\
x_1 + x_2 + x_3 &\leq 50{,}000 && (2.11)\\
x_1 &\geq 0 && (2.12)\\
x_2 &\geq 0 && (2.13)\\
x_3 &\geq 0 && (2.14)
\end{aligned}
$$

Usually, the nonnegative goals (2.12) through (2.14) (which are very common) are written simply as

$$x_1, x_2, x_3 \geq 0$$

Again, note that this baseline model contains only goals (i.e., there are no objectives to be maximized or minimized). Such a model often comes as a shock to those schooled in the traditional, single-objective mode, and they may find it difficult to resist the impulse to extract one of the foregoing goals (undoubtedly that of interest return) and convert it into a (maximization) objective. This impulse will be satisfied, as well as analyzed, in Part Two. However, at this point we will form the baseline model exactly as shown above—so as to reflect as closely as possible the problem as perceived by our decision maker.

Isolation of the rigid constraints. Our last step in the baseline model phase is simply to identify those goals that are absolute. This step should usually be accomplished with the assistance of the decision maker. For example, consider the goal of (2.6), the desire to achieve at least $5,000 per year in interest on our investments. In reviewing this goal we ask whether or not he could manage with less. Must the interest really equal or exceed $5,000? If not, then this goal is not absolute or a rigid constraint. Going through each of the goals as formulated in (2.6) through (2.14), we might then possibly find that goals (2.7), (2.9), and (2.11) through (2.14) are truly absolute or rigid.

Example 2-3: A Baseline Nonlinear Model

Although this book concentrates on the linear model, the formulation process for a nonlinear model is the same. In fact, often one does not discover that the model is either linear or nonlinear until the formulation process. Consider the following problem.

Postal regulations are placed on the size of parcels. For example, the length of a parcel might be restricted to $3\frac{1}{2}$ feet while its girth plus length might not be permitted to exceed 6 feet. Let us assume that a rectangular parcel is to be sent by mail wherein we wish to maximize the volume of the parcel while still satisfying postal regulations.

Determination of the decision variables. Since we have no control over postal regulations, our only control variables are the dimensions of the parcel. Thus, we let:

$$x_1 = \text{length, feet, of the parcel}$$
$$x_2 = \text{width, feet, of the parcel}$$
$$x_3 = \text{height, feet, of the parcel}$$

Formulation of the objectives and goals. The list of objectives and goals is rather short.

1. Maximize the volume of the parcel.
2. Satisfy postal regulations:
   a. Length less than $3\frac{1}{2}$ feet.
   b. Length plus girth less than 6 feet.
3. All variables must be nonnegative.

The formulation of these functions is then given as shown below.

The volume function is an objective since it has no right-hand side value. It is written simply as

$$\text{maximize } x_1 x_2 x_3$$

The restriction on length is

$$x_1 \leq 3\tfrac{1}{2}$$

and the restriction on girth is

$$x_1 + 2x_2 + 2x_3 \leq 6$$

Finally, the nonnegativity restrictions are

$$x_1, x_2, x_3 \geq 0$$

Combining these, we obtain the following nonlinear baseline model:

$$\text{maximize } x_1 x_2 x_3 \tag{2.15}$$
$$\text{satisfy: } x_1 \leq 3\tfrac{1}{2} \tag{2.16}$$
$$x_1 + 2x_2 + 2x_3 \leq 6 \tag{2.17}$$
$$x_1, x_2, x_3 \geq 0 \tag{2.18}$$

wherein we have one objective, (2.15), and three goals, (2.16), (2.17), and (2.18).

Isolation of absolute goals. The isolation of absolute goals is not difficult for this problem. The parcel cannot be sent unless it satisfies postal regulations and, further, the nonnegativity restrictions must hold. Consequently, all goals are absolute.

Example 2-4: A Baseline Model with Integer Variables

As stated earlier, regardless of whether or not the model is linear or nonlinear, the formulation process (as seen in Example 2-3) is the same. This also holds true regardless of whether the variables involved are all continuous, all discrete, or some combination (mixed). Our final example illustrates a model in which the variables are all to be integer-valued. Since the form of this model appears linear (ignoring the restrictions on the variables), it is commonly designated as a linear integer programming model.

Let us suppose that we must select scientific experiments for a space mission. A list of candidate experiments is shown in Table 2-2 together with estimates of their respective weights, volumes, power requirements, and "values." One value has been assigned to the scientific importance of the experiment (the higher the value, the more important), while another value has been given to express the political importance of the experiment (again, the higher this value, the more important it is from a political point of view). Limits on total weight, total volume, and total power consumption are 100 pounds, 200 cubic feet, and 1,000 watts, respectively.

Table 2-2. EXPERIMENT LIST AND SPECIFICATIONS

Experiment	*Weight*	*Volume*	*Power*	*Scientific Value*	*Political Value*
1	20	15	100	5	8
2	25	40	200	3	7
3	40	60	150	8	5
4	30	60	300	2	9
5	12	70	500	9	4

Determination of the decision variables. Since our control over politics, gravity, spacecraft design, and so forth is nil, our only decision variables are used to reflect the selection or nonselection of an experiment. Thus, we let:

$$x_1 = \begin{cases} 1 & \text{if experiment 1 is selected} \\ 0 & \text{otherwise} \end{cases}$$

$$x_2 = \begin{cases} 1 & \text{if experiment 2 is selected} \\ 0 & \text{otherwise} \end{cases}$$

$$x_3 = \begin{cases} 1 & \text{if experiment 3 is selected} \\ 0 & \text{otherwise} \end{cases}$$

$$x_4 = \begin{cases} 1 & \text{if experiment 4 is selected} \\ 0 & \text{otherwise} \end{cases}$$

$$x_5 = \begin{cases} 1 & \text{if experiment 5 is selected} \\ 0 & \text{otherwise} \end{cases}$$

Examining our definition of the decision variables, we see that they are not only restricted to integer values, they are even more specifically restricted to the values of *zero or one*. Such a problem is commonly known as a zero–one programming problem.

Formulation of the objectives and goals. Our performance list is assumed to be:

1. Maximize the total combined scientific value of the experiments selected.
2. Maximize the total combined political value of the experiments selected.
3. Do not exceed the limitations on total weight, volume, or power.
4. All variables must be nonnegative (which is actually a redundant objective in light of our variable definitions given above).

Mathematically, these objectives and goals may be written as shown below.

To maximize scientific value, we "simply" use the data from the scientific value column of Table 2-2 and write

$$\text{maximize } 5x_1 + 3x_2 + 8x_3 + 2x_4 + 9x_5$$

Similarly, for political value, we write

$$\text{maximize } 8x_1 + 7x_2 + 5x_3 + 9x_4 + 4x_5$$

The goals concerning weight, volume and power limits may also be written out using Table 2-2 and are:

$$\begin{aligned}
&\text{Weight:} && 20x_1 + 25x_2 + 40x_3 + 30x_4 + 12x_5 \leq 100 \\
&\text{Volume:} && 15x_1 + 40x_2 + 60x_3 + 60x_4 + 70x_5 \leq 200 \\
&\text{Power:} && 100x_1 + 200x_2 + 150x_3 + 300x_4 + 500x_5 \leq 1{,}000
\end{aligned}$$

The nonnegativity restrictions are

$$x_1, x_2, x_3, x_4, x_5 \geq 0$$

However, in addition, we must also specifically restrict our variables to zero–one values:

$$x_1, x_2, x_3, x_4, x_5 = 0, 1$$

Since these last goals dominate our nonnegativity restrictions, they may be dropped and our combined formulation is:

$$\begin{aligned}
&\text{maximize} && 5x_1 + 3x_2 + 8x_3 + 2x_4 + 9x_5 && (2.19) \\
&\text{maximize} && 8x_1 + 7x_2 + 5x_3 + 9x_4 + 4x_5 && (2.20) \\
&\text{satisfy:} && 20x_1 + 25x_2 + 40x_3 + 30x_4 + 12x_5 \leq 100 && (2.21) \\
& && 15x_1 + 40x_2 + 60x_3 + 60x_4 + 70x_5 \leq 200 && (2.22) \\
& && 100x_1 + 200x_2 + 150x_3 + 300x_4 + 500x_5 \leq 1{,}000 && (2.23) \\
& && x_1, x_2, x_3, x_4, x_5 = 0, 1 && (2.24)
\end{aligned}$$

Except for (2.24), the problem would appear to be a simple linear model. However, since the variables are zero–one, it is a linear zero–one model. Note that we have two objectives, (2.19) and (2.20), and all other functions are goals.

Isolation of absolute goals. We shall assume that the rigid constraints are associated with the weight, power, volume, and zero–one restrictions. Thus, goals (2.21) through (2.24) are absolute.

Model Validity

One of the questions that will, or at least should, enter the reader's mind is: How "good" is the model that has been developed? Does it really represent the actual system and respond to alternative policies as would the actual system? As we have already warned, the results obtained through the solution of a model are only as good as the model itself. Thus far we have only discussed the approach that should be taken so as to *develop* a model. It is believed that, if these steps are followed, the resulting model will be better than one developed without consideration given to such a systematic approach. However, even then, we may wonder about the validity of the model. Consequently, in this section we present some ideas on how one might satisfy himself or herself as to the validity of the model.

First, the reader must face the fact that the mathematical model of a real problem is never perfect (unless, perhaps, the problem is of a very trivial nature). Second, an accurate, absolute *measure* of validity simply does not exist. That is, we cannot speak with certainty of one model as having a validity of, say, 75 units while another has a validity of, say, 90 units. Validity is a multidimensional and highly subjective concept that just does not lend itself to such simple approaches. Consequently, when one speaks of model validity it should be recognized that this validity is an imperfect measure based very much on faith, a concept somewhat unnerving to those with an analytical bent and used to rigorous proofs and absolute definitions.

This difficulty in the measurement of model validity usually leads one to establish a *relative ranking* of models according to their perceived validity. That is, although we may be unable to assign an accurate measure of validity to a given model, we are generally comfortable with a comparison of the perceived validity between two models. If we believe that there is a distinguishable difference in validity between models, we may label one model as having a higher degree of validity than the other. The next question is: On what basis may the validity of models be compared? We suggest a validation procedure that consists of the following four phases:

1. An evaluation of model structure.
2. An evaluation of model logic.

3. An evaluation of the design and/or input data.
4. An evaluation of model response.

An Evaluation of Model Structure

As previously discussed, there are two basic approaches commonly used in model development. The first is to start with a model that contains all aspects and variables of the system. We then attempt to simplify this model (which is generally far too cumbersome for analysis) step by step to the point at which any further simplification would so distort the model as to be unacceptable. The second method, and by far the more preferable, is to begin with a preliminary, simplified model of the system. Such a model represents only the most basic factors and operations of the system. We next add detail to this model until we are satisfied that the responses of interest are accurately represented (obviously, a subjective phase in which faith in the model becomes of central importance). The first approach is both wasteful and time consuming, whereas the second approach is more systematic and logical and thus much more likely to lead to a model that is structurally correct. That is, the elements of the system and their place within the system are accurately represented.

An Evaluation of Model Logic

Our first evaluation examined the representation of those elements within the system that are believed to have a (significant) impact on the system's responses. We now determine the accuracy of the representation of the interrelationships and interactions between those elements. If the model logic truly reflects the system logic, the model will react to a stimulus (e.g., a change in policy) in the same manner as would the actual system. A common procedure for the evaluation of model logic is to stimulate the model with a representative range of inputs and observe the resultant model output.

Generally, it is not very important that the output response values of the model be of the same value as the actual system. Rather, it is the *relative differences* in the outputs that are of importance. For example, for a given policy, do profits rise or fall?

An Evaluation of the Design and/or Input Data

All too often, a great deal of work is performed in the development of a model, only to be negated by the use of poor or incorrect data. The data used in model development may be roughly classified into two types: (1) the design data or that information used to actually construct the model, and (2) the input data or data used to stimulate the system. Inattention to either may seriously degrade the validity of a model.

Data collection and verification may well be the most overlooked portion of model construction. Unfortunately, virtually all textbooks contribute to this problem, at least indirectly, by presenting the reader with a misleading impression of the data collection process. Textbook presentations are, and must be, simplified. One common simplification is that the data needed for model development and/or solution are usually presented directly to the reader in a list (as we did in Table 2-2), and thus the reader seldom, if ever, faces the problems involved in the actual collection of the data. Unfortunately, not only is the validity of the model dependent on these data, but, also, the process of data collection often consumes the major amount of time and resources when dealing with actual problems.

Consider, for a moment, the data shown in Table 2-2 for Example 2-3. It might at first seem that the data concerning experiment weights, volumes, and power requirements should be relatively easy to collect. This is not necessarily true. In fact, when the author built such a model (in the mid-1960s), several months were required to compile *valid* data on weight, volume, and power for some 150 experiments. Problems that slowed this data collection included:

1. Accurate measurements of the data on prototype experiments (earth-based) had not been made or even estimated by most of the scientists involved.
2. The prototype experiments did not necessarily reflect the experiment configuration to be used in an actual spaceflight.

Now consider the last two columns of Table 2-2. The "values" of each experiment in terms of scientific and political "importance" are simply presented for the model builder's use. The question that the reader should ask himself or herself, however, is: How are such values actually measured?

Let us now return to the question of how and when one obtains data. All too often data are collected *before* one has decided on the basic form of the model to be used. As a result, many of the data are useless in that they do not fit the requirements of the model finally developed. The amount of wasted effort may be considerable. What one should do is to *first* decide on the basic form of the model, construct a preliminary model, and *then* identify its specific data needs. One may then collect only those data that are actually needed to support the model.

The sources of data vary extensively based upon the particular situation. These sources may include historical records (if the system has been in existence), theoretical data (which are generally projected data for systems not yet in existence), and ongoing records (for systems in existence). If historical records are available, one must compare the system in its present (or proposed state) versus the state of the system at the time over which the historical records were collected. If no significant differences exist, confidence in the validity of the data is increased.

Although the use of ongoing records is usually a particularly attractive source of data, one often does not have the time to wait for the collection of these data.

An Evaluation of Model Response

The true validation of a model is often said to be reflected solely in its ability to predict the behavior of the system that has been modeled. Such a premise can be carried to the absurd. For example, some observers have noted a correlation between the state of the national economy and the hemlines on women's dresses. High hemlines have occurred simultaneously with a good state of the economy whereas low hemlines have occurred in times of depression or recession. The fact that such a "model" has happened to be an accurate one does not mean that it is valid. At any rate, the introduction of jeans and pants suits has made this a moot point.

Even if one accepts the premise of future verification, it is not always possible to wait for such results. Validation of model response must then be accomplished through the input of estimated, historical, or ongoing data. Again, however, the fact that the model performs "reasonably" with such inputs is not, by any means, an absolute guarantee of its validity or reaction to future data.

Although consideration of the four areas described above is by no means perfect nor completely objective, it does provide a practical means to consider validation on a relative basis and to compare the validity of models. If one can establish some degree of confidence in the structure, logic, data, and model response, we can establish some faith in the model.

Summary

In this chapter we have attempted to:

—Present an introduction to the modeling process
—Provide modeling guidelines
—Provide certain working definitions
—List and illustrate the steps of the procedure involved in the development of our preliminary model—the baseline model

Again, it is vital to realize that most traditional linear programming texts generally ignore this phase (i.e., the development of the baseline model) of the modeling process. In doing so, we believe that they tend to distort the impression, provided to the reader, of the modeling process that will actually be required to handle real problems. The reader with some previous exposure to traditional linear programming should, in particular, reflect on the difference between approaches.

REFERENCES

1. Ignizio, J. P. *Goal Programming and Extensions*. Lexington, Mass.: Heath (Lexington Books), 1976.

2. Ignizio, J. P., and Gupta, J. N. D. *Operations Research in Decision Making*. New York: Crane, Russak, 1975.

3. Rivett, P. *Principles of Model Building*. London: Wiley, 1972.

PROBLEMS

For Problems 2.1 through 2.5, list all the associated variables that you believe may be pertinent and then distinguish between those variables that may be controlled and those that may not.

2.1. The variables for a model that predicts the weight of livestock over a certain period of time.

2.2. The variables for a model that predicts the yield, in bushels, from a wheat field.

2.3. The variables for a model of voter response for a given presidential candidate.

2.4. The variables for an investment portfolio model.

2.5. The variables for a model for predicting the outcome of a football game.

2.6. Can you describe a systematic procedure that might be implemented so as to efficiently identify the variables within any given problem?

2.7. A new type of "hot dog" is now permitted by the government. The ingredients of this wiener include water, animal fat, soybean meal, animal lips, and seaweed. The prices of each of the ingredients are given in the accompanying table together with governmental regulations as to the maximum and minimum amounts permitted on a per pound basis. Develop a baseline model that reflects the desires to minimize costs while satisfying the regulations.

Ingredient	*Cost* ($/*lb*)	*Regulations*	
		Minimum (%)	*Maximum* (%)
Water	0.03	—	30
Animal fat	0.22	—	25
Soybean meal	0.38	10	30
Animal lips	0.50	20	—
Seaweed	0.35	5	20

2.8. In Problem 2.7, how might you mathematically model the objective to "make the hot dog tasty"?

2.9. In the coming term, you must select exactly 5 courses from 12 candidates. You wish to maximize, to your professional career, the "value" of these courses, but you still wish to achieve an expected grade-point average of 3.0 (A = 4.0) or better. Data on courses are given below. (The higher the "value," the better.) Develop the baseline model.

Course	*Expected Grade*	*Professional Value*
1	C	9
2	A	3
3	B	7
4	C	7
5	A	9
6	A	3
7	B	8
8	B	6
9	A	4
10	A	8
11	C	5
12	A	2

2.10. Reformulate Problem 2.9 if you decide that you also wish to maximize your grade-point average.

2.11. There are six new job openings at a small business. Six candidates have applied for the jobs (three men and three women), and each candidate has been tested on each job. The results of these tests are given in the accompanying table, where a cell entry designates the time required for a given candidate to perform a given job. It is desired to assign the six candidates to the jobs so as to:
(a) Achieve a *total* expected time of 30 hours or less.
(b) Assign at least one woman to either job 1 or 2.
Develop the baseline model for the problem and discuss.

	Job					
Candidate	1	2	3	4	5	6
Woman 1	8	7	5	1	3	10
Woman 2	4	7	4	7	3	8
Woman 3	6	4	3	2	3	3
Man 1	3	5	6	5	5	8
Man 2	5	3	6	8	3	6
Man 3	2	3	7	9	2	7

2.12. Remodel Problem 2.11 to select the six candidates in such a way as to achieve the minimum total time required to do the jobs.

2.13. Two recent graduates have decided to enter the new field of microcomputers. They intend to manufacture two types of microcomputers, Autoaccount and Scicomp. Because of the interest in microcomputers, they can (presently) sell all that they could possibly produce. However, they wish to size the production rate so as to satisfy various estimated limits with a small production crew. These include:

Assembly hours:	150 hours available per week
Test hours:	70 hours available per week

The Autoaccount requires 4 hours of assembly and 3 of testing, while Scicomp consumes 6 hours and 2 hours, respectively. Profit for the Autoaccount is estimated at $300 per unit; that of Scicomp is $450 per unit.

The partners wish to maximize profit while also maximizing the total *number* of microcomputers produced per week (to achieve market visibility). They do not wish to increase the assembly hours beyond 150, because of the lack of production equipment, but testing via some overtime would be permissible. Develop the baseline model.

2.14. Remodel Problem 2.13 if testing on overtime reduces profit on Autoaccount by $20 each and profit on Scicomp by $25 each. In addition, no more than 30 hours of overtime for testing is to be permitted.

2.15. A firm produces two microwave switches, switch *A* and switch *B*. The return per switch of switch *A* is $20, whereas that for *B* is $30. It has been determined that the repair cost, in dollars, for defective switches, is proportional to 10% of the product of the number of switches (i.e., of both types) made each day. At least 10 switches of each type must be produced each day. The firm wishes to obtain a daily profit of at least $5,000. Develop the baseline model.

2.16. An automotive firm produces three types of cars: a large luxury car, a medium-sized car, and a compact car. The gasoline mileage figures, predicted sales, and profit figures for each type of car is given in the accompanying table. Government regulations state that the average gasoline mileage for the company's entire line of cars should equal or exceed 30 miles per gallon (mpg). For every 1 mpg below 30, the company must pay a penalty of $200 per car produced. The firm wishes to maximize its profits and also minimize the mileage amount *under* 27 mpg (since not only must they pay a penalty but they will also receive bad publicity). Develop the baseline model for this problem.

Car	*Mileage (mpg)*	*Profit/Car*	*Market*
Large	18	$600	600,000
Medium	29	400	800,000
Compact	36	300	700,000

2.17. (a) Remodel Problem 2.16 if the penalty is $1,000,000 for every 1 mpg less than 30.

(b) Can you think of any other objectives the firm might actually have?

2.18. A refinery produces three grades (*A*, *B*, *C*) of gasolines from three different sources of crude oil (I, II, III). Any crude oil can be used to produce any of the gasolines as long as the following specifications are met:

Grade of Gasoline	*Specifications*	*Selling Price per Gallon*
A	Not less than 50% crude I, not more than 30% crude II	\$2.50
B	Not less than 35% crude I, not more than 45% crude II	2.20
C	Not more than 20% crude III	1.80

The maximum amount of crude oil available per period, and their costs, are:

Crude I: 10,000 gallons, \$2.60 cost/gallon
Crude II: 9,000 gallons, \$2.00 cost/gallon
Crude III: 3,000 gallons, \$1.20 cost/gallon

The oil refinery naturally wants to maximize profit. Formulate the baseline model.

2.19. The preliminary layout of communications network has been determined and is illustrated in the accompanying network diagram. The links of this network are depicted as branches, or arcs between each node, and the cost of transmitting a single message unit is indicated in parentheses beside each link. The nodes of the network (with the exception of nodes 1 and 5, which are simply transmission and reception terminals) actually represent repeater stations (whose purpose is to receive, amplify, error-check, and transmit any messages incident to the node), whose cost is a linear function of the (maximum) number of messages received, per second, at the station. Further, there are limits to the message units per second that may be transmitted via each link. We wish to determine the minimal-cost design of this network if the maximum (i.e., worst case) message load between terminals (nodes) 1 and 5 is to be 500,000 messages per second. The accompanying network diagram and tables provide a summary of the data to be used in constructing a model for this problem. Formulate the baseline model for this problem.

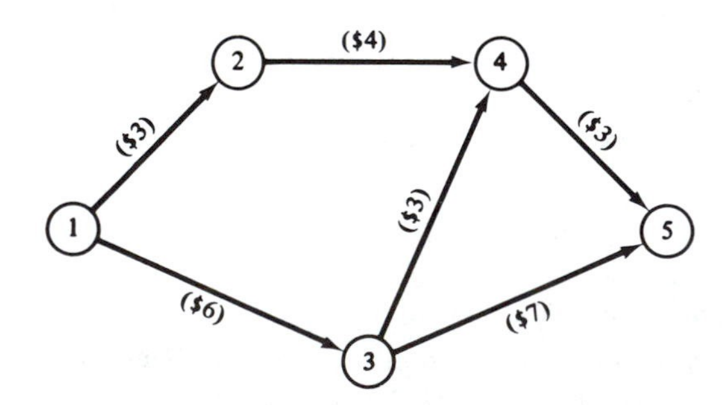

Figure 2P-19

Link	*Maximum Capacity (Messages/Second)*	*Transmission Cost per Message*
1–2	300,000	$3
1–3	400,000	6
2–4	400,000	4
3–4	100,000	3
3–5	300,000	7
4–5	300,000	3

Repeater Station, Node Number	*Cost per Messages/Second*
2	$2.5
3	2
4	3

2.20. Formulate Problem 2.19 in baseline-model format, if, in addition to minimizing total cost, you wish to:

(a) Balance the communication loading across all transmission links.
(b) Minimize the variation in repeater station sizing.
(c) Minimize the number of messages sent through link 3–4.

A REVIEW OF LINEAR ALGEBRA

The Need for a Review

In the chapters to follow, we introduce the simplex method for solving traditional linear programming problems and a modified simplex method for multiple-objective linear programming (or "goal programming") problems. Both techniques are based upon the concepts and tools of linear algebra. However, the specific concepts and tools used in linear programming are, for the most part, rather elementary and consequently our review of linear algebra need not be, and is not, an exhaustive survey of this area.

In this chapter we simply present a very brief look at those facets of linear algebra that are believed necessary for a full and better understanding of the material to be presented in succeeding chapters. Readers wishing further details or rigorous proofs are directed to the references [1–3], particularly to the texts by Nobel [2] and Hadley [1]. Readers satisfied that they have an adequate background in linear algebra may either skip or skim over this chapter.

The Model and Conditions

In Chapter 2 we introduced an approach to the construction of the unified baseline linear model. Recall that such a model may consist of various types of linear objectives, linear goals, and rigid linear constraints. Some goals or

3

constraints were equalities, some were inequalities, while objectives actually had no specific right-hand side. In this chapter we deal strictly with simultaneous linear *equations*. In later chapters we see how our linear models may be represented by such linear equations and thus the methods for dealing with simultaneous linear equations are also the methods used for dealing with linear programming problems. The study of simultaneous linear equations, the basic concepts underlying their analysis and solution, and their methods for representation, analysis, and solution all comprise what has come to be known as linear algebra.

The set of m simultaneous linear equations that will be of interest take the following form:

$$\left.\begin{aligned} a_{1,1}x_1 + \dots + a_{1,n}x_n &= b_1 \\ a_{2,1}x_1 + \dots + a_{2,n}x_n &= b_2 \\ \dots \quad a_{i,j}x_j \quad \dots &= b_i \\ a_{m,1}x_1 = \dots + a_{m,n}x_n &= b_m \end{aligned}\right\} \tag{3.1}$$

The coefficients $a_{i,j}$ are usually constants, as are the right-hand side values b_i. The $x_j, j = 1, \dots, n$, are the variables that must satisfy (if possible) the equations of (3.1)

The equations in (3.1) must also satisfy the following conditions:

1. Each term ($a_{i,j}x_j$) contains only a single variable.
2. Each variable (x_j) is raised only to the first power. (Further, no functions such as cos x, log x, or exp x are involved.)
3. The condition of additivity holds. That is, the total of the right-hand side (b_i) must be the sum of each term on the left-hand side (all the $a_{i,j}x_j$ terms).
4. The variables (x_j) must be divisible (i.e., continuous).

In the material that follows, we introduce a number of concepts that will allow us to represent more easily and to work with equations (3.1). These concepts include vectors, matrices, and determinants.

Vectors

A *vector* is a quantity having both magnitude and direction. A common physical example is force. We note that, to describe force completely, we must define both its magnitude (say, in pounds per square inch) *and* its direction (e.g., vertical, horizontal, etc.). A geometric representation of a vector is usually given by a line with an arrowhead at its end. The length represents magnitude and the arrowhead specifies the direction. It should be noted that a vector is thus most definitely not a number.

As noted in Chapter 2, vectors are printed in boldface type (e.g., **a**, **b**, **x**, **0**, **1**). As with numbers, we list a variety of operations that may be performed with vectors.

Generally, we will not deal much with the geometric concept of the vector (i.e., the line and arrowhead); rather, we use the following definition:

> **VECTOR** A vector is either a row or column of elements (numbers) wherein the position of each element is important. A vector has both magnitude and direction.

For example, the quantities of product 1, product 2, and product 3 might be 8, 12, and 5, respectively. Either a row or column vector may be used to summarize this information:

$$(8 \quad 12 \quad 5) \qquad \text{or} \qquad \begin{pmatrix} 8 \\ 12 \\ 5 \end{pmatrix}$$

Vector Addition

Vectors of the same type (i.e., either row or column) may be added together only if they have the same number of entries. Given two vectors, **a** and **b**, and given that both are either row or column vectors with the same number of entries, the addition of **a** and **b** is carried out by simply adding the elements of

a, position by position, to the elements of **b**. That is, letting $\mathbf{c} = \mathbf{a} + \mathbf{b}$, and with c_i being the element of **c** in its ith position (with a similar designation for a_i and b_i), we have

$$c_i = a_i + b_i \tag{3.2}$$

Example 3-1

A firm's sales figures are summarized by quarters. The sales over the past 4 years are listed in Table 3-1. We wish to determine the total quarterly sales figures in vector notation. We designate sales in year i by the column vector **s** wherein

$$\mathbf{s}_i = \text{sales vector for year } i$$

and

$$s_{i,j} = \text{sales in the } j\text{th quarter of year } i$$

Thus, in general,

$$\mathbf{s}_i = \begin{pmatrix} s_{i,1} \\ s_{i,2} \\ s_{i,3} \\ s_{i,4} \end{pmatrix}$$

and, specifically, using the data of Table 3-1, we have

$$\mathbf{s}_1 = \begin{pmatrix} 80 \\ 93 \\ 112 \\ 100 \end{pmatrix} \quad \mathbf{s}_2 = \begin{pmatrix} 72 \\ 77 \\ 68 \\ 85 \end{pmatrix} \quad \mathbf{s}_3 = \begin{pmatrix} 85 \\ 95 \\ 110 \\ 105 \end{pmatrix} \quad \mathbf{s}_4 = \begin{pmatrix} 88 \\ 99 \\ 115 \\ 110 \end{pmatrix}$$

Table 3-1. DATA FOR EXAMPLE 3-1

	Sales (Thousands of Dollars) in Quarter:			
Year	*1*	*2*	*3*	*4*
1	80	93	112	100
2	72	77	68	85
3	85	95	110	105
4	88	99	115	110

Consequently, the total of the sales, by quarter, for all 4 years is given by $\mathbf{t} = \mathbf{s}_1 + \mathbf{s}_2 + \mathbf{s}_3 + \mathbf{s}_4$, or

$$\mathbf{t} = \begin{pmatrix} 80 + 72 + 85 + 88 \\ 93 + 77 + 95 + 99 \\ 112 + 68 + 110 + 115 \\ 100 + 85 + 105 + 110 \end{pmatrix} = \begin{pmatrix} 325 \\ 364 \\ 405 \\ 400 \end{pmatrix}$$

Example 3-2

Given that

$$\mathbf{a} = \begin{pmatrix} 4 \\ 0 \\ 7 \end{pmatrix} \qquad \mathbf{b} = \begin{pmatrix} 5 \\ 9 \\ 1 \end{pmatrix} \qquad \mathbf{c} = (6 \quad 8 \quad 0) \qquad \mathbf{d} = \begin{pmatrix} 4 \\ 10 \\ 2 \\ 3 \end{pmatrix}$$

then

$$\mathbf{a} + \mathbf{b} = \begin{pmatrix} 9 \\ 9 \\ 8 \end{pmatrix}$$

$\mathbf{a} + \mathbf{c} =$ undefined (both vectors must be of same type)

$\mathbf{a} + \mathbf{d} =$ undefined (both vectors must have same number of elements)

Multiplication of a Vector by a Scalar

Numbers such as 3, 19, 37.5, and $\frac{2}{3}$ are scalars. Multiplication of a vector by a scalar is carried out by simply multiplying *each* element in the vector by the scalar. Thus, if α is the scalar and **a** is a row vector and **b** is a column vector, then

$$\alpha\mathbf{a} = (\alpha a_1, \ldots, \alpha a_n)$$

$$\alpha\mathbf{b} = \begin{pmatrix} \alpha b_1 \\ \cdot \\ \cdot \\ \cdot \\ \alpha b_m \end{pmatrix} \tag{3.3}$$

Example 3-3

Consider Example 3-1 again. If we wish to determine the average sales on a per quarter basis, we need to divide the total sales by $\frac{1}{4}$. That is, the average value for **t** is

$$\text{average } \mathbf{t} = \tfrac{1}{4}\begin{pmatrix} 325 \\ 364 \\ 405 \\ 400 \end{pmatrix} = \begin{pmatrix} 81.25 \\ 91 \\ 101.25 \\ 100 \end{pmatrix}$$

Vector Multiplication

Two vectors may be multiplied together only if they both have the same number of entries and one is a row vector while the other is a column vector. The result, often termed the *dot product*, is a *scalar*. By convention, the row vector is placed first and the column vector second. That is, $\mathbf{a} \cdot \mathbf{b}$, or $\mathbf{ab}$, infers that **a** is a row matrix and **b** a column matrix for the multiplication to be meaningful. To multiply the vectors together, corresponding entries are multiplied and the results added together. That is,

$$\mathbf{a} \cdot \mathbf{b} = \mathbf{ab} = \sum_{i=j=1}^{m} a_i b_j = \alpha \qquad \text{a scalar} \tag{3.4}$$

Example 3-4

Given that

$$\mathbf{a} = \begin{pmatrix} 3 \\ 0 \\ 7 \end{pmatrix} \qquad \mathbf{b} = (4 \quad 9 \quad 2) \qquad \mathbf{c} = \begin{pmatrix} -2 \\ 10 \\ 1 \end{pmatrix}$$

$$\mathbf{d} = (5 \quad 1 \quad 4 \quad 2) \qquad \mathbf{e} = (3 \quad -5)$$

find **ba**, **bc**, **da**, and **ea**.

$$\mathbf{ba} = (4 \quad 9 \quad 2)\begin{pmatrix} 3 \\ 0 \\ 7 \end{pmatrix} = 12 + 0 + 14 = 26$$

$$\mathbf{bc} = (4 \quad 9 \quad 2)\begin{pmatrix} -2 \\ 10 \\ 1 \end{pmatrix} = -8 + 90 + 2 = 84$$

$$\mathbf{da} = (5 \quad 1 \quad 4 \quad 2)\begin{pmatrix} 3 \\ 0 \\ 7 \end{pmatrix} = \text{undefined}$$

$$\mathbf{ea} = (3 \quad -5)\begin{pmatrix} 3 \\ 0 \\ 7 \end{pmatrix} = \text{undefined}$$

Special Vector Types

There are a number of vectors that occur so frequently that they have been given their own special names. These include:

UNIT VECTOR A unit vector has a 1 in the ith position and 0's elsewhere, such as

$$(0 \quad 1 \quad 0 \quad 0) \qquad \text{or} \qquad \begin{pmatrix} 1 \\ 0 \\ 0 \end{pmatrix}$$

NULL VECTOR The null vector, written **0** or ∅, is a vector having only 0's as its elements.

SUM VECTOR The sum vector, written **1**, has only 1's as its elements.

LINEAR DEPENDENCE AND INDEPENDENCE

The notion of linear dependence or independence plays an important part in the material that follows. We define these concepts next.

LINEAR DEPENDENCE A set of vectors $\mathbf{a}_1, \mathbf{a}_2, \ldots, \mathbf{a}_m$ is termed linearly dependent if there exist some scalars, α_i, which are not all zero such that

$$\alpha_1 \mathbf{a}_1 + \alpha_2 \mathbf{a}_2 + \cdots + \alpha_m \mathbf{a}_m = \mathbf{0} \tag{3.5}$$

LINEAR INDEPENDENCE If the only set of scalars, α_i, for which (3.5) holds is $\alpha_1, \alpha_2, \ldots, \alpha_m = 0$, the vectors are linearly independent.

Matrices

MATRIX A matrix is simply an ordered array of numbers; it has no numerical value. It may also be considered as an ordered array of either row or column vectors. As with vectors, the position of each element (number) within the matrix is important.

A vector may be considered as merely a matrix with only one row or column. Consequently, the vector operations previously discussed may be performed with matrices.

In this text we normally designate a matrix by a boldface capital letter. For example:

$$\mathbf{A} = \begin{pmatrix} 4 & 1 \\ 7 & 9 \end{pmatrix} \qquad \mathbf{B} = \begin{pmatrix} 8 & -4 \\ 3 & 0 \\ 7 & 5 \end{pmatrix} \qquad \mathbf{C} = \begin{pmatrix} c_{1,1} & c_{1,2} & c_{1,3} \\ c_{2,1} & c_{2,2} & c_{2,3} \end{pmatrix}$$

and if

$$\mathbf{c}_1 = (c_{1,1} \quad c_{1,2} \quad c_{1,3}) \qquad \text{and} \qquad \mathbf{c}_2 = (c_{2,1} \quad c_{2,2} \quad c_{2,3})$$

then

$$\mathbf{C} = \begin{pmatrix} \mathbf{c}_1 \\ \mathbf{c}_2 \end{pmatrix}$$

or if

$$\mathbf{b}_1 = \begin{pmatrix} \mathbf{b}_{1,1} \\ \mathbf{b}_{2,1} \\ \mathbf{b}_{3,1} \end{pmatrix} \qquad \text{and} \qquad \mathbf{b}_2 = \begin{pmatrix} \mathbf{b}_{1,2} \\ \mathbf{b}_{2,2} \\ \mathbf{b}_{3,2} \end{pmatrix}$$

then

$$\mathbf{B} = (\mathbf{b}_1 \quad \mathbf{b}_2)$$

The *order* of a matrix refers to the number of *rows and columns* the matrix contains. The rows are always given first. Consequently, a m by n matrix is one with exactly m rows and n columns. A *square* matrix has exactly the same number of rows as columns. Thus, $q \times q$, 3×3, 7×7, and so on, all represent orders of square matrices.

An element within a matrix is usually designated by a lowercase letter with two subscripts (often i and j). The first subscript refers to the row in which the element is found, whereas the second denotes the respective column. Thus, $a_{i,j}$ is the element in row i, column j of the matrix. Alternatively, we may specify an entire column or row in the matrix by a vector. That is, if $\mathbf{e}_1$, $\mathbf{e}_2$, and $\mathbf{e}_3$ are each column vectors with the same number of elements, we might represent $\mathbf{E}$ by

$$\mathbf{E} = (\mathbf{e}_1 \quad \mathbf{e}_2 \quad \mathbf{e}_3)$$

Two matrices, say $\mathbf{A}$ and $\mathbf{B}$, are equal (i.e., $\mathbf{A} = \mathbf{B}$) if and only if their corresponding elements are equal. This, of course, implies that they must also be of the same order. Consequently, if $a_{i,j}$ are the elements of $\mathbf{A}$ and $b_{i,j}$ are the elements of $\mathbf{B}$, then $\mathbf{A} = \mathbf{B}$ if and only if $a_{i,j} = b_{i,j}$ for all i and j. Examine the two matrices below. Matrix $\mathbf{C}$ equals matrix $\mathbf{D}$ as long as $x = y$.

$$\mathbf{C} = \begin{pmatrix} 3 & 7 & x \\ -1 & 2 & 1 \end{pmatrix} \qquad \mathbf{D} = \begin{pmatrix} 3 & 7 & y \\ -1 & 2 & 1 \end{pmatrix}$$

Matrix Addition

If two matrices are of the same order, they may be added. Otherwise, addition is undefined. To add matrix $\mathbf{A}$ to matrix $\mathbf{B}$, we simply add the elements in each corresponding position. Thus, if $\mathbf{C} = \mathbf{A} + \mathbf{B}$, then $c_{i,j} = a_{i,j} + b_{i,j}$, for every i and j. Matrix addition satisfies both the commutative law,

$$\mathbf{A} + \mathbf{B} = \mathbf{B} + \mathbf{A} \tag{3.6}$$

and the associative law,

$$\mathbf{A} + (\mathbf{B} + \mathbf{C}) = (\mathbf{A} + \mathbf{B}) + \mathbf{C} = \mathbf{A} + \mathbf{B} + \mathbf{C} \tag{3.7}$$

Example 3-6: Matrix Addition

Given that

$$\mathbf{X} = \begin{pmatrix} 7 & 1 & -2 \\ 3 & 3 & 0 \end{pmatrix} \qquad \mathbf{Y} = \begin{pmatrix} 2 & -1 & 4 \\ 1 & 1 & 9 \end{pmatrix} \qquad \mathbf{Z} = \begin{pmatrix} 2 & 1 \\ 7 & 3 \\ 9 & 2 \end{pmatrix}$$

then

$$\mathbf{X} + \mathbf{Y} = \begin{pmatrix} 9 & 0 & 2 \\ 4 & 4 & 9 \end{pmatrix} = \mathbf{Y} + \mathbf{X}$$

$\mathbf{X} + \mathbf{Z}$ is undefined

$\mathbf{Y} + \mathbf{Z}$ is undefined

$$\mathbf{X} - \mathbf{Y} = \begin{pmatrix} 5 & 2 & -6 \\ 2 & 2 & -9 \end{pmatrix}$$

Multiplication by a Scalar

As with vectors, if α is a scalar and **A** a matrix, the product α **A** is obtained by multiplying each element $(a_{i,j})$ of **A** by α. That is,

$$\alpha\mathbf{A} = \begin{pmatrix} \alpha a_{1,1} & \dots & \alpha a_{1,n} \\ \cdot & & \\ \cdot & & \\ \cdot & & \\ \alpha a_{m,1} & \dots & \alpha a_{m,n} \end{pmatrix} \tag{3.8}$$

Example 3-6: Multiplication of a Matrix by a Scalar

Given that

$$\alpha = 0.7 \qquad \beta = 3$$

$$\mathbf{A} = \begin{pmatrix} 8 & 3 \\ -1 & 2 \\ 7 & 1 \end{pmatrix} \qquad \mathbf{B} = \begin{pmatrix} 10 & 7 \\ -3 & 5 \end{pmatrix}$$

then

$$\alpha\mathbf{B} = \begin{pmatrix} 7 & 4.9 \\ -2.1 & 3.5 \end{pmatrix}$$

$$\beta\mathbf{A} = \begin{pmatrix} 24 & 9 \\ -3 & 6 \\ 21 & 3 \end{pmatrix}$$

Matrix Multiplication

Two matrices, **A** and **B**, may be multiplied together (i.e., **AB**) if and only if they are *conformable*, that is, only if the number of columns in **A** is equal to the number of rows in **B**. Thus, **AB** is defined for matrix **A** of order $m \times n$ and matrix **B** of order $p \times q$ if and only if $n = p$. If the two matrices *are* conformable, their multiplication produces the matrix **C** of order $m \times q$. That is, the number of rows in **C** equals the number of columns in **A**, and the number of columns in **C** equals the number of rows in **B**. Each element of **C** is given by

$$c_{i,j} = \sum_{k=1}^{t} a_{i,k} b_{k,j} \tag{3.9}$$

where t = number of rows in **B** or columns in **A**

$i = 1, \dots, m$

$j = 1, \dots, n$

Matrix multiplication satisfies the associative law,

$$(\mathbf{AB})\mathbf{C} = \mathbf{A}(\mathbf{BC}) = \mathbf{ABC} \tag{3.10}$$

(wherein the matrices are conformable) and the distributive law,

$$\mathbf{A}(\mathbf{B} + \mathbf{C}) = \mathbf{AB} + \mathbf{AC} \tag{3.11}$$

but does *not* satisfy the commutative law; that is, in *general*

$$\mathbf{AB} \neq \mathbf{BA} \tag{3.12}$$

Consequently, **AB** does not equal **BA** except under special conditions. Note that even if both **AB** and **BA** are defined, the resulting matrices need not be identical.

We may also perform matrix multiplication through the use of the multiplication rules previously defined for vectors. Recall that the product of a row and column vector is a scalar. If we wish to multiply two matrices **A** and **B** together, where

$$\mathbf{A} = \begin{pmatrix} \mathbf{a}_1 \\ \cdot \\ \cdot \\ \cdot \\ \mathbf{a}_m \end{pmatrix} \qquad \mathbf{B} = (\mathbf{b}_1 \ldots \mathbf{b}_q)$$

then $\mathbf{C} = \mathbf{AB}$ is a matrix of order $m \times q$ wherein

$$c_{i,j} = \mathbf{a}_i \mathbf{b}_j \tag{3.13}$$

Example 3-7: Matrix Multiplication

Given that

$$\mathbf{A} = \begin{pmatrix} 7 & 1 \\ 4 & -3 \\ 2 & 0 \end{pmatrix} \qquad \mathbf{B} = \begin{pmatrix} 2 & 1 & 7 \\ 0 & -1 & 4 \end{pmatrix}$$

$$\mathbf{C} = \begin{pmatrix} 1 & -1 & 3 \\ 2 & 2 & 3 \\ -1 & 4 & 7 \end{pmatrix} \qquad \mathbf{D} = \begin{pmatrix} 2 & 2 \\ 9 & 3 \end{pmatrix}$$

$$\mathbf{E} = \begin{pmatrix} 3 & 1 \\ 1 & -1 \end{pmatrix} \qquad \mathbf{F} = \begin{pmatrix} 2 & 0 \\ 1 & 1 \end{pmatrix}$$

then

$$\mathbf{AB} = \begin{pmatrix} 7 & 1 \\ 4 & -3 \\ 2 & 0 \end{pmatrix} \begin{pmatrix} 2 & 1 & 7 \\ 0 & -1 & 4 \end{pmatrix} = \begin{pmatrix} 14 & 6 & 53 \\ 8 & 7 & 16 \\ 4 & 2 & 14 \end{pmatrix}$$

CD is undefined

$$\mathbf{BC} = \begin{pmatrix} 2 & 1 & 7 \\ 0 & -1 & 4 \end{pmatrix} \begin{pmatrix} 1 & -1 & 3 \\ 2 & 2 & 3 \\ -1 & 4 & 7 \end{pmatrix} = \begin{pmatrix} -3 & 28 & 58 \\ -6 & 14 & 25 \end{pmatrix}$$

$$\mathbf{EF} = \begin{pmatrix} 3 & 1 \\ 1 & -1 \end{pmatrix} \begin{pmatrix} 2 & 0 \\ 1 & 1 \end{pmatrix} = \begin{pmatrix} 7 & 1 \\ 1 & -1 \end{pmatrix}$$

$$\mathbf{FE} = \begin{pmatrix} 2 & 0 \\ 1 & 1 \end{pmatrix} \begin{pmatrix} 3 & 1 \\ 1 & -1 \end{pmatrix} = \begin{pmatrix} 6 & 2 \\ 4 & 0 \end{pmatrix}$$

Note that, although both **EF** and **FE** are defined, they are not the same.

SPECIAL MATRICES

There are a number of special matrices that we shall encounter. These include the diagonal matrix, the identity matrix, the null or zero matrix, the matrix transpose, symmetric matrices, and augmented matrices.

DIAGONAL MATRIX The diagonal matrix is a square matrix (i.e., $m = n$) whose off-diagonal elements (those elements, $c_{i,j}$, where i and j are not the same) are all equal to zero.

Example 3-8: Diagonal Matrices

$$\mathbf{A} = \begin{pmatrix} 2 & 0 & 0 \\ 0 & 7 & 0 \\ 0 & 0 & -9 \end{pmatrix} \qquad \mathbf{B} = \begin{pmatrix} c_{1,1} & 0 & 0 \\ 0 & c_{2,2} & 0 \\ 0 & 0 & c_{3,3} \end{pmatrix}$$

IDENTITY MATRIX The identity matrix, also known as the unit matrix, is a special case of the diagonal matrix in which *all* diagonal elements are equal to unity. An identity matrix of order m is designated as either $\mathbf{I}_m$ or just $\mathbf{I}$. The identity matrix is the matrix algebra equivalent of the number 1.

Example 3-9: Identity Matrices

$$\mathbf{I}_2 = \begin{pmatrix} 1 & 0 \\ 0 & 1 \end{pmatrix} \qquad \mathbf{I}_4 = \begin{pmatrix} 1 & 0 & 0 & 0 \\ 0 & 1 & 0 & 0 \\ 0 & 0 & 1 & 0 \\ 0 & 0 & 0 & 1 \end{pmatrix}$$

NULL MATRIX A null matrix has all its elements equal to zero. The null matrix does not have to be square. This matrix is the matrix algebra equivalent of the number zero and is denoted by either $\mathbf{0}$ or $\varnothing$.

Example 3-10: Null Matrices

$$\varnothing = \begin{pmatrix} 0 \\ 0 \\ 0 \end{pmatrix} \qquad \varnothing = 0$$

$$\varnothing = \begin{pmatrix} 0 & 0 & 0 \\ 0 & 0 & 0 \\ 0 & 0 & 0 \end{pmatrix} \qquad \varnothing = \begin{pmatrix} 0 & 0 \\ 0 & 0 \\ 0 & 0 \end{pmatrix}$$

MATRIX TRANSPOSE The transpose of matrix $\mathbf{A}$, denoted as $\mathbf{A}^T$, is a reordering of the original matrix. To obtain the transpose of a matrix, we simply interchange the rows and columns, in order. That is, row 1 of the original matrix becomes column 1 of the transpose, row 2 of the original becomes column 2 of the transpose, and so on. Thus, if

$$\mathbf{A} = \begin{pmatrix} a_{1,1} & a_{1,2} & \cdots & a_{1,n} \\ a_{2,1} & a_{2,2} & \cdots & a_{2,n} \\ \cdot & & & \\ \cdot & & & \\ \cdot & & & \\ a_{m,1} & a_{m,2} & \cdots & a_{m,n} \end{pmatrix}$$

then

$$\mathbf{A}^T = \begin{pmatrix} a_{1,1} & a_{2,1} & \cdots & a_{m,1} \\ a_{1,2} & a_{2,2} & \cdots & a_{m,2} \\ \cdot & & & \\ \cdot & & & \\ \cdot & & & \\ a_{1,n} & a_{2,n} & \cdots & a_{m,n} \end{pmatrix}$$

or, in vector notation:

$$\mathbf{A} = \begin{pmatrix} \mathbf{a}_1 \\ \mathbf{a}_2 \\ \cdot \\ \cdot \\ \cdot \\ \mathbf{a}_m \end{pmatrix} \qquad \mathbf{A}^T = (\mathbf{a}_1^T \quad \mathbf{a}_2^T \quad \cdots \quad \mathbf{a}_m^T)$$

Example 3-11: Transposed Matrices

$$\mathbf{X} = \begin{pmatrix} 1 & 5 & 2 \\ 2 & 0 & 4 \end{pmatrix}$$

$$\mathbf{X}^T = \begin{pmatrix} 1 & 2 \\ 5 & 0 \\ 2 & 4 \end{pmatrix}$$

$$\mathbf{Y} = \begin{pmatrix} 2 & 0 & 0 \\ 0 & -4 & 0 \\ 0 & 0 & 7 \end{pmatrix}$$

$$\mathbf{Y}^T = \begin{pmatrix} 2 & 0 & 0 \\ 0 & -4 & 0 \\ 0 & 0 & 7 \end{pmatrix}$$

Note that the transpose of a diagonal matrix is unchanged:

$$\mathbf{Z} = (1 \quad 4 \quad 9) \qquad \mathbf{Z}^T = \begin{pmatrix} 1 \\ 4 \\ 9 \end{pmatrix}$$

SYMMETRIC MATRIX A symmetric matrix is one whose transpose and the matrix itself are equal. That is, $\mathbf{A} = \mathbf{A}^T$. Such a matrix must, obviously, be square, and $a_{i,j} = a_{j,i}$ for all i and j.

From the definition, it should be obvious that diagonal matrices and identity matrices are symmetric.

Example 3-12: Symmetric Matrices

$$\mathbf{A} = \begin{pmatrix} 1 & 2 & 3 \\ 2 & 6 & 4 \\ 3 & 4 & 9 \end{pmatrix} \qquad \mathbf{B} = \begin{pmatrix} 1 & 6 \\ 6 & 2 \end{pmatrix}$$

$$\mathbf{C} = \begin{pmatrix} -1 & 0 & 0 \\ 0 & 3 & 0 \\ 0 & 0 & 7 \end{pmatrix} \qquad \mathbf{I} = \begin{pmatrix} 1 & 0 \\ 0 & 1 \end{pmatrix}$$

AUGMENTED MATRIX An augmented matrix is one in which rows or columns of another matrix, of appropriate order, are appended to the original matrix.

If **B** is augmented (on the right) to **A**, the resulting matrix is denoted as either $(\mathbf{A}, \mathbf{B})$ or $(\mathbf{A} | \mathbf{B})$. All the rules of matrix algebra apply equally as well to the augmented matrix.

Example 3-13: Augmented Matrices

If

$$\mathbf{X} = \begin{pmatrix} 6 & 6 \\ 5 & 5 \end{pmatrix} \qquad \mathbf{Y} = \begin{pmatrix} 3 \\ 1 \end{pmatrix} \qquad \mathbf{I} = \begin{pmatrix} 1 & 0 \\ 0 & 1 \end{pmatrix}$$

then

$$(\mathbf{X} | \mathbf{Y}) = \left(\begin{array}{cc|c} 6 & 6 & 3 \\ 5 & 5 & 1 \end{array}\right) \quad \text{or} \quad \begin{pmatrix} 6 & 6 & 3 \\ 5 & 5 & 1 \end{pmatrix}$$

$$(\mathbf{X} | \mathbf{I}) = \left(\begin{array}{cc|cc} 6 & 6 & 1 & 0 \\ 5 & 5 & 0 & 1 \end{array}\right) \quad \text{or} \quad \begin{pmatrix} 6 & 6 & 1 & 0 \\ 5 & 5 & 0 & 1 \end{pmatrix}$$

$$(\mathbf{Y} | \mathbf{I}) = \left(\begin{array}{c|cc} 3 & 1 & 0 \\ 1 & 0 & 1 \end{array}\right) \quad \text{or} \quad \begin{pmatrix} 3 & 1 & 0 \\ 1 & 0 & 1 \end{pmatrix}$$

Determinants

DETERMINANT Every square matrix (and *only* square matrices) has a number associated with it. This number is called the determinant of the matrix. Given a square matrix **A**, the determinant of **A** is denoted as $|\mathbf{A}|$. Since the rows and columns of a determinant are equal, we may specify the order of the determinant or, actually, its associated matrix by simply the number of rows or columns. For example, a third-order determinant has three rows and three columns.

By definition, a first-order determinant is always equal to its single element; that is,

$$|c_{1,1}| = c_{1,1} \qquad \text{or} \qquad |-3| = -3$$

and the value of a second-order determinant is given as

$$\begin{vmatrix} a_{1,1} & a_{1,2} \\ a_{2,1} & a_{2,2} \end{vmatrix} = a_{1,1}a_{2,2} - a_{1,2}a_{2,1} \tag{3.14}$$

The diagonal of a determinant from left to right is called the *principal diagonal*, where, for each element $a_{i,j}$, $i = j$. Thus, in a determinant $|\mathbf{A}|$ of order three, the principal diagonal would consist of elements $a_{1,1}$, $a_{2,2}$, and $a_{3,3}$. The other diagonal (i.e., from right to left) is denoted as the *secondary diagonal*. Again, for our third-order determinant $|\mathbf{A}|$, the secondary diagonal would contain elements $a_{1,3}$, $a_{2,2}$, and $a_{3,1}$.

Every element of a determinant, except for the special case of a first-order determinant, has an associated *minor*. To determine the minor of an element, we simply cross out the row and column associated with that element. The remaining elements represent a determinant, of order one less than the original, that is the minor for that element. If the element is given as $a_{i,j}$, we omit row i and column j to obtain the minor denoted as $|\mathbf{A}_{i,j}|$.

The *cofactor* of an element is simply its minor with the sign $(-1)^{i+j}$ assigned to it. That is, the cofactor of $a_{i,j}$ is $(-1)^{i+j}|\mathbf{A}_{i,j}|$.

Example 3-14: Cofactors

Given the determinant

$$|\mathbf{A}| = \begin{vmatrix} 7 & -1 & 0 \\ 3 & 2 & 1 \\ 8 & 1 & -4 \end{vmatrix}$$

the cofactor for $a_{2,1}$ is

$$(-1)^{2+1}|\mathbf{A}_{2,1}| = (-1)\begin{vmatrix} -1 & 0 \\ 1 & -4 \end{vmatrix} = -4$$

and the cofactor for $a_{3,3}$ is

$$(-1)^{3+3}|\mathbf{A}_{3,3}| = (+1)\begin{vmatrix} 7 & -1 \\ 3 & 2 \end{vmatrix} = 17$$

Thus far we have only discussed and illustrated how one may find the value for a determinant of order one or two. In general, most determinants will be of considerably higher order. We shall present just one of several ways to obtain the value for the determinant of any order. This involves the use of cofactors and an expansion of the determinant by any given row or column.

The *value* of a determinant may be found by adding the products of each element, $a_{i,j}$, of any given row or column, by its respective cofactor. This may be written as

$$|\mathbf{A}| = \sum_{j=1}^{n} a_{i,j}(-1)^{i+j}|\mathbf{A}_{i,j}| \tag{3.15}$$

for any row i, or

$$|\mathbf{A}| = \sum_{i=1}^{m} a_{i,j}(-1)^{i+j}|\mathbf{A}_{i,j}| \tag{3.16}$$

for any column j.

Example 3-15: Values of Determinants

Given that

$$|\mathbf{A}| = \begin{vmatrix} a_{1,1} & a_{1,2} & a_{1,3} \\ a_{2,1} & a_{2,2} & a_{2,3} \\ a_{3,1} & a_{3,2} & a_{3,3} \end{vmatrix}$$

$$|\mathbf{B}| = \begin{vmatrix} 1 & 4 & 3 \\ 2 & 0 & 2 \\ 1 & 3 & 5 \end{vmatrix} \qquad |\mathbf{I}| = \begin{vmatrix} 1 & 0 & 0 \\ 0 & 1 & 0 \\ 0 & 0 & 1 \end{vmatrix}$$

a. Expanding $|\mathbf{A}|$ by row 2, we obtain

$$\begin{aligned} |\mathbf{A}| &= a_{2,1}(-1)^{2+1}|\mathbf{A}_{2,1}| + a_{2,2}(-1)^{2+2}|\mathbf{A}_{2,2}| + a_{2,3}(-1)^{2+3}|\mathbf{A}_{2,3}| \\ &= -a_{2,1}|\mathbf{A}_{2,1}| + a_{2,2}|\mathbf{A}_{2,2}| - a_{2,3}|\mathbf{A}_{2,3}| \end{aligned}$$

b. Expanding $|\mathbf{B}|$ by column 3, we get

$$\begin{aligned} |\mathbf{B}| &= 3(-1)^{1+3}\begin{vmatrix} 2 & 0 \\ 1 & 3 \end{vmatrix} + 2(-1)^{2+3}\begin{vmatrix} 1 & 4 \\ 1 & 3 \end{vmatrix} + 5(-1)^{3+3}\begin{vmatrix} 1 & 4 \\ 2 & 0 \end{vmatrix} \\ &= 3(6) - 2(-1) + 5(-8) = -20 \end{aligned}$$

c. Expanding $|\mathbf{I}|$ by row 1:

$$\begin{aligned} |\mathbf{I}| &= 1(-1)^{1+1}\begin{vmatrix} 1 & 0 \\ 0 & 1 \end{vmatrix} + 0(-1)^{1+2}\begin{vmatrix} 0 & 0 \\ 0 & 1 \end{vmatrix} + 0(-1)^{1+3}\begin{vmatrix} 0 & 1 \\ 0 & 0 \end{vmatrix} \\ &= 1 \end{aligned}$$

Notice carefully that if a determinant of order greater than three is to be evaluated, we would have to perform a *series* of expansions. Unfortunately, such an approach is quite unwieldy for large problems.

The expansion of determinants may often be simplified considerably by utilizing the following six properties:

1. In the following statements, the word "row" may be replaced by the word "column" (and vice versa) without affecting the validity of the statement.
2. If one complete row of a determinant is all zero, the value of the determinant is zero.
3. If two rows have elements that are proportional to one another, the value of the determinant is zero.
4. If two rows of a determinant are interchanged, the value of the new determinant is equal to the negative of the value of the old determinant.
5. Elements of any row may be multiplied by a nonzero constant if, in turn, the entire determinant (i.e., outside the determinant) is multiplied by the reciprocal of the constant.
6. To the elements of any row, you may add a constant times the corresponding elements of any other row without changing the value of the determinant.

Using some of our previous definitions, we may now define the *adjoint of a matrix*. If $\mathbf{A}$ is a square matrix, the adjoint of $\mathbf{A}$ (designated as $\mathbf{A}^\alpha$) may be found by the following process:

1. Replace each element $a_{i,j}$ of $\mathbf{A}$ by its cofactor (i.e., by $(-1)^{i+j}|\mathbf{A}_{i,j}|$).
2. Take the transpose of the matrix of cofactors, found in step 1.
3. The resulting matrix is $\mathbf{A}^\alpha$, the adjoint of $\mathbf{A}$. $\mathbf{A}^\alpha$ may be described mathematically as follows.

Let $\gamma_{i,j}$ be the cofactor for element $a_{i,j}$; then

$$\mathbf{A}^\alpha = \begin{pmatrix} \gamma_{1,1} & \gamma_{2,1} & \cdots & \gamma_{m,1} \\ \gamma_{1,2} & \gamma_{2,2} & \cdots & \gamma_{m,2} \\ \cdot & & & \cdot \\ \cdot & & & \cdot \\ \cdot & & & \cdot \\ \gamma_{1,n} & \gamma_{2,n} & \cdots & \gamma_{m,n} \end{pmatrix} \tag{3.17}$$

where, since $\mathbf{A}$ is square, $m = n$.

Matrix Inverse and Rank

Having set forth the concept and some properties of determinants, we may now present two additional, very important matrix concepts: the inverse of a matrix and the rank of a matrix.

The *inverse* of a matrix, say **A**, is denoted by $\mathbf{A}^{-1}$. Only *square* matrices have inverses. Further, for the inverse to exist, the square matrix must be *nonsingular*, wherein a nonsingular matrix is one whose determinant does *not* equal zero. Conservely, the determinant of a *singular* matrix *does* equal zero. Every nonsingular matrix has an inverse, and only nonsingular matrices have inverses. Given that **A** is nonsingular, its inverse may be found by

$$\mathbf{A}^{-1} = \frac{1}{|\mathbf{A}|}\mathbf{A}^{\alpha} \tag{3.18}$$

Example 3-16: Computation of Matrix Inverses

Given that

$$\mathbf{C} = \begin{pmatrix} c_{1,1} & c_{1,2} \\ c_{2,1} & c_{2,2} \end{pmatrix} \qquad \mathbf{A} = \begin{pmatrix} 2 & 0 \\ 1 & 3 \end{pmatrix}$$

then $\mathbf{C}^{-1}$ and $\mathbf{A}^{-1}$ are found as follows:

a. $\mathbf{C}^{-1} = \dfrac{1}{|\mathbf{C}|}\mathbf{C}^{\alpha}$

and, for a 2×2 matrix, the value of the determinant is simply $c_{1,1}c_{2,2} - c_{1,2}c_{2,1} = |\mathbf{C}|$. Next, we find the adjoint of **C**:

$$\mathbf{C}^{\alpha} = \begin{pmatrix} |\mathbf{C}_{1,1}| & -|\mathbf{C}_{2,1}| \\ -|\mathbf{C}_{1,2}| & |\mathbf{C}_{2,2}| \end{pmatrix}$$

Thus, $\mathbf{C}^{-1}$ is

$$\mathbf{C}^{-1} = \frac{1}{c_{1,1}c_{2,2} - c_{1,2}c_{2,1}} \begin{pmatrix} c_{2,2} & -c_{1,2} \\ -c_{2,1} & c_{1,1} \end{pmatrix}$$

and this is then the *general* formula for the inverse of any nonsingular 2×2 matrix.

b. From (a), $\mathbf{A}^{-1}$ is simply

$$\mathbf{A}^{-1} = \frac{1}{6-0}\begin{pmatrix} 3 & 0 \\ -1 & 2 \end{pmatrix} = \begin{pmatrix} \frac{1}{2} & 0 \\ -\frac{1}{6} & \frac{1}{3} \end{pmatrix}$$

The inverse of a matrix, say **A**, may also be defined as follows:

MATRIX INVERSE Given a square matrix **A**, if there exists a square matrix **B** such that

$$\mathbf{AB} = \mathbf{BA} = \mathbf{I} \tag{3.19}$$

then **B** is termed the inverse of **A**.

Going back to Example 3-16(b), we may multiply **A** by $\mathbf{A}^{-1}$ to obtain

$$\begin{pmatrix} 2 & 0 \\ 1 & 3 \end{pmatrix}\begin{pmatrix} \frac{1}{2} & 0 \\ -\frac{1}{6} & \frac{1}{3} \end{pmatrix} = \begin{pmatrix} 1 & 0 \\ 0 & 1 \end{pmatrix} = \mathbf{I}$$

Some of the more useful properties of the inverse include:

1. The inverse of a nonsingular matrix is unique.
2. Given that **A** and **B** are of the same order and nonsingular, then

a. $$(\mathbf{AB})^{-1} = \mathbf{B}^{-1}\mathbf{A}^{-1} \tag{3.20}$$

b. $$(\mathbf{A}^{-1})^{-1} = \mathbf{A} \tag{3.21}$$

c. $$(\mathbf{A}^{T})^{-1} = (\mathbf{A}^{-1})^{T} \tag{3.22}$$

It should be stressed that (3-18) is not the only way by which the inverse may be computed. Another, rather common, approach is to augment the square matrix **A** with the identity matrix **I** and then, by performing elementary row operations, we may develop $\mathbf{A}^{-1}$. This is probably best explained by an example.

Example 3-17: Alternative Method for the Inverse

We shall use the same **A** matrix as in Example 3-16. Thus, augmenting a second-order identity matrix to the right of **A**, we have

$$\left(\begin{array}{cc|cc} 2 & 0 & 1 & 0 \\ 1 & 3 & 0 & 1 \end{array}\right)$$

Our objective then is, by row operations on $(\mathbf{A}|\mathbf{I})$, to form the identity matrix on the *left*. This will result in $\mathbf{A}^{-1}$ on the right, or $(\mathbf{I}|\mathbf{A}^{-1})$. First, we may divide row 1 by 2 to obtain

$$\left(\begin{array}{cc|cc} 1 & 0 & \frac{1}{2} & 0 \\ 1 & 3 & 0 & 1 \end{array}\right)$$

Next, let us subtract row 1 from row 2.

$$\left(\begin{array}{cc|cc} 1 & 0 & \frac{1}{2} & 0 \\ 0 & 3 & -\frac{1}{2} & 1 \end{array}\right)$$

Finally, dividing row 2 by the number 3, we have

$$\left(\begin{array}{cc|cc} 1 & 0 & \frac{1}{2} & 0 \\ 0 & 1 & -\frac{1}{6} & \frac{1}{3} \end{array}\right)$$

Notice that the inverse of **A** is now to the right of the identity matrix.

MATRIX RANK The rank of a matrix, **A**, is designated by $r(\mathbf{A})$ and is equal to the number of linearly independent columns (or, alternatively, the number of linearly independent rows) of **A**. **A** does not have to be square.

One way to determine this rank is by determining the order of the largest nonsingular determinant that may be formed from **A**.

Example 3-18: The Rank of a Matrix

$$\mathbf{A} = \begin{pmatrix} 2 & 1 & 2 & 3 \\ 1 & 3 & 1 & 9 \\ 1 & 1 & 1 & 3 \end{pmatrix}$$

We may form *four* third-order determinants from **A**:

$$\begin{vmatrix} 2 & 1 & 2 \\ 1 & 3 & 1 \\ 1 & 1 & 1 \end{vmatrix} \quad \begin{vmatrix} 2 & 2 & 3 \\ 1 & 1 & 9 \\ 1 & 1 & 3 \end{vmatrix} \quad \begin{vmatrix} 2 & 1 & 3 \\ 1 & 3 & 9 \\ 1 & 1 & 3 \end{vmatrix} \quad \begin{vmatrix} 1 & 2 & 3 \\ 3 & 1 & 9 \\ 1 & 1 & 3 \end{vmatrix}$$

However, all of these are singular (i.e., their determinant value is zero).

We next form the second-order determinants of **A** (there are 18 such determinants). However, we do not have to form all of these because we see that the determinant given by

$$\begin{pmatrix} a_{1,1} & a_{1,2} \\ a_{2,1} & a_{2,2} \end{pmatrix} \neq 0$$

and thus the rank of **A** is 2. That is, $r(\mathbf{A}) = 2$.

The Solution of Simultaneous Linear Equations

Probably the best known use of matrices and determinants is their employment in the solution of simultaneous linear equations. The use of matrices and vectors permits a shorthand, concise means of expressing the problem while the use of determinants (and other of the concepts previously mentioned) provides the tools for a procedural approach to their solution.

Repeating the form of the set of simultaneous linear equations given in (3.1), we have

$$\begin{aligned} a_{1,1}x_1 + \cdots + a_{1,n}x_n &= b_1 \\ a_{2,1}x_1 + \cdots + a_{2,n}x_n &= b_2 \\ &\vdots \\ a_{m,1}x_1 + \cdots + a_{m,n}x_n &= b_m \end{aligned}$$

These equations may be written in concise matrix form as simply $\mathbf{Ax} = \mathbf{b}$, where

$$\mathbf{A} = \begin{pmatrix} a_{1,1} & \cdots & a_{1,n} \\ a_{2,1} & \cdots & a_{2,n} \\ \cdot & & \\ \cdot & & \\ \cdot & & \\ a_{m,1} & \cdots & a_{m,n} \end{pmatrix} \qquad \mathbf{x} = \begin{pmatrix} x_1 \\ x_2 \\ \cdot \\ \cdot \\ \cdot \\ x_n \end{pmatrix} \qquad \mathbf{b} = \begin{pmatrix} b_1 \\ b_2 \\ \cdot \\ \cdot \\ \cdot \\ b_m \end{pmatrix}$$

The matrix **A** is known as the coefficient matrix, the matrix (actually column vector) **x** is the variable or program vector, while the matrix (again, actually a column vector) **b** is the constant or right-hand side vector.

The set of equations $\mathbf{Ax} = \mathbf{b}$ has either no solution, a unique solution, or an infinite number of solutions. To determine which case exists, the use of the concept of matrix rank is required. These conditions and their identification are given below.

No Solution to $\mathbf{Ax} = \mathbf{b}$

A system of simultaneous linear equations has *no* solution if that system is *not consistent.* For a system to be consistent or inconsistent, the following definitions apply.

CONSISTENT SYSTEM A system of simultaneous linear equations is said to be consistent if the rank of the coefficient matrix, $r(\mathbf{A})$, is equal to the rank of the coefficient matrix augmented with the constant (right-hand side) vector, $r(\mathbf{A}\,|\,\mathbf{b})$.

INCONSISTENT SYSTEM A system of simultaneous linear equations is said to be inconsistent if the rank of the coefficient matrix, $r(\mathbf{A})$, is *not* equal to the rank of the coefficient matrix augmented with the constant vector, $r(\mathbf{A}\,|\,\mathbf{b})$. An inconsistent system has no solution.

Notice that $r(\mathbf{A}\,|\,\mathbf{b})$ cannot possibly be less than $r(\mathbf{A})$, nor can it be more than one greater. Thus, more precisely:

$$\text{if } r(\mathbf{A}\,|\,\mathbf{b}) > r(\mathbf{A}) \qquad \text{the system is inconsistent} \tag{3.23}$$

and

$$\text{if } r(\mathbf{A}\,|\,\mathbf{b}) = r(\mathbf{A}) \qquad \text{the system is consistent} \tag{3.24}$$

Example 3-19: A System with No Solution

Consider the following system:

$$\begin{aligned} 2x_1 + x_2 + 2x_3 &= 6 \\ x_1 + 3x_2 + x_3 &= 9 \\ x_1 + x_2 + x_3 &= 3 \end{aligned}$$

Thus,

$$\mathbf{A} = \begin{pmatrix} 2 & 1 & 2 \\ 1 & 3 & 1 \\ 1 & 1 & 1 \end{pmatrix} \qquad \mathbf{x} = \begin{pmatrix} x_1 \\ x_2 \\ x_3 \end{pmatrix} \qquad \mathbf{b} = \begin{pmatrix} 6 \\ 9 \\ 3 \end{pmatrix}$$

$$(\mathbf{A}\,|\,\mathbf{b}) = \left(\begin{array}{ccc|c} 2 & 1 & 2 & 6 \\ 1 & 3 & 1 & 9 \\ 1 & 1 & 1 & 3 \end{array}\right)$$

The rank of **A** is two, because this is the order of the largest nonvanishing determinant that may be formed from **A**. However, the rank of $(\mathbf{A}|\mathbf{b})$ is three (e.g., form the determinant for columns 1, 2, and 4). Thus, the system is inconsistent and has no solution.

A Unique Solution to $\mathbf{Ax} = \mathbf{b}$

If a system of simultaneous linear equations is consistent and the coefficient matrix **A** is nonsingular (and thus square), a *unique* solution exists. This could also be stated as:

UNIQUE SOLUTION If **A** is a square matrix and $r(\mathbf{A}) = n$ (where n is the number of variables), a unique solution exists.

There are numerous methods used to solve for a unique solution, including Cramer's rule and Gaussian reduction [1, 2]. We shall now present Cramer's rule, although it should be noted that it is not a very (computationally) efficient approach.

We let

$|\mathbf{A}|$ represent the determinant of **A**

$|\mathbf{A}|_j$ represent the determinant of the matrix in which the jth column of **A** is replaced by the **b** vector

Then

$$x_j = \frac{|\mathbf{A}|_j}{|\mathbf{A}|} \qquad \text{for all } j \tag{3.25}$$

Equation (3.25) summarizes Cramer's rule.

Example 3-20: Employment of Cramer's Rule

We now use Cramer's rule to solve the following set of simultaneous linear equations. Notice that **A** is square and its rank is equal to the number of variables, and thus a unique solution does exist.

$$\begin{aligned} 2x_1 + x_2 + 2x_3 &= 6 \\ 2x_1 + 3x_2 + x_3 &= 9 \\ x_1 + x_2 + x_3 &= 3 \end{aligned}$$

$$\mathbf{A} = \begin{pmatrix} 2 & 1 & 2 \\ 2 & 3 & 1 \\ 1 & 1 & 1 \end{pmatrix} \qquad \mathbf{x} = \begin{pmatrix} x_1 \\ x_2 \\ x_3 \end{pmatrix} \qquad \mathbf{b} = \begin{pmatrix} 6 \\ 9 \\ 3 \end{pmatrix}$$

Using (3.26), we obtain x_1, x_2, and x_3:

$$x_1 = \frac{|\mathbf{A}|_1}{|\mathbf{A}|} = \frac{\begin{vmatrix} 6 & 1 & 2 \\ 9 & 3 & 1 \\ 3 & 1 & 1 \end{vmatrix}}{\begin{vmatrix} 2 & 1 & 2 \\ 2 & 3 & 1 \\ 1 & 1 & 1 \end{vmatrix}} = \frac{6}{1} = 6$$

$$x_2 = \frac{|\mathbf{A}|_2}{|\mathbf{A}|} = \frac{\begin{vmatrix} 2 & 6 & 2 \\ 2 & 9 & 1 \\ 1 & 3 & 1 \end{vmatrix}}{1} = \frac{0}{1} = 0$$

$$x_3 = \frac{|\mathbf{A}|_3}{|\mathbf{A}|} = \frac{\begin{vmatrix} 2 & 1 & 6 \\ 2 & 3 & 9 \\ 1 & 1 & 3 \end{vmatrix}}{1} = \frac{-3}{1} = -3$$

Another approach to the solution of simultaneous linear equations for a unique solution is by means of the inverse. Notice that

$$\mathbf{Ax} = \mathbf{b}$$

and that

$$\mathbf{A}^{-1}\mathbf{A} = \mathbf{I}$$

Thus, if we premultiply $\mathbf{Ax} = \mathbf{b}$ by $\mathbf{A}^{-1}$, we obtain

$$\mathbf{A}^{-1}\mathbf{Ax} = \mathbf{A}^{-1}\mathbf{b}$$

or

$$\mathbf{Ix} = \mathbf{A}^{-1}\mathbf{b}$$

and since $\mathbf{Ix} = \mathbf{x}$, we have

$$\mathbf{x} = \mathbf{A}^{-1}\mathbf{b} \tag{3.26}$$

Example 3-21: Alternative Approach

We now solve the problem given in Example 3-20 by use of (3.26):

$$\mathbf{x} = \mathbf{A}^{-1}\mathbf{b}$$

$\mathbf{A}^{-1}$ may be found through the use of (3.18) and is

$$\mathbf{A}^{-1} = \begin{pmatrix} 2 & 1 & -5 \\ -1 & 0 & 2 \\ -1 & -1 & 4 \end{pmatrix}$$

Thus,

$$\mathbf{x} = \begin{pmatrix} 2 & 1 & -5 \\ -1 & 0 & 2 \\ -1 & -1 & 4 \end{pmatrix} \begin{pmatrix} 6 \\ 9 \\ 3 \end{pmatrix} = \begin{pmatrix} 6 \\ 0 \\ -3 \end{pmatrix}$$

which is the same solution as obtained through the use of Cramer's rule, as it should be.

An Infinite Number of Solutions to $\mathbf{Ax} = \mathbf{b}$

The final case to be considered is actually the one of most interest because it reflects the problem class that is typically encountered in linear programming. This is the case of a set of simultaneous linear equations for which an *infinite* number of solutions exist.

INFINITE NUMBER OF SOLUTIONS An infinite number of solutions will exist for a set of simultaneous linear equations wherein (1) the system is consistent, and (2) the rank of **A**, or $(\mathbf{A}\,|\,\mathbf{b})$ is *less* than the number of variables, n, that is, a consistent system in which $r(\mathbf{A}) < n$ has an infinite number of solutions.

Example 3-22: An Infinite Number of Solutions

Consider the following system:

$$\begin{aligned} 3x_1 + x_2 - x_3 &= 8 \\ x_1 + x_2 + x_3 &= 4 \end{aligned}$$

This system is consistent since $r(\mathbf{A}) = r(\mathbf{A}\,|\,\mathbf{b})$, where

$$\mathbf{A} = \begin{pmatrix} 3 & 1 & -1 \\ 1 & 1 & 1 \end{pmatrix} \qquad \mathbf{b} = \begin{pmatrix} 8 \\ 4 \end{pmatrix}$$

That is,

$$r(\mathbf{A}) = 2 \qquad \text{and} \qquad r(\mathbf{A}\,|\,\mathbf{b}) = 2$$

Also, the number of variables ($n = 3$) is greater than the rank of the system (2). For such a system, we may choose r equations and find r of the variables in terms of the remaining $n - r$ variables. That is,

$$\begin{aligned} 3x_1 + x_2 &= 8 + x_3 \\ x_1 + x_2 &= 4 - x_3 \end{aligned}$$

Solving these two equations for x_1 and x_2 only, we obtain

$$\begin{aligned} x_1 &= 2 + x_3 \\ x_2 &= 2 - 2x_3 \end{aligned}$$

Since we may assign *any* real value to x_3, the number of values for x_1 and x_2 are infinite.

As mentioned, the type of problem most usually encountered in linear programming is one in which an infinite number of solutions (i.e., decision variable values) do exist. However, as we discover later, only a finite number of these infinite values will really be of interest. This finite set of values is called the *basic solution set.*

Basic Solutions and the Basis

The concept of a basis (matrix) and of basic solutions is fundamental to the development and understanding of linear programming. We shall summarize and discuss these ideas herein without the use of highly formalized definitions. Those desiring more rigor and/or details are directed to the references [1, 2].

BASIC SOLUTION Given a system, $\mathbf{Ax} = \mathbf{b}$, of m simultaneous linear equations in n unknowns, wherein the number of solutions are infinite and $r(\mathbf{A}) = m$ $(m < n)$, we may select any $m \times m$ *nonsingular* submatrix from $\mathbf{A}$ and set the remaining $n - m$ variables to zero. The solution to the resulting system (of m variables and m equations) is called a basic solution and is denoted as $\mathbf{x}_B$.

BASIC VARIABLES The m variables associated with the basic solution are termed the basic variables and they may take on negative, positive, or zero values.

NONBASIC VARIABLES The $n - m$ variables that were *set* to zero are termed the nonbasic variables.

BASIS (MATRIX) The $m \times m$ *nonsingular* matrix selected above is termed the basis, or basis matrix, and is denoted as $\mathbf{B}$. Since it is nonsingular, it must contain a set of linearly independent columns (or rows).

If, for the system $\mathbf{Ax} = \mathbf{b}$, $n - m$ variables are set to zero and the resulting matrix is a basis, our resulting system may be expressed as:

$$\mathbf{Bx}_B = \mathbf{b}$$

or

$$\mathbf{x}_B = \mathbf{B}^{-1}\mathbf{b} \tag{3.27}$$

where $\mathbf{B}$ = basis matrix

$\mathbf{b}$ = original right-hand side vector

$\mathbf{x}_B$ = basic solution vector

Example 3-23: Finding the Basic Solutions

We shall first return to the problem given in Example 3-22. It was already shown that this problem has an infinite number of solutions. Let us now examine the *basic* solutions to

$$3x_1 + x_2 - x_3 = 8$$
$$x_1 + x_2 + x_3 = 4$$

Setting x_1 to zero, the resulting $m \times m$ (2×2) matrix from $\mathbf{A}$ is

$$\begin{pmatrix} 1 & -1 \\ 1 & 1 \end{pmatrix}$$

Since this matrix is nonsingular (i.e., the determinant does not equal zero), it may serve as a basis. Thus,

$$\mathbf{B} = \begin{pmatrix} 1 & -1 \\ 1 & 1 \end{pmatrix}$$

and, from (3.27),

$$\mathbf{x}_B = \mathbf{B}^{-1}\mathbf{b}$$

$$\mathbf{B}^{-1} = \tfrac{1}{2}\begin{pmatrix} 1 & 1 \\ -1 & 1 \end{pmatrix} \qquad \mathbf{b} = \begin{pmatrix} 8 \\ 4 \end{pmatrix}$$

and is

$$\mathbf{x}_B = \begin{pmatrix} \frac{1}{2} & \frac{1}{2} \\ -\frac{1}{2} & \frac{1}{2} \end{pmatrix}\begin{pmatrix} 8 \\ 4 \end{pmatrix} = \begin{pmatrix} 6 \\ -2 \end{pmatrix}$$

or

$$x_2 = 6 \qquad x_3 = -2$$

Note *very carefully* the interpretation of $\mathbf{x}_B$. Since the *first column* of $\mathbf{B}$ is associated with x_2, the *first value* in $\mathbf{x}_B$ is associated with x_2. The *second column* of $\mathbf{B}$ is associated with x_3, and thus the *second value* in $\mathbf{x}_B$ is the value associated with x_3. This basis matrix column/decision vector row relationship must be observed if one is to correctly interpret the results given by (3.27).

We shall now attempt to find *another* basic solution to the foregoing problem. This time, we shall set $x_2 = 0$. The resulting $m \times m$ matrix from $\mathbf{A}$ is

$$\begin{pmatrix} 3 & -1 \\ 1 & 1 \end{pmatrix}$$

and since it is nonsingular, it may be used as a basis matrix (in which the first column is associated with x_1 and the second with x_3). Thus, letting $\mathbf{B}$ be

$$\mathbf{B} = \begin{pmatrix} 3 & -1 \\ 1 & 1 \end{pmatrix}$$

and

$$\mathbf{B}^{-1} = \tfrac{1}{4}\begin{pmatrix} 1 & 1 \\ -1 & 3 \end{pmatrix} = \begin{pmatrix} \frac{1}{4} & \frac{1}{4} \\ -\frac{1}{4} & \frac{3}{4} \end{pmatrix}$$

then $\mathbf{x}_B$ is

$$\mathbf{x}_B = \begin{pmatrix} \frac{1}{4} & \frac{1}{4} \\ -\frac{1}{4} & \frac{3}{4} \end{pmatrix}\begin{pmatrix} 8 \\ 4 \end{pmatrix} = \begin{pmatrix} 3 \\ 1 \end{pmatrix}$$

or

$$x_1 = 3 \qquad x_3 = 1$$

Next, we set x_3 to zero and solve. The resulting $m \times m$ matrix is nonsingular; thus,

$$\mathbf{B} = \begin{pmatrix} 3 & 1 \\ 1 & 1 \end{pmatrix} \qquad \mathbf{B}^{-1} = \tfrac{1}{2}\begin{pmatrix} 1 & -1 \\ -1 & 3 \end{pmatrix}$$

$$\mathbf{x}_B = \begin{pmatrix} \frac{1}{2} & -\frac{1}{2} \\ -\frac{1}{2} & \frac{3}{2} \end{pmatrix}\begin{pmatrix} 8 \\ 4 \end{pmatrix} = \begin{pmatrix} 2 \\ 2 \end{pmatrix}$$

or

$$x_1 = 2 \qquad x_2 = 2$$

We have now examined all the basic solutions to this problem. This is because there are no other $m \times m$ (i.e., 2×2) basis matrices contained within **A**. We summarize our results in Table 3-2.

Table 3-2. SUMMARY OF RESULTS FOR EXAMPLE 3-23

Basic Vector	*Basic Variables*	*Nonbasic Variables*
$\mathbf{x}_B = \begin{pmatrix} 6 \\ -2 \end{pmatrix}$ or $x_2 = 6, x_3 = -2$	x_2, x_3	x_1
$\mathbf{x}_B = \begin{pmatrix} 3 \\ 1 \end{pmatrix}$ or $x_1 = 3, x_3 = 1$	x_1, x_3	x_2
$\mathbf{x}_B = \begin{pmatrix} 2 \\ 2 \end{pmatrix}$ or $x_1 = 2, x_2 = 2$	x_1, x_2	x_3

From our results in Example 3-23, we see that the number of basic solutions is limited. The *maximum* possible is equal to the number of combinations of the variables (n) taken m at a time. That is,

$$\frac{n!}{m!(n-m)!}$$

is the *maximum* number of basic solutions that may exist. In general, not *all* of these will exist, as will be seen in the next example.

Example 3-24: Finding the Basic Solutions

We shall now find the basic solutions to

$$\begin{aligned} 4x_1 + 2x_2 + x_3 &= 7 \\ -x_1 + 4x_2 + 2x_3 &= 14 \end{aligned}$$

Setting x_1 to zero, our resulting $m \times m$ matrix is

$$\begin{pmatrix} 2 & 1 \\ 4 & 2 \end{pmatrix}$$

This matrix, however, is *singular* and thus *cannot* be a basis (this is immediately obvious because the columns are not linearly independent—and, in fact, neither are the rows). As a consequence, we set x_2 to zero and obtain a resulting $m \times m$ matrix as

$$\begin{pmatrix} 4 & 1 \\ -1 & 2 \end{pmatrix}$$

This matrix can be a basis and so

$$\mathbf{x}_B = \mathbf{B}^{-1}\mathbf{b} = \begin{pmatrix} \frac{2}{9} & -\frac{1}{9} \\ \frac{1}{9} & \frac{4}{9} \end{pmatrix} \begin{pmatrix} 7 \\ 14 \end{pmatrix} = \begin{pmatrix} 0 \\ \frac{63}{9} \end{pmatrix}$$

or

$$x_1 = 0 \qquad x_3 = \tfrac{63}{9} = 7$$

(Notice particularly that x_1, although *not* a nonbasic variable, still has a value of zero in this basic solution.)

Finally, we set x_3 to zero and, since the resulting $m \times m$ matrix is nonsingular, it may serve as a basis.

$$\mathbf{x}_B = \begin{pmatrix} \frac{2}{9} & -\frac{1}{9} \\ \frac{1}{18} & \frac{2}{9} \end{pmatrix} \begin{pmatrix} 7 \\ 14 \end{pmatrix} = \begin{pmatrix} 0 \\ \frac{63}{18} \end{pmatrix}$$

or

$$x_1 = 0 \qquad x_2 = \tfrac{7}{2}$$

and again, a basic variable has a value of zero.

Example 3-24 serves to illustrate two factors. First, not every 2×2 matrix formed from **A** is a basis, and second, it is quite possible for a *basic variable* to have a value of zero. We have a special designation for such a basic solution; it is termed a *degenerate* solution.

DEGENERATE SOLUTION A basic solution to $\mathbf{Ax} = \mathbf{b}$ is degenerate if one or more of the basic variables vanish.

Convex Sets

When we move on (in the following chapters) to formulating and solving linear programming problems, we shall see that a primary operation will be the derivation of the basic solutions to a set of simultaneous linear equations. More specifically, we shall be looking only at *certain* basic solutions.

In all the problems that we face, the linear equations involved "form" a certain type of region or set. This region either does not exist (i.e., is empty) *or*, more often, this region is of a certain special class: a convex set.

> **CONVEX SET** A set (region) is convex if, for *any* two points (say $\mathbf{x}_1$ and $\mathbf{x}_2$) in that set, the line segment joining these two points lies entirely within the set.

A point itself is, by convention, also a convex set. Figure 3-1 depicts some examples of convex and nonconvex sets.

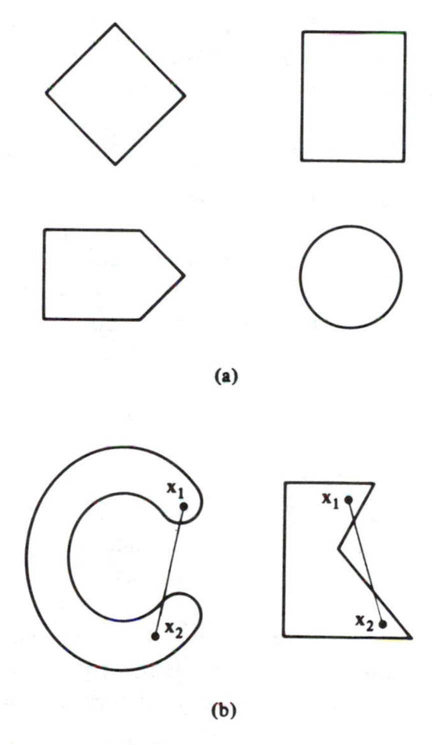

Figure 3-1. Convex and nonconvex sets: (a) convex sets; (b) nonconvex sets.

Notice, from Figure 3-1, that, in most cases at least, a convex set has "points" on its exterior boundaries. These "points," in linear models, occur at the intersection of the straight lines that form the boundaries of the convex set. Such points are called "extreme points" and are very important.

> **EXTREME POINT** A point, $\mathbf{x}$, is termed an extreme point of a given convex set if and only if $\mathbf{x}$ does not lie on the line between two other distinct points of the convex set.

We may also define the concept of *adjacent extreme points.*

ADJACENT EXTREME POINTS Two distinct extreme points, say $\mathbf{x}_1$ and $\mathbf{x}_2$ are adjacent if the line segment joining them is the edge of the convex set.

We see in later chapters that our technique for solving linear programming problems involves the determination of these extreme points and a "movement" between adjacent extreme points. Two other important properties of convex sets that will be used to construct our solution technique are summarized below. They will be discussed in more detail in later chapters.

1. Each linear equation of the linear model forms a convex set.
2. The intersection of two convex sets is also convex (and thus the set of simultaneous linear equations will form a convex set).

The Product Form of the Inverse

Our last topic for discussion in this chapter is known as the product form of the inverse and is a concept that is fundamental to the study of a modification of simplex known as revised simplex. As we shall soon learn, the simplex method of linear programming typically involves the following steps:

1. A basic matrix is selected from among a set of simultaneous linear equations.
2. The basic solution (actually, a restricted basic solution) is evaluated and a decision is made as to whether or not to examine a new basic solution.
3. If a new basic solution is to be examined, it is (usually) obtained by changing *one column* of the previous basis matrix.

On occasion, we wish to calculate the inverse of a basis matrix, $\mathbf{B}_c$, that is different by only one column from a basis matrix, $\mathbf{B}$, whose inverse is known. The product form of the inverse allows us to determine this new inverse in an efficient manner.

First, let us consider the following definitions:

$\mathbf{B}$ is the original basic matrix.
$\mathbf{B}_c$ is the new basis matrix, which is identical to $\mathbf{B}$ *except* for column r.
$\mathbf{c}$ is the rth column of matrix $\mathbf{B}_c$, the only column different from those in $\mathbf{B}$.

$$\mathbf{e} = \mathbf{B}^{-1}\mathbf{c} \tag{3.28}$$

$$\boldsymbol{\eta} = \left(-\frac{e_1}{e_r}, \ldots, -\frac{e_{r-1}}{e_r}, \frac{1}{e_r}, -\frac{e_{r+1}}{e_r}, \ldots, -\frac{e_J}{e_r}\right) \tag{3.29}$$

where e_j is the jth component of $\mathbf{e}$ as computed in (3.28) and J is the total number of elements in the column vector **e**. Thus,

$$\mathbf{B}_c^{-1} = \xi_r \mathbf{B}^{-1} \tag{3.30}$$

where $\mathbf{B}_c^{-1}$ = inverse of $\mathbf{B}_c$

$\mathbf{B}^{-1}$ = inverse of the previous matrix

ξ_r = identity matrix with its rth column replaced by η^T

Example 3-25: Using the Product Form of the Inverse

We now use (3.30) to illustrate the computation of the inverse of a basis matrix that differs by only a single column from another basis matrix (whose inverse is known).

Consider the two matrices shown below. Both are nonsingular and differ by only one column, the second. The inverse of **B** is given and we wish to find the inverse of $\mathbf{B}_c$.

$$\mathbf{B} = \begin{pmatrix} 2 & 1 & 2 \\ 2 & 3 & 1 \\ 1 & 1 & 1 \end{pmatrix} \quad \mathbf{B}_c = \begin{pmatrix} 2 & 2 & 2 \\ 2 & 4 & 1 \\ 1 & 0 & 1 \end{pmatrix} \quad \mathbf{B}^{-1} = \begin{pmatrix} 2 & 1 & -5 \\ -1 & 0 & 2 \\ -1 & -1 & 4 \end{pmatrix}$$

We first compute **e** from (3.28), where

$$\mathbf{c} = \begin{pmatrix} 2 \\ 4 \\ 0 \end{pmatrix}$$

(i.e., the second column in $\mathbf{B}_c$)

$$\mathbf{e} = \mathbf{B}^{-1}\mathbf{c} = \begin{pmatrix} 2 & 1 & -5 \\ -1 & 0 & 2 \\ -1 & -1 & 4 \end{pmatrix}\begin{pmatrix} 2 \\ 4 \\ 0 \end{pmatrix} = \begin{pmatrix} 8 \\ -2 \\ -6 \end{pmatrix}$$

Next, from (3.29) we establish η:

$$\boldsymbol{\eta} = \left(-\frac{8}{-2} \quad \frac{1}{-2} \quad -\frac{-6}{-2}\right) = (4 \quad -\tfrac{1}{2} \quad -3)$$

Thus,

$$\xi_2 = \begin{pmatrix} 1 & 4 & 0 \\ 0 & -\frac{1}{2} & 0 \\ 0 & -3 & 1 \end{pmatrix}$$

and from (3.30),

$$\mathbf{B}_c^{-1} = \begin{pmatrix} 1 & 4 & 0 \\ 0 & -\frac{1}{2} & 0 \\ 0 & 3 & 1 \end{pmatrix}\begin{pmatrix} 2 & 1 & -5 \\ -1 & 0 & 2 \\ -1 & -1 & 4 \end{pmatrix} = \begin{pmatrix} 6 & 1 & 3 \\ \frac{1}{2} & 0 & -1 \\ 2 & -1 & -2 \end{pmatrix}$$

Summary

In this chapter we have attempted to review those aspects of elementary linear algebra believed most fundamental to the field of linear programming. It is strongly suggested that, in addition to reading this chapter, the problems at the end of the chapter be attempted. It is only through the solution of such problems that the reader can truly evaluate his or her understanding of the material presented—material that forms the foundation of the methodology to be employed in later chapters.

We shall now summarize the material presented:

—The chapter is devoted to the study of simultaneous linear equations.

—Vectors and matrices are used as a convenient shorthand representation of a system of simultaneous linear equations.

—Certain operations may be performed with vectors, including vector addition, the multiplication of a vector by a scalar, and the multiplication of vectors.

—A set of vectors, $\mathbf{a}_1, \mathbf{a}_2, \ldots, \mathbf{a}_m$ are linearly independent if

$$\alpha_1\mathbf{a}_1 + \alpha_2\mathbf{a}_2 + \ldots + \alpha_m\mathbf{a}_m = \mathbf{0}$$

only when $\alpha_1, \alpha_2, \ldots, \alpha_m = 0$.

—A matrix, which consists of a set of column vectors (or row vectors) has several special classes of matrices to be considered, including diagonal matrices, identity matrices, null matrices, the matrix transpose, symmetric matrices, and augmented matrices.

—Certain operations may be performed with matrices, including matrix addition, the multiplication of a matrix by a scalar, and the multiplication of matrices. The laws governing these operations (and the operations on vectors, which are simply a special subclass of matrices) include:

Addition

$$\mathbf{A} + \mathbf{B} = \mathbf{B} + \mathbf{A} \qquad \text{(commutative law)}$$

$$\mathbf{A} + (\mathbf{B} + \mathbf{C}) = (\mathbf{A} + \mathbf{B}) + \mathbf{C} = \mathbf{A} + \mathbf{B} + \mathbf{C} \qquad \text{(associative law)}$$

$$\mathbf{A} + \boldsymbol{\varnothing} = \mathbf{A}$$

Multiplication

$$(\mathbf{AB})\mathbf{C} = \mathbf{A}(\mathbf{BC}) = \mathbf{ABC} \qquad \text{(associative law)}$$

$$\mathbf{A}(\mathbf{B} + \mathbf{C}) = \mathbf{AB} + \mathbf{AC} \qquad \text{(distributive law)}$$

In general, $\mathbf{AB} \neq \mathbf{BA}$.

$$\mathbf{A}\boldsymbol{\alpha} = \boldsymbol{\alpha}\mathbf{A} = \mathbf{A}$$

$$\mathbf{A}\boldsymbol{\varnothing} = \boldsymbol{\varnothing}\mathbf{A} = \boldsymbol{\varnothing}$$

—Determinants exist only for square matrices and are the number associated with that matrix.
—Definitions necessary to the understanding of determinants include diagonals, minors, cofactors, singularity, nonsingularity, and the adjoint matrix.
—The inverse of a matrix exists only for square matrices that are nonsingular.
—The order of the largest nonvanishing determinant that may be formed from the elements of a matrix is equal to its rank.
—Simultaneous linear equations either have no solution, a unique solution, or an infinite number of solutions.
—A basis matrix is a nonsingular matrix formed from the (independent) columns of a set of simultaneous linear equations.
—If **B** is the basis matrix, the associated basic solution is

$$\mathbf{x}_B = \mathbf{B}^{-1}\mathbf{b}$$

—The maximum number of basic solutions possible is

$$\frac{n!}{m!\,(n-m)!}$$

—A set is convex if, for any two points in the set, $\mathbf{x}_1$ and $\mathbf{x}_2$, the line segment joining these points lies entirely within the set.
—Some important concepts associated with convex sets are extreme points, adjacent extreme points, and the intersection of convex sets.
—The product form of the inverse may be used to find the inverse of a (basis) matrix which differs by only one column from a basis matrix with a known inverse.

REFERENCES

1. HADLEY, G. *Linear Algebra.* Reading, Mass.: Addison-Wesley, 1973.
2. NOBLE, B. *Applied Linear Algebra.* Englewood Cliffs, N.J.: Prentice-Hall, 1969.
3. PENNINGTON, R. H. *Introductory Computer Methods and Numerical Analysis.* Toronto: Macmillan, 1970.

PROBLEMS

3.1. Are the following sets of vectors linearly independent?

(a) $(1 \quad -9 \quad 3)$, $(4 \quad 7 \quad 2)$

(b) $(3 \quad 1 \quad 4)$, $(6 \quad 2 \quad 8)$

(c) $(1 \quad 2 \quad 3)$, $(3 \quad 0 \quad -2)$, $(3 \quad 2 \quad 1)$

(d) $(4 \quad 1 \quad 0 \quad 5)$, $(12 \quad 2 \quad 0 \quad 15)$

3.2. Calculate the products **AB** and **BA** for each of the following pairs of matrices:

(a) $\mathbf{A} = \begin{pmatrix} 3 & 1 \\ 4 & 2 \\ 0 & -3 \end{pmatrix}$ $\quad \mathbf{B} = \begin{pmatrix} 4 & 1 \\ -3 & 2 \end{pmatrix}$

(b) $\mathbf{A} = \begin{pmatrix} 4 & 1 & 3 \\ 0 & 2 & 7 \end{pmatrix}$ $\quad \mathbf{B} = \begin{pmatrix} 1 & 1 & 3 \\ 1 & 2 & 3 \end{pmatrix}$

(c) $\mathbf{A} = \begin{pmatrix} 1 & 0 & 0 \\ 0 & 1 & 0 \\ 0 & 0 & 1 \end{pmatrix}$ $\quad \mathbf{B} = \begin{pmatrix} 4 & 2 \\ 2 & 1 \\ 7 & 1 \end{pmatrix}$

3.3. Determine the inverse for:

$$\mathbf{A} = \begin{pmatrix} 1 & 3 \\ 7 & -2 \end{pmatrix} \qquad \mathbf{B} = \begin{pmatrix} 2 & 1 & 1 \\ 5 & -3 & 7 \\ 4 & 2 & 2 \end{pmatrix}$$

$$\mathbf{C} = \begin{pmatrix} 3 & 2 & 1 \\ 1 & 2 & 3 \\ 0 & 7 & 7 \end{pmatrix} \qquad \mathbf{D} = \begin{pmatrix} 1 & 0 \\ 0 & 1 \end{pmatrix}$$

3.4. Find the rank for:

$$\mathbf{A} = \begin{pmatrix} 5 & -3 & 1 \\ 10 & 6 & 2 \\ 5 & 4 & 1 \end{pmatrix} \qquad \mathbf{B} = \begin{pmatrix} 4 & 3 \\ 1 & 1 \end{pmatrix}$$

$$\mathbf{C} = \begin{pmatrix} 3 & 1 & 7 & 2 \\ 4 & 8 & 0 & 3 \end{pmatrix} \qquad \mathbf{D} = \begin{pmatrix} 4 \\ 3 \end{pmatrix}$$

3.5. Solve the following sets of simultaneous linear equations:

(a) $2x_1 - x_2 = 6$
$\quad x_1 + x_2 = 3$

(b) $4x_1 + 2x_2 + x_3 = 10$
$\quad x_1 + x_2 + x_3 = 6$

(c) $3x_1 + x_2 = 10$
$\quad 6x_1 + 2x_2 = 30$

(d) $5x_1 + 3x_2 + x_3 = 8$
$\quad x_1 + x_3 = 4$
$\quad x_1 - x_2 + 4x_3 = 18$

3.6. Find all the basic solutions for:

$$3x_1 + 4x_2 + x_3 + 2x_4 + 4x_5 = 10$$
$$2x_1 + 2x_2 + 3x_3 + 5x_4 + 2x_5 = 20$$

3.7. List all the basis matrices from Problem 3.6 that provide a solution in which all variables are nonnegative. For each of these basis matrices, list the associated basic and nonbasic variables.

3.8. Can you find the solution (values of x_1 and x_2) to the following problem? Find x_1 and x_2 to maximize $3x_1 + 7x_2$ where

$$x_1 + x_2 = 10$$
$$4x_1 - x_2 = 25$$

Discuss your results.

part two

SINGLE-OBJECTIVE LINEAR PROGRAMMING

THE SINGLE-OBJECTIVE LINEAR PROGRAMMING MODEL

Chapter Objective

In Chapter 2 we discussed and illustrated the development of the preliminary, baseline mathematical model. The general mathematical form of the baseline linear model is repeated below.

Find $\boldsymbol{x} = (x_1, x_2, \ldots, x_n)$ so as to

$$\text{maximize:} \quad a_{r,1}x_1 + a_{r,2}x_2 + \cdots + a_{r,n}x_n \qquad \text{for all } r \tag{4.1}$$

$$\text{minimize:} \quad a_{s,1}x_1 + a_{s,2}x_2 + \cdots + a_{s,n}x_n \qquad \text{for all } s \tag{4.2}$$

$$\text{satisfy:} \quad a_{t,1}x_1 + a_{t,2}x_2 + \cdots + a_{t,n}x_n \ (*)\ b_t \qquad \text{for all } t \tag{4.3}$$

where

$$x_1, x_2, \ldots, x_n \geq 0 \tag{4.4}$$

and in (4.3) the expression (*) is either a type I inequality ($\leq$), or a type II inequality ($\geq$), or an equality ($=$) for each t.

There are then r maximization objectives, s minimization objectives, and t goals (some of which may be absolute, or rigid constraints). The goals in set (4.4) are known as the nonnegativity conditions on the decision variables. As mentioned earlier, although there are cases in which variables may take on nega-

4

tive values, we shall, according to conventional practice, assume that such restrictions exist and are, in fact, rigid.[a]

Unfortunately, although the baseline model is the closest that we shall get to truly representing the actual system, we do not have any effective means for solving such a model. This is why we transform the baseline model into either the single- or multiple-objective form. Our purpose in this chapter is thus to convert the baseline model of (4.1) through (4.4) into the following traditional single-objective linear programming format:

Find $\mathbf{x} = (x_1, x_2, \ldots, x_n)$ so as to maximize (or minimize)

$$z = c_1x_1 + c_2x_2 + \cdots + c_nx_n \tag{4.5}$$

subject to the *absolute* satisfaction of the following rigid constraints:

$$\left.\begin{array}{l} a_{1,1}x_1 + a_{1,2}x_2 + \cdots + a_{1,n}x_n \; (*) \; b_1 \\ a_{2,1}x_1 + a_{2,2}x_2 + \cdots + a_{2,n}x_n \; (*) \; b_2 \\ \cdot \\ \cdot \\ \cdot \\ a_{m,1}x_1 + a_{m,2}x_2 + \cdots + a_{m,n}x_n \; (*) \; b_m \end{array}\right\} \tag{4.6}$$

[a]If a variable (say x_7) is actually unrestricted, we may replace it by the difference between two restricted variables. That is, let $x_7 = u - v$, where $u, v \geq 0$.

where

$$x_1, x_2, \ldots, x_n \geq 0 \tag{4.7}$$

and (*) indicates either $=$, $\leq$, or $\geq$ signs.

We should instantly take note of the far more specific form of the traditional model. Only a single objective is allowed and all other objectives and goals must be considered as rigid constraints—whether they actually are or not. This single-objective model will offer the advantage of easy solution by the powerful simplex method—but only at a possible degradation in the validity of the model.

Why Use the Single-Objective Model?

In Chapters 1 and 2 some of the deficiencies of the single-objective linear programming model were mentioned. This was not done to discourage the use of this model. Rather, our purpose has been to make the reader aware of the scope and limitations of this model so that he or she may better understand and evaluate any solutions derived. *All* models have, to different degrees, such limitations and, in fact, without a number of simplifying assumptions, we would be unable to solve most problems.

Another question that, hopefully, has entered the reader's mind is: Why use the single-objective model when multiple-objective models exist? This is a legitimate query.

The primary reason we use the single-objective model is because, historically, it was the first to be developed and thus it has received considerably more exposure, been put to more use, and is generally considered to be at a relatively high level of refinement. Thus, the implication is that a simple, well-tested tool is available and we may well be inclined to fit our problem to this model despite the assumptions required.

Single-objective linear programming is a tool that has been refined and polished by many investigators over three decades. The multiple-objective linear programming approach is less well known and has, until recently, received far less exposure and utilization. Consequently, it is not surprising that many investigators feel more comfortable with the choice of single-objective linear programming.

Another "advantage" of the single-objective model is that, according to common practice, the baseline model development is bypassed completely. One picks a single maximization or minimization type of objective and then considers all other goals and objectives to be rigid constraints. Such an approach certainly minimizes the decisions to be made and hastens the model development process, but only at the expense of a possible loss in model understanding and validity.

We shall leave the choice of tools (i.e., single- or multiple-objective linear programming) to the user. Our main concern is that, regardless of the choice made, the user should always be aware of all the assumptions inherent in his or her choice—and their potential impact on the resultant solution. In the past this concern has sometimes been ignored and some rather irresponsible and even ridiculous solutions have resulted.

The Conversion Process

The general approach that we shall employ to convert the baseline linear model into the conventional single-objective linear programming model is relatively straightforward. Of major importance, however, is a reflection upon those assumptions that are used to perform such a conversion.

The steps to be followed may be summarized as follows:

1. Establish two simplifying assumptions:
   a. Only a *single* objective or goal is to be optimized (i.e., maximized or minimized).
   b. All remaining functions, whether objectives or nonrigid goals, are to be considered as rigid constraints of either the equality or inequality form.
2. Select the single objective or goal that we shall optimize. Note that, in the case of the selection of a goal, we disregard the right-hand side or aspiration level so as to convert it into an objective.
3. Convert all remaining objectives into goals by assigning aspiration levels (i.e., right-hand side values) and treat all of these goals, as well as all others, as rigid constraints.

Next, we elaborate on step 3, prior to presenting some examples of the conversion process. The objectives in (4.1) and (4.2) are of the maximization and minimization form, respectively. As such, no right-hand side values were specified. However, to convert these objectives into rigid constraints, we must first convert them into goals. The right-hand side constants for these goals should then reflect an *aspired level* which we would like to have the left-hand side achieve. Unfortunately, a poor choice of this aspired to value can have a severe impact on the model, as we shall see later.

Maximization objectives must be converted into type II ($\geq$) inequalities. Thus,

$$\text{maximize } a_{r,1}x_1 + a_{r,2}x_2 + \cdots + a_{r,n}x_n$$

is replaced by

$$a_{r,1}x_1 + a_{r,2}x_2 + \cdots + a_{r,n}x_n \geq b_r$$

Minimization objectives, on the other hand, are changed into type I ($\leq$) inequalities. Thus,

$$\text{minimize } a_{s,1}x_1 + a_{s,2}x_2 + \cdots + a_{s,n}x_n$$

is replaced by

$$a_{s,1}x_1 + a_{s,2}x_2 + \cdots + a_{s,n}x_n \leq b_s$$

We are now ready to present a few illustrative examples of the conversion process.

Conversion Examples

Two brief examples will be presented to clarify the general steps of the conversion process as stated in the preceding section.

Example 4-1: First Conversion Example

We shall assume that the following baseline model has been constructed: Find x_1, x_2, and x_3 so as to

$$\text{maximize} \quad 3x_1 - 2x_2 + 4x_3 \tag{4.8}$$

$$\text{maximize} \quad x_1 + 4x_2 + x_3 \tag{4.9}$$

$$\text{minimize} \quad 5x_1 + 2x_2 + 2x_3 \tag{4.10}$$

$$\text{satisfy} \quad x_1 + x_2 + x_3 \leq 38 \tag{4.11}$$

$$2x_1 + 6x_2 + x_3 \leq 100 \tag{4.12}$$

$$x_1 + x_2 \geq 7 \tag{4.13}$$

$$x_1, x_2, x_3 \geq 0 \tag{4.14}$$

Objective (4.8) is a profit objective expressed in dollars. Objective (4.9) is a market-share objective expressed in units of market shares. Objective (4.10) reflects the amounts of a certain pollutant (as yet unregulated and thus no upper limit exists) generated in the production of the products x_1, x_2, and x_3. We shall further assume that the market-share objective is ranked highest among these three objectives and that all goals, (4.11) through (4.13), are absolute.

Given the foregoing information, we may immediately consider objective (4.9) as the single objective of our traditional linear programming model. We must now convert objectives (4.8) and (4.10) into rigid constraints. Confering with our decision maker, we determine that she would like to achieve a profit level per period of 800 units while keeping the units per period of pollution to less than 100. Our resultant single-objective linear programming formulation is:

Find x_1, x_2, and x_3 to

$$\text{maximize } z = x_1 + 4x_2 + x_3 \tag{4.15}$$

subject to (the absolute satisfaction of)

$$3x_1 - 2x_2 + 4x_3 \geq 800 \tag{4.16}$$

$$5x_1 + 2x_2 + 2x_3 \leq 100 \tag{4.17}$$

$$x_1 + x_2 + x_3 \leq 38 \tag{4.18}$$

$$2x_1 + 6x_2 + x_3 \leq 100 \tag{4.19}$$

$$x_1 + x_2 \geq 7 \tag{4.20}$$

$$x_1, x_2, x_3 \geq 0 \tag{4.21}$$

Unfortunately, our single-objective model does not have a solution. (We shall discuss this problem in more detail later.) This is because the aspired level of profit (800 units) in constraint (4.16) has been set too high, and thus there are no values of x_1, x_2, and x_3 that will satisfy all the constraints.

Another potential problem with the foregoing single-objective model is that we may be unsure as to which objective, that is, (4.8), (4.9), or (4.10), should really be the single objective. Obviously, we could try all three and thus have three models to solve but, even then, this is no assurance that the resultant solution is actually optimal for the baseline model.

Example 4-2: Second Conversion Example

In this problem we assume that the decision maker has a plant that produces products *A*, *B*, and *C*. The profit per unit is \$4, \$6, and \$3 for *A*, *B*, and *C*, respectively. The decision maker would like to achieve a weekly profit of \$1,000. Two resources are consumed in the manufacture of these products, as shown in Table 4-1. Only a limited amount of these resources are available each week.

Table 4-1. RESOURCE USAGE

Resource	*Amount of Resource Consumed per Unit of Product*			*Total Amount of Resource Available per Week (units)*
	A	*B*	*C*	
I	8	10	9	2,100
II	4	2	6	1,450

The baseline model of this problem is given below. Let

x_1 = units per week of product *A* produced

x_2 = units per week of product *B* produced

x_3 = units per week of product *C* produced

Find x_1, x_2, and x_3 so as to satisfy

$$\text{(Profit goal)} \quad 4x_1 + 6x_2 + 3x_3 \geq 1{,}000 \tag{4.22}$$

$$\text{(Resource I goal)} \quad 8x_1 + 10x_2 + 9x_3 \leq 2{,}100 \tag{4.23}$$

$$\text{(Resource II goal)} \quad 4x_1 + 2x_2 + 6x_3 \leq 1{,}450 \tag{4.24}$$

$$x_1, x_2, x_3 \geq 0 \tag{4.25}$$

Note that we have no maximization or minimization objectives in this model. Assuming that goals (4.23) and (4.24) are absolute, we may then consider the remaining nonabsolute goal (4.22) to be our single objective for the traditional model. However, notice that it is an inequality rather than a maximization or minimization objective. Since our single objective must be in the maximization or minimization form, we simply disregard the right-hand side (the desired profit level) and instead, maximize the left-hand side. The resultant single-objective model is then:

Find x_1, x_2, and x_3 so as to

$$\text{maximize } z = 4x_1 + 6x_2 + 3x_3 \tag{4.26}$$

subject to

$$8x_1 + 10x_2 + 9x_3 \leq 2{,}100 \tag{4.27}$$

$$4x_1 + 2x_2 + 6x_3 \leq 1{,}450 \tag{4.28}$$

$$x_1, x_2, x_3 \geq 0 \tag{4.29}$$

Notice that our decision maker would have been satisfied with a profit of $1,000 per week, but now that we attempt to maximize profit, we could (and actually can) achieve an even higher profit level. In this example, our decision maker actually had his sights set too low with regard to profit.

Summary

Since the baseline model is not in a form that may readily and efficiently be analyzed, we must transform it into either the single- or multiple-objective linear programming formulation. In this chapter we have stated and illustrated the steps required for conversion to the traditional model. However, although the conversion process is relatively straightforward, there is no guarantee that the resulting single-objective model will be a totally valid representation of the original baseline model. We generally accept this condition in order to use a computationally efficient tool (i.e., simplex) to solve and analyze our converted model.

REFERENCE

1. IGNIZIO, J. P. "A Baseline Model for Mathematical Programming," Working Paper, Pennsylvania State University, 1978.

PROBLEMS

In Problems 4.1 through 4.10, convert the baseline model into a conventional, *single*-objective linear programming model. Discuss any assumptions that you employ.

4.1. The model of Problem 2.8.

4.2. The model of Problem 2.9.

4.3. The model of Problem 2.10.

4.4. The model of Problem 2.11.

4.5. The model of Problem 2.12.

4.6. The model of Problem 2.13.

4.7. The model of Problem 2.14.

4.8. The model of Problem 2.15. Do you detect any difficulties? Why?

4.9. The model of Problem 2.16.

4.10. The model of Problem 2.17(a).

4.11. Discuss the role that a firm and sound understanding of the problem and its environment must play in both the development of the baseline model and the conversion to a "good" single-objective representation.

GRAPHICAL SOLUTION AND INTERPRETATION

Introduction

Those readers anxious to finally solve a linear programming problem will get their chance in this chapter. We present a methodology for both problem solution and interpretation, but it is *not* the simplex method that we have referred to previously. Rather, we restrict the linear programming problems of this chapter to such a size (two or three variables) that they may be solved *graphically*.

PURPOSE

It would be unusual to ever discover a problem in the real world that consisted of no more than three variables. Consequently, it should not be inferred that graphical analysis is a practical approach to linear programming. It is, however, a superb teaching tool and visual aid in linear programming. Two particular benefits are that:

1. It provides the reader with the "flavor" of the process that we shall ultimately employ in the solution of larger linear programming problems.
2. It presents the reader with a clear, visual definition of many of the terms and concepts that we utilize throughout the text.

As a consequence, the graphical approach is a valuable adjunct to the plausible reasoning approach used in this text.

5

The Graphical Simplex Procedure

The mechanics of the graphical procedure are brief and straightforward. We first establish the problem in terms of the single-objective model. Next, we plot all rigid constraints on a Cartesian coordinate system in which each axis represents one decision variable. If the constraint is an equality, it will plot out as a straight line in two dimensions (i.e., two variables) or as a plane in three dimensions (i.e., three variables). However, if the constraint is an *inequality*, it defines a *region* that is bounded by the straight line (or plane) obtained when the function is considered an equality. To clarify this, consider the following example.

Example 5-1: The Graphical Representation of Constraints

Consider the following constraints:

$$3x_1 + 2x_2 = 12 \tag{5.1}$$

$$x_1 + x_2 \leq 8 \tag{5.2}$$

$$4x_1 + x_2 \geq 10 \tag{5.3}$$

$$2x_1 - 3x_2 \leq 12 \tag{5.4}$$

Constraint (5.1) is an equality and thus is plotted simply as a straight line in Figure

5-1(a). Any point on the line AB satisfies this constraint. However, since we *normally* consider the decision variables in a linear programming problem to be constrained to be nonnegative, we only consider the portion of the line (or region, if it were an inequality) that lies within the first quadrant [i.e., $x_1 \geq 0$, $x_2 \geq 0$ in Figure 5-1(a)], that is, the segment CD.

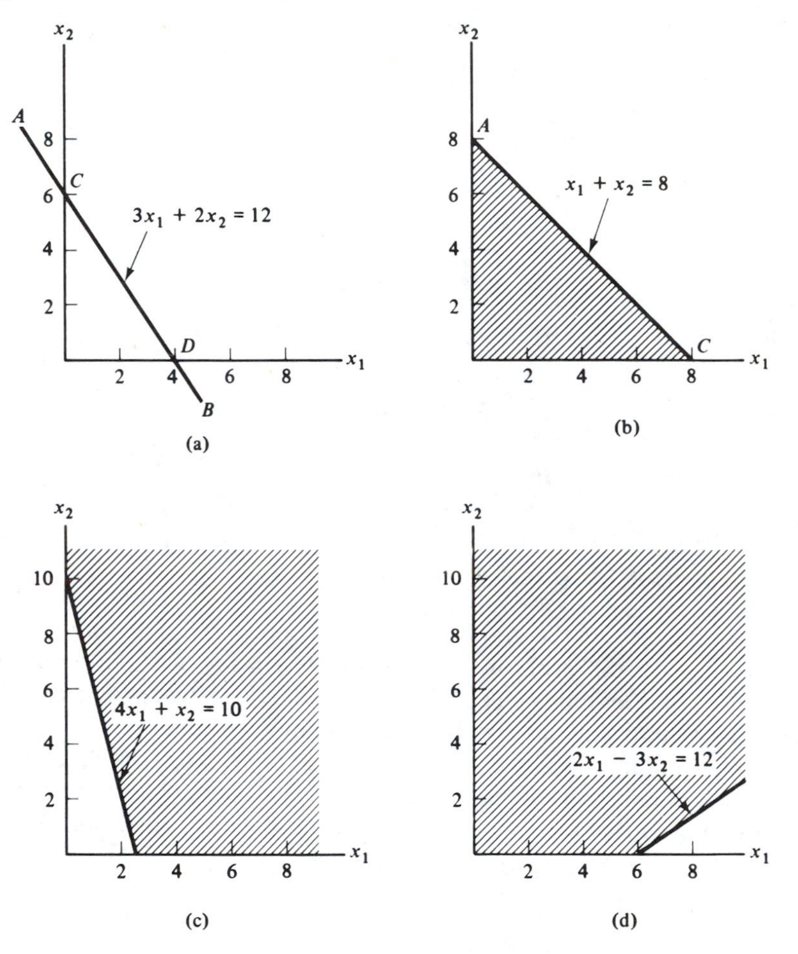

Figure 5-1. Graphical representation of Example 5-1.

Constraint (5.2) is an inequality (type I) and thus describes a region bounded by the straight line defined by considering (5-2) to be an equality, that is, $x_1 + x_2 = 8$. The crosshatched region in Figure 5-1(b) describes all the x_1, x_2 values that satisfy (5.2) *and* are nonnegative. Notice that this includes not only the crosshatched region but also its boundaries (the lines AB, AC, and BC).

Constraint (5.3) is a type II constraint and thus also describes a region having, as a boundary, the line $4x_1 + x_2 = 10$. However, carefully note [in Figure 5-1(c)] that the crosshatched region containing the points satisfying the constraint is not closed. That is, we can increase x_1, x_2 or both to infinity and still satisfy the constraint. Constraint

(5.4) also describes [in Figure 5-1(d)] an unbounded region having, as one boundary, the line $2x_1 - 3x_2 = 12$.

Before proceeding further, the reader should be confident that he or she understands just how the foregoing regions were defined and plotted.

After plotting *all* the constraints associated with the model, we must next consider the maximization or minimization of the single-objective function over the final region as simultaneously defined by these constraints. If the constraints actually describe a region that does exist, this region will take on a very special form: it will be a *convex* region. There will either be only a single point in a convex region (recall from Chapter 3 that a point is itself a convex region) and thus the solution to the problem is trivial, or, more likely, there will be an *infinite* number of points within this region. In the latter case, the optimal solution to the problem is one (or possibly more) of these infinite points. Fortunately, it has been proved (as we discuss in Chapter 6) that, if there is an optimal solution to a linear programming problem, at least one of the extreme points of the convex region (as defined by the constraints) will be optimal. The impact of this statement is that, rather than having to investigate the *infinite* number of solutions within the convex region or on its boundaries, we need only investigate the *finite* number of extreme points of the set. This observation was one of the key discoveries in linear programming.

We are now in a position to state the steps involved in the graphical solution of a linear programming problem:

Step 1. *Define the coordinate system.* Sketch the axes of the coordinate system and associate, with each axis, a specific decision variable.

Step 2. *Plot the constraints.* Establish the line (in the case of an equality) or region (in the case of an inequality) associated with each constraint.

Step 3. *Identify the resultant solution space.* The *intersection* of all the regions in step 2 determines the convex solution space in which all points simultaneously satisfy all the problem constraints. In the event that there is no such intersection (i.e., no resultant convex set), no solution exists that will satisfy *all* the constraints. In this case, the problem is termed *infeasible.* If there is a solution space, go to step 4.

Step 4. *Identify the candidate set of optimal solution points.* Each extreme point of the convex set defined by the constraints is a possible candidate for the optimal solution to the problem. List these points (i.e., their decision variable values or coordinates).

Step 5. *Identify the optimal solution(s).* For each extreme point in step 4, determine the associated objective function value. The point(s) that maximizes (or minimizes) this value is optimal.

These steps are best illustrated by means of a number of example problems. The problems in the next section have been selected to both demonstrate the implementation of the foregoing steps and also to help clarify a number of very important terms and concepts of linear programming.

Examples of the Graphical Approach

Example 5-2: A Maximization Problem

Find x_1 and x_2 so as to

$$\text{maximize } z = 5x_1 + 2x_2$$

$$\text{subject to}$$

$$x_1 + x_2 \leq 8$$

$$4x_1 + x_2 \leq 12$$

$$x_1, x_2 \geq 0$$

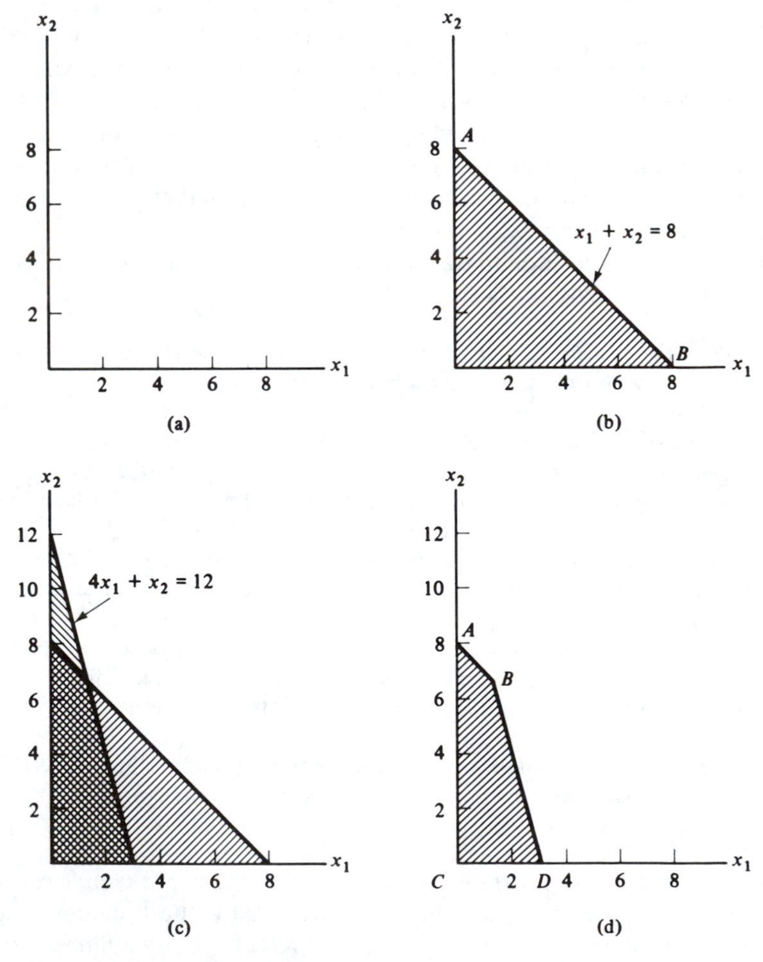

Figure 5-2. Graphs for Example 5-2.

Labeling one axis as x_1 and the other as x_2, we establish our coordinate system as shown in Figure 5-2(a). Next, each constraint is plotted. Considering the first constraint ($x_1 + x_2 \leq 8$) initially, we plot its boundaries by setting the left-hand side equal to the right (i.e., $x_1 + x_2 = 8$). The line from A to B in Figure 5-2(b) results. The region that this line bounds is shown crosshatched in the figure. Our next step is to repeat this process with our second constraint ($4x_1 + x_2 \leq 12$). This is shown in Figure 5-2(c), where the region satisfying this constraint (and the nonnegativity conditions on x_1 and x_2) is shaded. The intersection of both regions is shown in Figure 5-2(d) and its extreme points, candidates for optimal programs, are labeled A, B, C, and D.

Our final step is to evaluate the objective function ($5x_1 + 2x_2$) at points A, B, C, and D to determine which one(s) is optimal. This is accomplished in Table 5-1. (Notice that point B occurs at the intersection of $x_1 + x_2 = 8$ and $4x_1 + x_2 = 12$, or $x_1 = \frac{4}{3}$ and $x_2 = \frac{20}{3}$.) Since point B gives the maximum value (20) for the objective, it is the optimal program.

Table 5-1. CANDIDATE SOLUTIONS FOR EXAMPLE 5-2

Extreme Point	*Coordinates* (x_1, x_2)	*Objective Function Value* ($5x_1 + 2x_2$)
A	0, 8	16
B	$\frac{4}{3}, \frac{20}{3}$	20
C	0, 0	0
D	3, 0	15

We describe this result as

$$\mathbf{x}^* = \left(\tfrac{4}{3}, \tfrac{20}{3}\right) \quad \text{and} \quad z^* = 20$$

Example 5-3: An Alternative Approach

A slightly different approach to the solution of a graphical linear programming problem is achieved through the use of the "isoprofit" (if maximizing) or "isocost" (if minimizing) objective. Our steps will remain identical, through step 4, to those previously presented. However, step 5, the identification of the optimal solution, is conducted through a strictly graphical approach. The easiest way to explain this approach is through an actual example, and we use the previous example and Figure 5-2(d) to accomplish this. The convex solution set defined by the constraints of Example 5-2 is shown in Figure 5-2(d).

Our next step will be to plot out several isoprofit lines. We do this by setting the objective ($5x_1 + 2x_2$) *equal* to a number of arbitrary profit values (assuming that the measure of this objective is in terms of profit).

a. $5x_1 + 2x_2 = 0$
b. $5x_1 + 2x_2 = 10$
c. $5x_1 + 2x_2 = 15$
d. $5x_1 + 2x_2 = 30$

The lines determined by these equalities are shown in Figure 5-3.

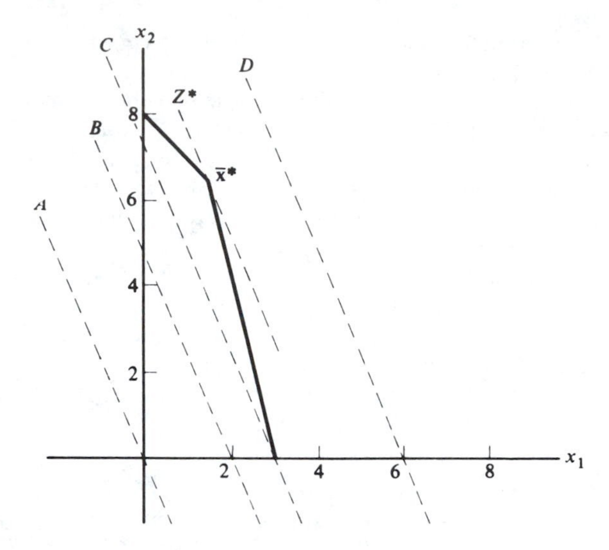

Figure 5-3. Plot of isoprofit lines.

Notice, in this figure, that any point on, for example, the line C gives an objective function value of 15. Further, as the value of the objective rises, the isoprofit lines move farther away from the origin, while always remaining parallel to one another. This observation leads us to a technique for performing step 5 graphically. We simply plot *one* isoprofit line. If the line is within the solution space, we move it (parallel to its original position) until it leaves the solution space. *Usually*, we move the line away from the origin to maximize and toward the origin to minimize. However, as will be illustrated in Example 5-6, this is *not* always the case. The last point (or line, or surface) that the isoprofit line touches immediately prior to leaving the solution space is the optimal solution. This is line Z^* in Figure 5-3, which touches point $\mathbf{x}^* = (\frac{4}{3}, \frac{20}{3})$ last as it leaves the region of feasible solutions.

Example 5-4: A Minimization Problem

Find x_1 and x_2 so as to

$$\text{minimize } z = 2x_1 + 8x_2$$
$$\text{subject to}$$
$$x_1 + x_2 \geq 9$$
$$3x_1 + x_2 \geq 12$$
$$x_1, x_2 \geq 0$$

We have plotted both constraints, and the convex solution space that they form, in Figure 5-4. The crosshatched region is the solution space and notice that it is not a closed set. That is, either x_1 or x_2 or both may go to infinity. However, since we are minimizing and since the coefficients of both x_1 and x_2 in z are positive, infinite values of the decision variables are obviously not of interest. We need only consider points A, B, and C.

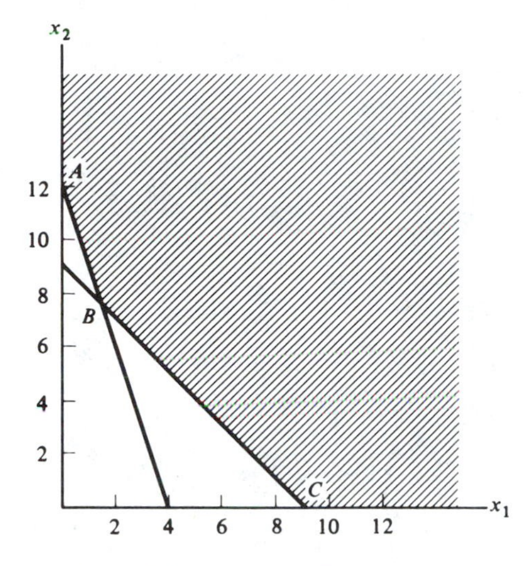

Figure 5-4. Graph for Example 5-4.

Before proceeding further, let us stress that, for some problems, our "solution" may exist at an infinite value of one or more decision variables. This will signify what is termed an *unbounded* problem and we shall discuss such a problem in the example to follow.

Evaluating the value of z for points A, B, and C in Figure 5-4, we find that, at point A ($x_1 = 0, x_2 = 12$), $z = 96$. At point B ($x_1 = \frac{3}{2}, x_2 = \frac{15}{2}$), $z = 63$. At point C ($x_1 = 9, x_2 = 0$), $z = 18$. Thus, point C is the point that minimizes z. Thus, $\mathbf{x}^* = (9, 0)$ and $z^* = 18$.

In our next example we present an illustration of a situation that is highly unlikely to ever truly exist: an *unbounded* problem. That is, if our objective is to be maximized, we may increase it to infinity and, if it is to be minimized, to minus infinity. Such a situation would infer, for example, unbounded profits and, although the oil-producing countries seem close to the achievement of such a status, there are virtually always restrictions (such as limited resources, market saturation, etc.) that make such cases impossible. However, it *is* likely that the reader will encounter a number of *models* (as opposed to real-life situations) that will be unbounded. What this generally means is that certain restrictions were overlooked and, in such a case, it is a signal to revise the model.

Example 5-5: An Unbounded Model

Find x_1 and x_2 so as to

$$\text{minimize } z = -4x_1 + 2x_2$$

$$\text{subject to}$$

$$x_1 + x_2 \geq 9$$

$$3x_1 + x_2 \geq 12$$

$$x_1, x_2 \geq 0$$

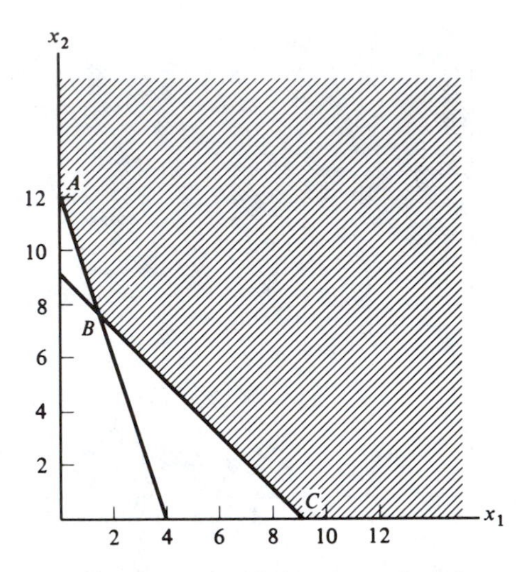

Figure 5-5. Graph for Example 5-5.

We have plotted our constraints in Figure 5-5. The crosshatched area is the set of all solutions. As in Example 5-4, this solution set is not closed, and, in fact, is identical to the space of Example 5-4.

Since z has at least one variable with a negative coefficient, we must examine not only points A, B, and C but, also the point associated with setting the variable (or variables in the general case) with the negative coefficient to infinity. That is, we must also examine $x_1 = \infty$, $x_2 = 0$. Rather obviously, this latter point with a z value of $-\infty$ is "optimal" and the problem is seen to be unbounded.

Example 5-6: Other Unbounded Models

We now present two other cases of unbounded problems. The first is:

a. Find x_1, x_2 so as to

$$\text{maximize } z = 3x_1 + 7x_2$$

$$\text{subject to}$$

$$x_1 + x_2 \geq 10$$

$$4x_1 - x_2 \geq 12$$

$$x_1, x_2 \geq 0$$

This problem is graphed in Figure 5-6, wherein the crosshatched region is the solution set. Since we are maximizing z, we must consider setting variables to infinity which have a positive coefficient in z. Thus, our points to consider are A ($x_1 = \frac{22}{5}$, $x_2 = \frac{28}{5}$), B ($x_1 = 10$, $x_2 = 0$), and C ($x_1 = \infty$, $x_2 = \infty$). Again, point C is obviously "optimal" and the problem is unbounded.

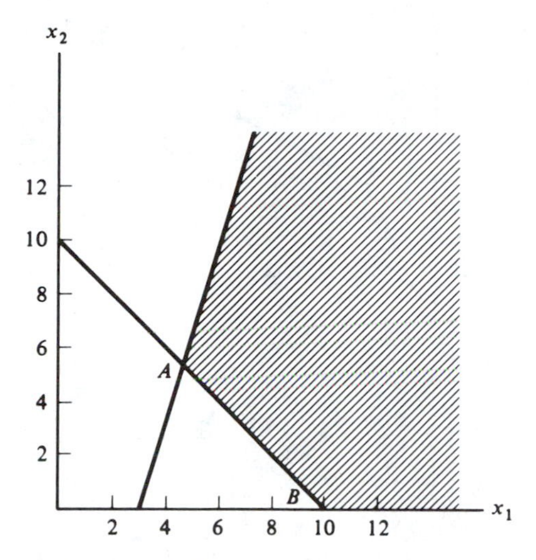

Figure 5-6. Graph for Example 5-6(a).

We now examine:

b. Find x_1 and x_2 so as to

$$\text{minimize } z = x_1 - 2x_2$$
$$\text{subject to}$$
$$-2x_1 + x_2 \leq 8$$
$$-0.5x_1 + x_2 \leq 12$$
$$x_1, x_2 \geq 0$$

This second problem is quite unusual and demonstrates not only an unusual case of unboundedness but, also, a counterexample to our earlier "intuitive" deduction as to the movement of the isocost line. Figure 5-7 provides the graphical representation of the problem together with several isocost lines. Notice carefully that the minimum value of z is *finite*. It occurs along the *entire* line from point A, along the $z = -24$ line, to infinity. However, although z is finite, the *program is unbounded*, since any x_1, x_2 values on on the $z = -24$ line, which extends to *infinity*, are solutions.

As our second point, look carefully at the set of isocost lines. We see that z decreases as these lines move *away* from the origin. Thus, any conclusion that we always move toward the origin for minimization (or away for maximization) is obviously not true in general.

There is actually a third interesting characteristic that is exhibited in this problem. This is the fact that the optimal program occurs along a *line* rather than at a single point as in our previous examples. We discuss this facet in more detail in the next example.

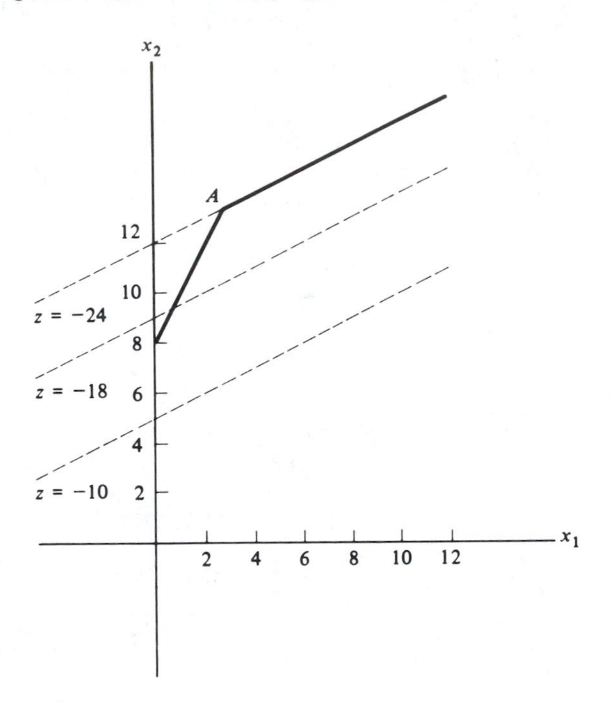

Figure 5-7. Graph for Example 5-6(b).

Example 5-7: Alternative Optimal Solutions

Find x_1 and x_2 so as to

$$\text{maximize } z = 10x_1 + 6x_2$$
$$\text{subject to}$$
$$5x_1 + 3x_2 \leq 30$$
$$x_1 + 2x_2 \leq 18$$
$$x_1, x_2 \geq 0$$

This problem is plotted in Figure 5-8. Table 5-2 lists the evaluation of the four extreme points *A*, *B*, *C*, and *D*.

Notice that *both* points *B* and *C* have z values of 60. Both extreme points are thus optimal and no unique optimal solution exists for this problem. Such a problem is said to have alternative optimal solutions. There are, in fact, an *infinite* number of values of x_1, x_2 that give a value of $z = 60$ and are thus optimal. *Any* point on the line between *B* and *C* (the reader can check this if he or she so wishes) is an optimal program.

Another way to identify a problem with alternative optimal solutions is by means of the isoprofit line. Notice that our objective function is parallel to the line *BC* (or the first constraint), and thus the last "point" it touches as it moves away from the origin is actually an entire line segment. Consequently, if any problem has an alternative optimal extreme point (or points), it has an infinite number of optimal programs.

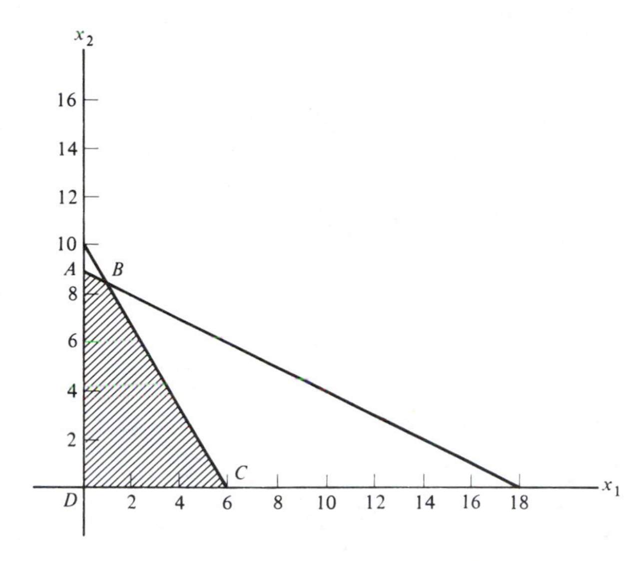

Figure 5-8. Graph for Example 5-7.

Table 5-2. CANDIDATE SOLUTIONS FOR EXAMPLE 5-7

Extreme Point	*Coordinates* (x_1, x_2)	*Objective Function Value* $(10x_1 + 6x_2)$
A	$x_1 = 0,\ x_2 = 9$	54
B	$x_1 = \frac{6}{7},\ x_2 = \frac{60}{7}$	60
C	$x_1 = 6,\ x_2 = 0$	60
D	$x_1 = 0,\ x_2 = 0$	0

Example 5-8: Redundant Constraints

Find x_1 and x_2 so as to

$$\text{maximize } z = 6x_1 + 12x_2$$

$$\text{subject to}$$

$$x_1 + 2x_2 \leq 10$$

$$2x_1 - 5x_2 \leq 20$$

$$x_1 + x_2 \leq 15$$

$$x_1, x_2 \geq 0$$

The problem is plotted in Figure 5-9. Notice that the only constraints that serve to define the final convex solution set are the first ($x_1 + 2x_2 \leq 10$) and the nonnegativity conditions ($x_1, x_2 \geq 0$). The second and third constraints play no part in either defining the solution set or in determining the optimal solution. Such constraints are termed *redundant*.

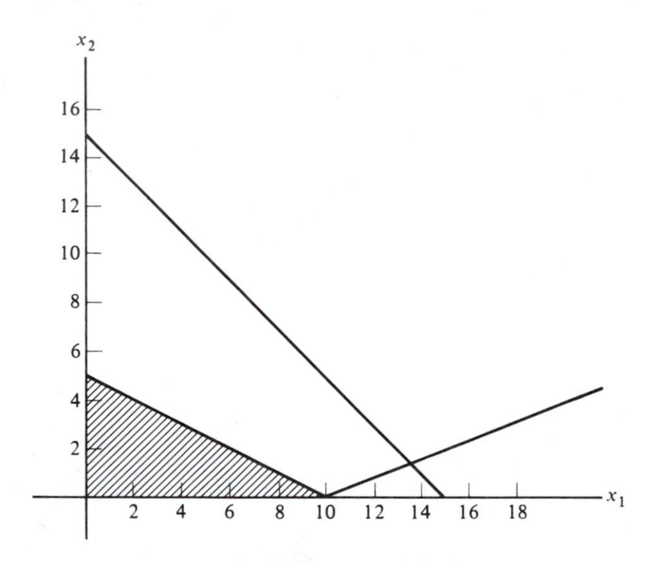

Figure 5-9. Graph for Example 5-8.

Example 5-9 An Infeasible Problem

For this example let us consider a specific problem that is developed first through the baseline model. Let us assume that our employer, a small-appliance manufacturer, wishes to have us analyze the production program of a particular plant. This plant manufactures only two products, a popcorn popper and a deluxe popcorn popper. The per unit profit for the regular and deluxe poppers are \$3 and \$5, respectively. The manufacture of a popper, whether of the deluxe or regular type, takes place in two main process areas. The processing time, per popper, is listed in Table 5-3.

Table 5-3. THE PANDORA POPCORN POPPER PROCESSING PLANT'S PERFORMANCE

	Processing Time (Minutes) for:		*Daily Processing Time Available without Overtime (Minutes)*
Process	*Regular*	*Deluxe*	
A	5	10	400
B	6	8	440

The philosophy of the company president, Pedro Pandora, has always been to minimize overtime within his plants. He has always been opposed to such practices.

Since popcorn poppers are a new venture for Pandora Manufacturing, Mr. Pandora has stated that he would like to see a large number of poppers produced per day (i.e., the combined total) so as to increase the company's visibility.

A possible baseline model for this problem is given below. Let

x_1 = number of regular poppers produced each day

x_2 = number of deluxe poppers produced each day

Find x_1 and x_2 so as to:

maximize profit: $3x_1 + 5x_1$ (5.5)

maximize units produced per day: $x_1 + x_2$ (5.6)

satisfy time limits on process A: $5x_1 + 10x_2 \leq 400$ (5.7)

satisfy time limits on process B: $6x_1 + 8x_2 \leq 440$ and $x_1, x_2 \geq 0$ (5.8)

After some deliberation, it is decided that goals (5.7) and (5.8) will be considered absolute and that the profit objective (5.5) shall be used as our single objective. This means that objective (5.6) must be considered a rigid constraint and, further, a right-hand side value must be assigned.

Our discussions with Mr. Pandora lead us to conclude that he would be happy with a combined production of 100 or more units per day. Thus, (5.6) is rewritten as: $x_1 + x_2 \geq 100$. The resultant single-objective model is then:

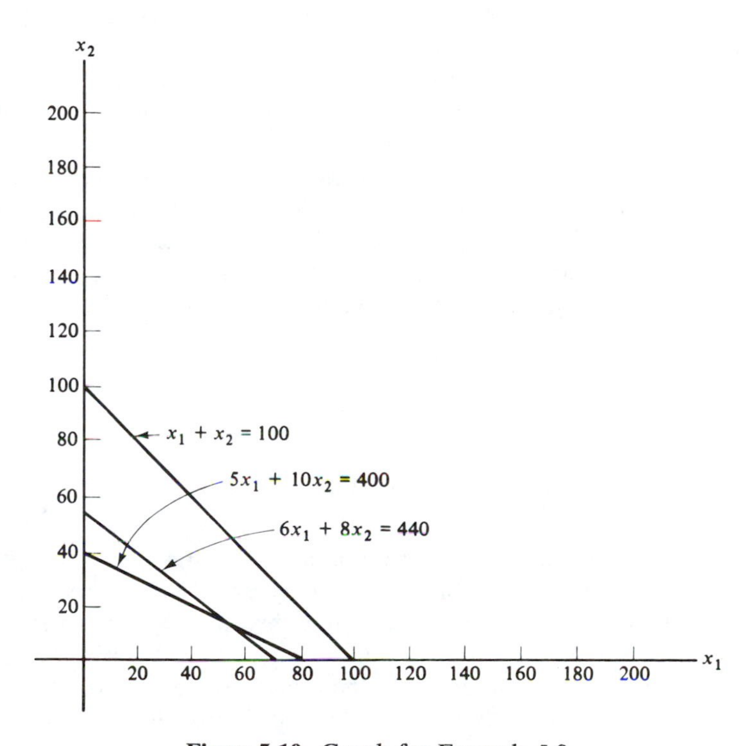

Figure 5-10. Graph for Example 5-9.

Find x_1 and x_2 so as to

$$\text{maximize } z = 3x_1 + 5x_2 \tag{5.9}$$

subject to

$$x_1 + x_2 \geq 100 \tag{5.10}$$

$$5x_1 + 10x_2 \leq 400 \tag{5.11}$$

$$6x_1 + 8x_2 \leq 440 \tag{5.12}$$

$$x_1, x_2 \geq 0$$

The three constraints have been plotted in Figure 5-10. There is, however, *no* intersection of the three regions defined by these constraints (in the first quadrant). Thus, there are *no* values of x_1 and x_2 that can satisfy all the constraints. As we mentioned earlier, such a problem is termed *infeasible.*

Now, in some texts, one is left with the impression that, if a problem is infeasible, we cannot solve it. This is true from a rigid, mathematical point of view. However, it should be obvious that, if we value our job at Pandora Manufacturing, we would not tell our president: "Sorry, there is no solution to your problem." Anyone naïve enough to do such a thing would deserve the consequences.

What we *should* do is to reexamine our model, particularly the conversion process that was used to transform the baseline model. Obviously, we either set our sights too high, with regard to the right-hand side of constraint (5.10), or else something is going to have to "give"—most probably our desire not to allow overtime. Another possibility is that all three constraints must "give." Whatever the case, we need to find a solution to this problem that comes as close to satisfying all our goals as possible. The conclusion that there is no "feasible" solution is one that is normally unwarranted in actual practice.

Some Definitions and Their Graphical Counterparts

As mentioned earlier, linear programming is a relatively easy subject. However, some students invariably seem to have difficulty with the topic—not because it is beyond their abilities, but simply because they fail to master the "jargon" that is employed. This is both unfortunate and unnecessary.

In this section we discuss some of the more common terms and concepts in linear programming. This discussion will be aided through the use of graphical representations. As such, each term and concept is not rigorously defined but, rather, is *described* in such a manner as to make the idea understandable.

> **FEASIBLE SOLUTION** The set of feasible solutions to a linear programming problem form a convex set for which all points in this set satisfy all the constraints and nonnegativity conditions.

If a feasible solution set exists and is not trivial (i.e., a point), it contains an infinite number of solutions to the set of constraints. Our desire, of course, is to find one (or more) of these points that optimizes our objective function.

> **INFEASIBLE SOLUTION** Any point that does *not* satisfy all the constraints and nonnegativity conditions is termed infeasible.

Consider Figure 5-11. The crosshatched area is the region satisfying all constraints and thus represents the set of feasible solutions. The entire region not included in this set is the infeasible region.

> **BASIC SOLUTION** A basic solution was defined mathematically in Chapter 3. Graphically, a basic solution may be identified as the intersection of two (or more) constraints. Thus, in Figure 5-11, points *A* through *G* are *all* basic solutions.

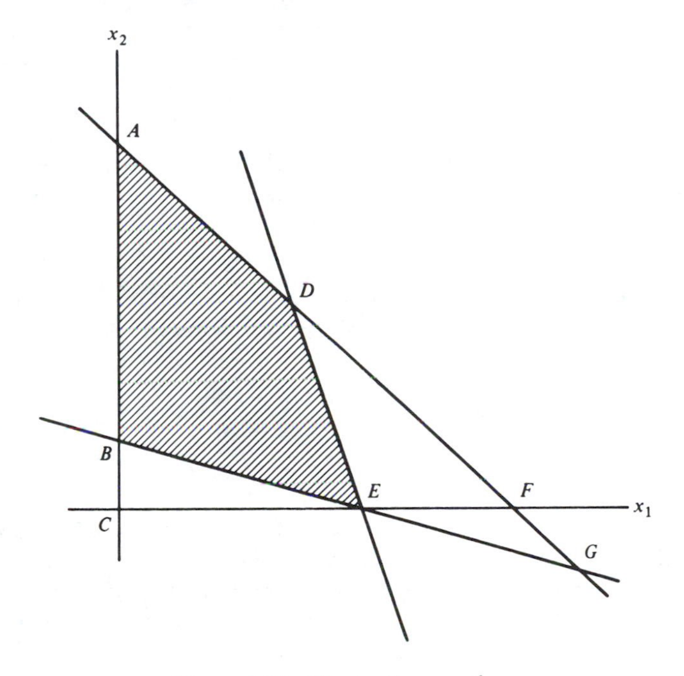

Figure 5-11. Illustrative graph.

From this definition of basic solutions, it should be obvious that we may subdivide basic solutions into two mutually exclusive sets: those that are basic *feasible* solutions and those that are basic *infeasible* solutions. In Figure 5-11, points A, B, D, and E are basic feasible, whereas points C, F, and G are basic infeasible.

EXTREME POINTS The extreme points of the convex set of feasible solutions are equivalent to the set of basic feasible solutions. That is, an extreme point and basic feasible solution are one and the same.
ADJACENT EXTREME POINTS (OR ADJACENT BASIC FEASIBLE SOLUTION) Two extreme points are adjacent if the line segment joining them is on edge of the convex set. Thus, in Figure 5-11, extreme point D is adjacent to both A and E.
OPTIMAL SOLUTION The optimal solution (the optimal values of $\mathbf{x}$ and z) to a linear programming problem, if it exists, will occur at one or more extreme points.

Notice carefully in the definition of optimal solutions that, although this optimal solution exists at one or more extreme points, the definition does not rule out the possibility of an optimal solution existing at other than an extreme point. In fact, a (feasible) linear programming problem will either have a unique solution (i.e., a single optimal extreme point as in Example 5-2) or an infinite number of solutions (i.e., several extreme points plus the edge or surface between them as in Example 5-7). For the second case we say that we have *alternative optimal solutions.*

ALTERNATIVE OPTIMAL SOLUTIONS If more than one extreme point is optimal, the entire edge or surface of the convex set between these extreme points is also optimal.

We now conclude our discussion of linear programming terminology with a description of unbounded and infeasible "problems." It should, of course, be realized that it is extremely unlikely that a real *problem* would be either unbounded or infeasible. However, quite often our imperfect *models* of such problems do exhibit such characteristics—but usually only because of our failure to accurately describe the actual problem.

UNBOUNDED PROBLEM (i.e., MODEL) Usually, an unbounded problem is one in which the objective may be increased or decreased without limit. Recall, however [see Example 5-6(b)], that some problems may have a finite objective value but have unbounded programs. In either case, such an occurrence usually means that some restriction(s) has been overlooked.

INFEASIBLE PROBLEM (i.e., MODEL) An infeasible problem is said to have no solution. That is, the constraint set does not have any common intersection. What this usually means is that one or more of the constraints should not be truly rigid. That is, they should "give."

Summary

This chapter was directed toward two areas:

1. How to graphically solve two- or three-variable linear programming problems.
2. How to associate graphical results with certain linear programming terminology.

The ideas and notions presented here will serve to ease the task of mastering the simplex method, which is developed in the chapters that immediately follow.

PROBLEMS

Graph and solve the following problems and identify (a) their extreme points, (b) optimal solutions, and (c) any other features of interest (i.e., unboundedness, infeasibility, alternative optimal solutions, etc.).

5.1. Maximize $z = 5x_1 + x_2$ subject to

$$\begin{aligned} x_1 + x_2 &\leq 12 \\ 4x_1 + x_2 &\leq 20 \\ x_1, x_2 &\geq 0 \end{aligned}$$

5.2. Maximize $z = 5x_1 + x_2$ subject to

$$\begin{aligned} x_1 + x_2 &\leq 12 \\ 4x_1 - x_2 &\leq 20 \\ x_1, x_2 &\geq 0 \end{aligned}$$

5.3. Minimize $z = 3x_1 - x_2$ subject to

$$\begin{aligned} x_1 + 2x_2 &\geq 12 \\ 4x_1 - x_2 &\leq 20 \\ 3x_1 + 6x_2 &\geq 36 \\ x_1, x_2 &\geq 0 \end{aligned}$$

5.4. Same as Problem 5.3, except that x_1 and x_2 are *unrestricted.*

5.5. Maximize $z = x_1 + x_2$ subject to

$$x_1 + x_2 \leq 10$$
$$x_1 + x_2 \geq 10$$
$$x_1, x_2 \geq 0$$

5.6. Maximize $z = 5x_1 + x_2$ subject to

$$3x_1 + x_2 \geq 12$$
$$x_2 \geq 7$$
$$x_1, x_2 \geq 0$$

5.7. Minimize $5x_1 - x_2$ subject to

$$x_1 + 3x_2 = 7$$
$$3x_1 - x_2 = 0$$
$$x_1, x_2 \geq 0$$

5.8. Maximize $4x_1 + 2x_2$ subject to

$$x_1 + x_2 \leq 7$$
$$3x_1 + 2x_2 \geq 30$$
$$x_1, x_2 \geq 0$$

5.9. Minimize $3x_1 - 4x_2$ subject to

$$-3x_1 + x_2 \leq 9$$
$$-9x_1 + 12x_2 \leq 21$$
$$x_1, x_2 \geq 0$$

5.10. The Wheeler Dealer Company produces two types of automobile phones for customers from the OPEC nations. The Midas Touch phone is 14-carat gold; the less flamboyant Silver Sheik is silver and jewel-encrusted. Orders for the phones are far outrunning production capabilities and the main concern of the firm is the optimal production policy.

The Midas Touch requires 2 hours each in production and 1 hour of testing; each of the Silver Sheiks requires 1.5 hours in production and 1 hour of test. A total of 100 worker-hours of production time and 120 worker-hours of testing time is available each week. Profit per phone is $2,000 for each Midas Touch and $1,700 each for the Silver Sheik.

(a) Let us first assume that the firm wishes to minimize the total number of units produced (so as to reduce handling problems, etc.) while achieving a weekly profit of at least $150,000. Graph and solve this problem as a single-objective model and discuss the results.

(b) Now assume that we seek to maximize weekly profit and disregard the number of phones produced. Graph this problem and discuss the results.

(c) Discuss the amount of time available for test and production. Could either amount be reduced (and by how much) without affecting the profit found in part (b)?

5.11. Graph the following problem and solve for the optimal program:

$$\text{maximize } z = x_1 + 1.5x_2$$

subject to

$$10x_1 + 10x_2 \leq 166$$

$$100x_1 \geq 580$$

$$x_1, x_2 \geq 0$$

5.12. Solve Problem 5.11 graphically if both x_1 and x_2 must be integers. Discuss your approach and results.

5.13. What is the impact of simply rounding off the continuous solution (found in Problem 5.11) to the nearest integer value? What happens if the solution found in Problem 5.10 is rounded off?

THE FOUNDATIONS OF THE SIMPLEX METHOD

Introduction

The purpose of this chapter is to introduce the reader to the rationale underlying the simplex method of linear programming. With some understanding of what simplex is, how it works, and why it works, one is better equipped to take advantage of the method's potential and to more clearly understand the meaning and implications of the results.

For the most part, the major theorems of linear programming and the simplex method are simply presented and discussed. These ideas are then reinforced by illustrative examples in this and succeeding chapters. However, we present a geometric derivation of what is perhaps the key theorem in linear programming: the extreme point theorem.

Those readers who are not interested in these basic details and wish to skip this material should at least read the chapter summary before going on to Chapter 7.

The review of linear algebra presented in Chapter 3 serves as a basis for much of the material that is discussed in this chapter. We now proceed to a narrative summary of the simplex method for the single-objective linear programming model.

6

The Procedure: A Narrative Summary

As we have noted previously, and as we prove later, if an optimal solution exists to a linear programming problem, the constraints of that problem form a convex set. In turn, this set (except in the trivial case of a single point) contains an infinite number of solutions to the constraint set, whereas all points lying outside the convex set violate (i.e., do not satisfy) at least one of the problem constraints.

Now, if one had to search the entire set of solutions for that one solution that optimized the objective, the solution of linear programming problems would be an impossible dream. However, as we shall see, none of the points in the interior of the convex set need be examined, as the optimum must lie on the boundaries of the set. This still entails an infinite number of points to be searched, and thus our next step will be to prove that, if an optimal solution exists, one (or more) of the extreme points of the convex set will be optimal.

The concepts just discussed are collectively termed, by this author, the *simplex filter*. Notice that, with each step of our discussion, we filter out more and more of the solution space until, finally, we have isolated a *finite* number of candidate solution points: the extreme points of the convex set.

Unfortunately, the fact that we have reduced our search to a finite set of points is still of little *practical* value. This is because of the enormous number of extreme points that typically exist in a problem of moderate to large size. For example, a problem with just 20 variables and 10 constraints may have as many as 184,756 extreme points. The simplex method provides a means for alleviating this difficulty by conducting an efficient, *systematic* search process which quarantees convergence in a finite number of steps. (In doing so, it also detects alternative optimal solutions, unboundedness, and infeasibility.)

The general steps of the simplex process are then:

1. Begin the search at an extreme point (i.e., a basic feasible solution).
2. Determine if movement to an adjacent extreme point can improve on the optimization of the objective function. If not, the present solution is optimal. If, however, improvement is possible, we proceed to the next step.
3. Move to the adjacent extreme point which offers (or, perhaps, *appears* to offer) the most improvement in the objective function.
4. Continue steps 2 and 3 until the optimal solution is found or until it can be shown that the problem is either unbounded or infeasible.

We term step 2 the *simplex index*. That is, in this step we obtain an *indication* as to whether or not improvement is possible. In actuality, with the exception of this step, the majority of simplex operations simply involve solution of simultaneous linear equations.

In Chapter 3 we defined, geometrically, an adjacent extreme point. Recall that it is simply an extreme point that may be reached by movement along one edge of the convex set. Algebraically, the movement is accomplished by a one-for-one exchange of a basic variable for a nonbasic variable. The accomplishment of this exchange, which we discuss later, is also a straightforward process.

Before leaving this discussion, there are three points that need emphasis. First, there are a wide number of variations on the simplex method and the one we have discussed is of a general, and rather elementary form. Second, the first step of the simplex method was to begin the search at a basic feasible solution. This is a highly restrictive rule to which we shall initially adhere. Later, however, we show how one may avoid this restriction, with an increase in solution efficiency, by starting our search at *either* a basic feasible solution *or* a basic *infeasible* solution. Although it may not now be apparent to the reader, the increased flexibility of this latter approach is significant. Our third, and last, point involves the form of the linear programming constraints with which we shall deal. Linear programming constraints are either inequalities (of type I or II) or equalities; however, when applying the simplex method we restrict our attention to equalities. Such a restriction allows us to apply the basic tools of linear algebra for the solution of simultaneous linear equations. This implies that we must convert all inequality constraints into equality constraints. The approach typically taken in the conversion of linear programming constraints into a form suitable for (elementary) simplex is the topic of the next section.

Slack, Surplus, and Artificial Variables

It is, in general, far easier to deal with equations than with inequalities. In this section we provide a few simple rules that allow one to convert any inequality into an equation by means of introducing some additional variables into the formulation.

For simplicity in procedure, we shall assume that, regardless of the form of the inequality, the right-hand side is nonnegative (i.e., $b_i \geq 0$). Thus, if the constraint initially has a negative b_i, we multiply the entire constraint by -1 and reverse the direction of the inequality (thus, a type I becomes a type II, and vice versa, whereas an equality remains unchanged).

Constraint Conversion

Consider first an inequality of the type I form. A typical constraint of this class, say constraint r, may be written as

$$\sum_{j=1}^{n} a_{r,j}x_j \leq b_r \tag{6.1}$$

We introduce a new variable, $s_r \geq 0$, termed the *slack* or *negative deviation* variable, so that

$$\sum_{j=1}^{n} a_{r,j}x_j + s_r = b_r \tag{6.2}$$

That is,

$$s_r = b_r - \sum_{j=1}^{n} a_{r,j}x_j \tag{6.3}$$

In words, s_r is the (nonnegative) difference between the right-hand side constant and the original left-hand side and, thus, it "takes up the slack." Physically, it often represents the amount of resource $r(b_r)$ that is unused or idle. From a mathematical view, it allows us to express our type I inequality in the more convenient equality format.

Next, consider a typical type II inequality, say inequality t.

$$\sum_{j=1}^{n} a_{t,j}x_j \geq b_t \tag{6.4}$$

Multiplying this entire constraint by -1, we have

$$-\sum_{j=1}^{n} a_{t,j}x_j \leq -b_t \tag{6.5}$$

Since we know that a type I inequality may be converted to an equation by the introduction of a slack variable, let us introduce variable s_t in (6.5) and then multiply the result by -1, to obtain

$$\sum_{j=1}^{n} a_{t,j}x_j - s_t = b_t \tag{6.6}$$

Note that, in this case, s_t is

$$s_t = \sum_{j=1}^{n} a_{t,j}x_j - b_t \tag{6.7}$$

That is, it represents the (nonnegative) difference between the left-hand side and the right-hand side of the type II inequality. Such a variable is termed a *surplus* or *positive deviation* variable. Physically, the surplus variable represents the amount by which we *exceed* the right-hand side, and thus "surplus" seems to be a fairly appropriate term.

Example 6-1

Consider the following two inequalities. Convert these into equations and find a basic feasible solution (see Chapter 3).

$$4x_1 + x_2 \leq 20 \tag{6.8}$$

$$x_1 + 4x_2 \leq 40 \tag{6.9}$$

Add slack variable s_1 to (6.8) and s_2 to (6.9), and we then obtain

$$4x_1 + x_2 + s_1 = 20 \tag{6.10}$$

$$x_1 + 4x_2 + s_2 = 40 \tag{6.11}$$

Notice that a basic feasible solution may be seen by inspection, that is,

$$s_1 = 20 \qquad s_2 = 40$$

Physically, this basic feasible solution suggests that all of resources 1 and 2 are idle and, thus, corresponds to doing nothing (i.e., $x_1 = x_2 = 0$).

Example 6-2

Let us try to repeat what was done in Example 6-1 with the following two constraints:

$$x_1 + 3x_2 \leq 10 \tag{6.12}$$

$$5x_1 + 3x_2 \geq 12 \tag{6.13}$$

Adding slack variable s_1 to (6.12) and subtracting surplus variable s_2 from (6.13), we have

$$x_1 + 3x_2 + s_1 = 10 \tag{6.14}$$

$$5x_1 + 3x_2 - s_2 = 12 \tag{6.15}$$

Unfortunately, we cannot (as we did in Example 6-1) immediately find a basic feasible solution by inspection through the use of our additional variables (i.e., s_1 and s_2). This is because the sign of s_2 is *negative*. Setting x_1 and x_2 to zero would force $s_2 = -12$ and, since *all* variables in the linear programming problem must be nonnegative, this is not permitted.

The dilemma of Example 6-2 may be circumvented via several ways. The first approach, and the most common, that we shall discuss is by means of the employment of an *artificial variable*. At this point, some readers are undoubtably saying to themselves: "artificial variable?" Aren't the slack and surplus variables artificial? No; the slack and surplus variables represent physically meaningful parameters and thus should not be termed artificial. On the other hand, the artificial variable we are about to introduce does *not* have any *physical* significance; it is truly an artificial device employed so that we may obtain an immediately recognizable basic feasible solution. Once this purpose has been achieved, we shall, in the simplex algorithm, attempt to drive the artificial variables out of the basis (i.e., nonbasic).

An artificial variable will be represented by q_i. We add artificial variables to *both* type II inequality and *equality* constraints. Thus, a type II inequality of the form shown in (6.4) is transformed into an equality by means of subtracting a surplus variable and adding an artificial variable as follows:

$$\sum_{j=1}^{n} a_{t,j}x_j - s_t + q_t = b_t \tag{6.16}$$

An *equation*, say constraint k, is *also* converted before employing the (elementary) simplex algorithm. Thus,

$$\sum_{j=1}^{n} a_{k,j}x_j = b_k \tag{6.17}$$

is replaced by

$$\sum_{j=1}^{n} a_{k,j}x_j + q_k = b_k \tag{6.18}$$

An additional example should clarify the preceding discussion.

Example 6-3

Given the following constraints, convert them into a form suitable for the simplex algorithm.

$$4x_1 - 2x_2 \leq 28 \tag{6.19}$$

$$x_1 + x_2 \geq 5 \tag{6.20}$$

$$4x_1 + x_2 = 16 \tag{6.21}$$

We first add a slack variable, s_1, to the left-hand side of (6.19). Next, a surplus variable, s_2, is subtracted from the left-hand side of (6.20) while an artificial variable, q_1, is added. Finally, an artificial variable, q_2, is added to the left-hand side of (6.21). We then have

$$4x_1 - 2x_2 + s_1 = 28 \tag{6.22}$$

$$x_1 + x_2 - s_2 + q_1 = 5 \tag{6.23}$$

$$4x_1 + x_2 + q_2 = 16 \tag{6.24}$$

Notice that an immediately obvious basic feasible solution is

$$s_1 = 28, \qquad q_1 = 5, \qquad \text{and} \qquad q_2 = 16$$

When one employs slack, surplus, and artificial variables to convert constraints (as we shall do for the next few chapters), the result is that a basic feasible solution is immediately found which consists of slacks and/or artificials. This fact allows for one to begin the steps of the simplex procedure with a basic feasible solution found with minimal effort.

The Impact on the Objective Function

The choice of decision variables (i.e., the x_j's) directly affects the value of the objective function. This holds true as well with the slack, surplus, and artificial variables. As a result, each variable introduced in the constraint conversion process should also be introduced, with a proper coefficient, into the objective function. To make this process considerably easier, we shall consider only objective functions of the *maximization* form. This in no way eliminates the consideration of minimization-type objectives for, if a function z is to be minimized, the maximization of $-z$ will accomplish the same end. Thus, when dealing with a minimization objective, we simply multiply the objective by -1 and maximize it. This should be done *prior* to the introduction of slacks, surplus, and artificial variables.

Thus, given a maximizing objective z, r surplus and slack variables, and t artificial variables, the modified objective is:

$$\text{maximize } z = \sum_{j=1}^{n} c_j x_j + \sum_{k=1}^{r} c_k s_k + \sum_{p=1}^{t} c_p q_p \tag{6.25}$$

The first term, $\sum_{j=1}^{n} c_j x_j$ is simply the original objective function. The second term corresponds to the impact of the slack and surplus variables while the third represents the contribution of the artificial. The question that still remains is: What are the values for c_k and c_p? We shall answer this by first looking at the coefficients of the slack and surplus variables (c_k). Now, if there is a cost or profit associated with idle resources or a surplus, the values of c_k should, accordingly, reflect this value. However, in most textbooks it is usually assumed that these costs or profits are zero and, as a result, one normally sees c_k given a value of zero. This is convenient and speeds the formulation process, but it is a dangerous habit to fall into since, in actual practice, there is almost always some nonzero value associated with a slack or surplus.

Now consider the coefficients of the artificial variables (c_p). Since they are truly artificial, one cannot associate with them an analogous *physical* cost or profit. Rather, we should realize that they have been introduced merely as a mathematical convenience—so as to start the simplex process—and we wish to rid ourselves of these variables as soon as possible. With this intent in mind, we shall see that it is reasonable to assign an extremely large, negative value to each c_p—thus making these artificial variables extremely costly to include in the basis (i.e., at a positive value). What is commonly done is to assign to each artificial a coefficient of $-M$, where M is some very large number in comparison to the other objective function coefficients.

Example 6-4

Given the following linear programming model, convert both the constraints and objective function into the form required by the simplex algorithm.

$$\text{Minimize } z = 7x_1 - 3x_2 + 5x_3 \text{ (cost, in dollars)}$$

$$\text{subject to}$$

$$\begin{aligned} x_1 + x_2 + x_3 &\geq 9 \\ 3x_1 + 2x_2 + x_3 &\leq 12 \\ x_1, x_2, x_3 &\geq 0 \end{aligned}$$

We shall assume that the cost of a surplus unit in the first constraint is zero while the cost of slack in the second constraint is \$1.5 per unit. The resultant model is:

$$\text{maximize } z' = -7x_1 + 3x_2 - 5x_3 + 0s_1 - Mq_1 - 1.5s_2$$

$$\text{subject to}$$

$$\begin{aligned} x_1 + x_2 + x_3 - s_1 + q_1 &= 9 \\ 3x_1 + 2x_2 + x_3 + s_2 &= 12 \\ x_1, x_2, x_3, s_1, s_2, q_1 &\geq 0 \end{aligned}$$

Notice that the original objective, z, was to minimize *cost*. Thus, the objective $z' = -z$ must be to maximize the negative of cost, which is, of course, profit. Also, since the cost of a slack resource unit is \$1.5, it is a *negative* contribution to the profit objective.

Quite often, authors do not distinguish between a decision variable (x_j) or slack, surplus, or artificial variable and the foregoing model would be written:

$$\text{maximize } z' = -7x_1 + 3x_2 - 5x_3 + 0x_4 - Mx_5 - 1.5x_6$$

subject to

$$\begin{aligned} x_1 + x_2 + x_3 - x_4 + x_5 \qquad &= 9 \\ 3x_1 + 2x_2 + x_3 \qquad + x_6 &= 12 \\ \mathbf{x} \geq \mathbf{0} \end{aligned}$$

where

$$x_4 = s_1$$
$$x_5 = q_1$$
$$x_6 = s_2$$

The latter equivalent formulation will be easier to deal with in the sections that follow.

Definitions and Notation

The material in the next two sections (and in the chapters that follow) will be made easier if we introduce, at this point, some simplifying notation and definitions. First, we generally let the column vector $\mathbf{x}$ represent *all* the variables in the converted model (i.e., the model in which the appropriate slack, surplus, and artificial variables have been introduced). Thus, $\mathbf{x}$ may include some slack, surplus, and artificial variables. It is of order $n \times 1$.

We may then write the converted linear programming model as:

$$\text{maximize } z = \mathbf{cx} \tag{6.26}$$

subject to

$$\mathbf{Ax} = \mathbf{b} \tag{6.27}$$

$$\mathbf{x} \geq \mathbf{0} \tag{6.28}$$

where $\mathbf{c}$ = row vector containing the coefficients of the variables in $\mathbf{x}$ (order $1 \times n$)

$\mathbf{A}$ = $m \times n$ matrix of the coefficients of the (converted) constraints

$\mathbf{b}$ = right-hand-side column vector (order $m \times 1$)

We further denote the jth column of matrix $\mathbf{A}$ as $\mathbf{a}_j$. The basis matrix formed by including m linearly independent columns from $\mathbf{A}$ is denoted as $\mathbf{B}$. $\mathbf{B}$ is then a $m \times m$ nonsingular matrix. The columns of $\mathbf{B}$ (not to be confused with the right-hand-side vector, $\mathbf{b}$) are termed $\mathbf{b}_1, \mathbf{b}_2, \ldots, \mathbf{b}_m$.

Example 6-5

Consider the following linear programming model, in which x_4 and x_5 are slack variables.

$$\text{Maximize } z = 5x_1 + 7x_2 + x_3 + 0x_4 + 0x_5$$

$$\text{subject to}$$

$$\begin{aligned} x_1 + 3x_2 - x_3 + x_4 \qquad &= 12 \\ 5x_1 + 6x_2 \qquad\qquad + x_5 &= 24 \end{aligned}$$

$$\mathbf{x} \geq \mathbf{0}$$

This problem may be rewritten as

$$\text{maximize } z = \mathbf{cx}$$

$$\text{subject to}$$

$$\mathbf{Ax} = \mathbf{b}$$

$$\mathbf{x} \geq \mathbf{0}$$

where

$$\mathbf{c} = (5 \quad 7 \quad 1 \quad 0 \quad 0)$$

$$\mathbf{x} = \begin{pmatrix} x_1 \\ x_2 \\ x_3 \\ x_4 \\ x_5 \end{pmatrix}$$

$$\mathbf{A} = \begin{pmatrix} 1 & 3 & -1 & 1 & 0 \\ 5 & 6 & 0 & 0 & 1 \end{pmatrix}$$

$$\mathbf{b} = \begin{pmatrix} 12 \\ 24 \end{pmatrix}$$

Next, let us form a basis matrix from **A**. We use the third and first columns, *in that order*, to form

$$\mathbf{B} = \begin{pmatrix} -1 & 1 \\ 0 & 5 \end{pmatrix}$$

Since the matrix **B** is nonsingular, it is a suitable basis. Notice that

$$\mathbf{b}_1 = \mathbf{a}_3$$

$$\mathbf{b}_2 = \mathbf{a}_1$$

for this basis.

Any basis matrix **B** determines a basic solution to $\mathbf{Ax} = \mathbf{b}$. However, such a basic solution may be either feasible (all $x_j \geq 0$) or infeasible (one or more $x_j < 0$). The basic solution given by $\mathbf{B}$ is

$$\mathbf{x}_B = \mathbf{B}^{-1}\mathbf{b} \tag{6.29}$$

where

$$\mathbf{x}_B = \begin{pmatrix} x_{B,1} \\ x_{B,2} \\ \cdot \\ \cdot \\ \cdot \\ x_{B,m} \end{pmatrix}$$

Example 6-6

The basis of Example 6-5 may now be used to determine a basic solution

$$\mathbf{B} = \begin{pmatrix} -1 & 1 \\ 0 & 5 \end{pmatrix}$$

Thus,

$$\mathbf{B}^{-1} = \begin{pmatrix} -1 & \frac{1}{5} \\ 0 & \frac{1}{5} \end{pmatrix}$$

$$\mathbf{x}_B = \begin{pmatrix} -1 & \frac{1}{5} \\ 0 & \frac{1}{5} \end{pmatrix} \begin{pmatrix} 12 \\ 24 \end{pmatrix} = \begin{pmatrix} -\frac{36}{5} \\ \frac{24}{5} \end{pmatrix}$$

Interpreting our solution, we see that

$$x_{B,1} = -\tfrac{36}{5}$$
$$x_{B,2} = \tfrac{24}{5}$$

but recall that $\mathbf{a}_3$ is the first column of $\mathbf{B}$ and $\mathbf{a}_1$ is the second column. Thus,

$$x_3 = x_{B,1} = -\tfrac{36}{5}$$
$$x_1 = x_{B,2} = \tfrac{24}{5}$$

Our solution is basic *infeasible* because of the sign on x_3.

The basic solution, $\mathbf{x}_B$, is associated also with the m-component row vector, denoted as $\mathbf{c}_B$. The vector $\mathbf{c}_B$, in turn, is the vector containing the objective function coefficients of the basic variables $x_{B,1}, x_{B,2}, \ldots, x_{B,m}$. That is, $c_{B,i}$ is the coefficient, from the objective function, of variable $x_{B,i}$. In Example 6-6, $\mathbf{c}_B$ would have as its first element the coefficient associated with $x_{B,1} = x_3$ and thus $c_{B,1} = 1$. The next element is the coefficient associated with $x_{B,2} = x_1$ or $c_{B,2} = 5$. Thus,

$$\mathbf{c}_B = (1 \quad 5)$$

Given a *basic feasible solution*, the value of z (the objective function) may be found by

$$z = \mathbf{c}_B \mathbf{x}_B \tag{6.30}$$

where

$$\mathbf{c}_B = (c_{B,1} \quad c_{B,2} \quad \ldots \quad c_{B,m})$$

Example 6-7

Returning to Example 6-5, let us select columns 4 and 5 (in that order) from **A** as our basis. Thus,

$$\mathbf{B} = \begin{pmatrix} 1 & 0 \\ 0 & 1 \end{pmatrix}$$

$$\mathbf{x}_B = \mathbf{B}^{-1}\mathbf{b} = \begin{pmatrix} 1 & 0 \\ 0 & 1 \end{pmatrix}\begin{pmatrix} 12 \\ 24 \end{pmatrix} = \begin{pmatrix} 12 \\ 24 \end{pmatrix}$$

$$x_{B,1} = x_4 = 12$$

$$x_{B,2} = x_5 = 24$$

Since $\mathbf{x}_B$ is basic feasible, we find z as

$$z = \mathbf{c}_B \mathbf{x}_B = (0 \quad 0)\begin{pmatrix} 12 \\ 24 \end{pmatrix} = 0$$

which is reasonable since the basic feasible solution consists solely of slack variables.

Any column, $\mathbf{a}_j$, of the matrix **A** may be written as a linear combination of the columns in **B**. That is,

$$\begin{aligned} \mathbf{a}_j &= y_{1,j}\mathbf{b}_1 + \cdots + y_{m,j}\mathbf{b}_m \\ &= \sum_{i=1}^{m} y_{i,j}\mathbf{b}_i \\ &= \mathbf{B}\mathbf{y}_j \end{aligned} \tag{6.31}$$

This may also be written (as it most often is) as

$$\mathbf{y}_j = \mathbf{B}^{-1}\mathbf{a}_j \tag{6.32}$$

where

$$\mathbf{y}_j = \begin{pmatrix} y_{1,j} \\ \cdot \\ \cdot \\ \cdot \\ y_{i,j} \\ \cdot \\ \cdot \\ \cdot \\ y_{m,j} \end{pmatrix}$$

Each $y_{i,j}$ is a scalar in which the subscript i refers to the associated column in $\mathbf{B}$ and the subscript j refers to the vector $\mathbf{a}_j$. Thus, $y_{3,7}$ refers to the scalar multiplier of the column $\mathbf{b}_3$ in $\mathbf{B}$ and $\mathbf{a}_j = \mathbf{a}_7$. Notice in particular that if $\mathbf{B}$ is an identity matrix, then $\mathbf{y}_j$ and $\mathbf{a}_j$ are identical.

Our last definition in this section concerns the scalar parameter z_j, which has, as we shall see later, a significant economic interpretation. At this point we simply state that

$$\begin{aligned} z_j &= y_{1,j}c_{B,1} + \cdots + y_{m,j}c_{B,m} \\ &= \sum_{i=1}^{m} y_{i,j}c_{B,i} \\ &= \mathbf{c}_B\mathbf{y}_j \end{aligned} \tag{6.33}$$

There is a z_j associated with every column $\mathbf{a}_j$ of matrix $\mathbf{A}$. As the basis, $\mathbf{B}$, is changed, these z_j values in turn will change.

The following example should help to tie together all the concepts presented in this section.

Example 6-8

$$\text{Maximize } z = 3x_1 + x_2 + 2x_3$$

$$\text{subject to}$$

$$x_1 + 2x_2 + 4x_3 \leq 18$$

$$3x_1 + 2x_2 + 12x_3 \leq 54$$

$$\mathbf{x} \geq \mathbf{0}$$

Adding slack variable x_4 to the first constraint and slack variable x_5 to the second, and assuming a zero cost for each, we have

$$\text{maximize } z = \mathbf{cx}$$

$$\text{subject to}$$

$$\mathbf{Ax} = \mathbf{b}$$

$$\mathbf{x} \geq \mathbf{0}$$

where

$$\mathbf{c} = (3 \quad 1 \quad 2 \quad 0 \quad 0)$$

$$\mathbf{x} = \begin{pmatrix} x_1 \\ x_2 \\ x_3 \\ x_4 \\ x_5 \end{pmatrix}$$

$$\mathbf{A} = \begin{pmatrix} 1 & 2 & 4 & 1 & 0 \\ 3 & 2 & 12 & 0 & 1 \end{pmatrix}$$

$$\mathbf{b} = \begin{pmatrix} 18 \\ 54 \end{pmatrix}$$

First, notice that, if the vectors $\mathbf{a}_1$ and $\mathbf{a}_3$ were to be used for a basis, that the matrix $\mathbf{B}$ would be singular. The easiest choice of vectors would be those belonging to the slack variables (i.e., $\mathbf{a}_4$ and $\mathbf{a}_5$). Normally, these would be the initial set of vectors selected. However, to add more information to this example, let us select $\mathbf{a}_2$ and $\mathbf{a}_3$ (in that order). The resultant matrix, $\mathbf{B}$, as shown below, is nonsingular and thus suitable for a basis.

$$\mathbf{B} = \begin{pmatrix} 2 & 4 \\ 2 & 12 \end{pmatrix} = (\mathbf{a}_2 \quad \mathbf{a}_3) = (\mathbf{b}_1 \quad \mathbf{b}_2)$$

Using (6.30) to find $\mathbf{x}_B$, we have

$$\mathbf{x}_B = \mathbf{B}^{-1}\mathbf{b} = \begin{pmatrix} \frac{3}{4} & -\frac{1}{4} \\ -\frac{1}{8} & \frac{1}{8} \end{pmatrix} \begin{pmatrix} 18 \\ 54 \end{pmatrix} = \begin{pmatrix} 0 \\ \frac{9}{2} \end{pmatrix}$$

Since $x_{B,1}$ and $x_{B,2}$ are both nonnegative, this is a basic feasible solution. Recall from Chapter 3 that this is also a *degenerate* solution. The value of z corresponding to this solution is then

$$z = \mathbf{c}_B\mathbf{x}_B = (c_{B,1} \quad c_{B,2}) \begin{pmatrix} x_{B,1} \\ x_{B,2} \end{pmatrix} = (1 \quad 2) \begin{pmatrix} 0 \\ \frac{9}{2} \end{pmatrix} = 9$$

Next, let us find the values of $\mathbf{y}_1$ and z_1.

$$\mathbf{y}_1 = \mathbf{B}^{-1}\mathbf{a}_1 = \begin{pmatrix} \frac{3}{4} & -\frac{1}{4} \\ -\frac{1}{8} & \frac{1}{8} \end{pmatrix} \begin{pmatrix} 1 \\ 3 \end{pmatrix} = \begin{pmatrix} 0 \\ \frac{1}{4} \end{pmatrix}$$

$$z_1 = \mathbf{c}_B\mathbf{y}_1 = (1 \quad 2) \begin{pmatrix} 0 \\ \frac{1}{4} \end{pmatrix} = \frac{1}{2}$$

The Extreme Point Theorem: A Geometric Derivation [2]

Next, we derive the extreme point theorem. Our derivation, for convenience in both narrative and visual presentation, is restricted to two dimensions (i.e., two variables). However, the general n-dimension case is easily arrived at via induction.

The extreme point theorem allows us to eliminate all but extreme points from consideration in our search for the optimal solution. Our first step in its derivation will be to prove that the constraint set of a linear programming problem forms a convex solution space.

THE PARAMETRIC EQUATIONS OF A LINE Consider a plane with coordinates x_1 and x_2 and the linear function of the form $f(x_1, x_2) = ax_1 + bx_2$, where a and b are not both zero. If $Y(y_1, y_2)$ and $Z(z_1, z_2)$ are two distinct points on this plane, the *parametric equations* of the line through Y and Z are

$$\left.\begin{aligned} x_1 &= (1-\alpha)y_1 + \alpha z_1 \\ x_2 &= (1-\alpha)y_2 + \alpha z_2 \end{aligned}\right\} \tag{6.34}$$

where α is any real number.

CLOSED HALF-PLANE A linear constraint, $ax_1 + bx_2 \leq c$, defines a closed half-plane in which the boundary line corresponds to equality in the constraint.

From the latter definition, we note that every linear programming constraint forms a closed half-plane. Let us now show that a closed half-plane is a convex region.

Theorem 6-1

The closed half-plane, $ax_1 + bx_2 \leq c$, is a convex region.

Proof

Recall from Chapter 3 that a convex region is one in which, for any two points, the line segment drawn between them lies entirely in the region. Let us then define two points, Y and Z, that belong to the closed half-plane $ax_1 + bx_2 \leq c$. Thus, the coordinates of $Y(y_1, y_2)$ and of $Z(z_1, z_2)$ satisfy

$$ay_1 + by_2 \leq c$$
$$az_1 + bz_2 \leq c$$

Points Y and Z are indicated in Figure 6-1. Also, as shown in Figure 6-1, we designate the point $U\,(u_1, u_2)$ to be *any* point on the line segment connecting Y and Z. The parametric equations of the line segment YZ may be written as

$$\begin{aligned} au_1 + bu_2 &= a[(1-\alpha)y_1 + \alpha z_1] + b[(1-\alpha)y_2 + \alpha z_2] \\ &= (1-\alpha)(ay_1 + by_2) + \alpha(az_1 + bz_2) \end{aligned}$$

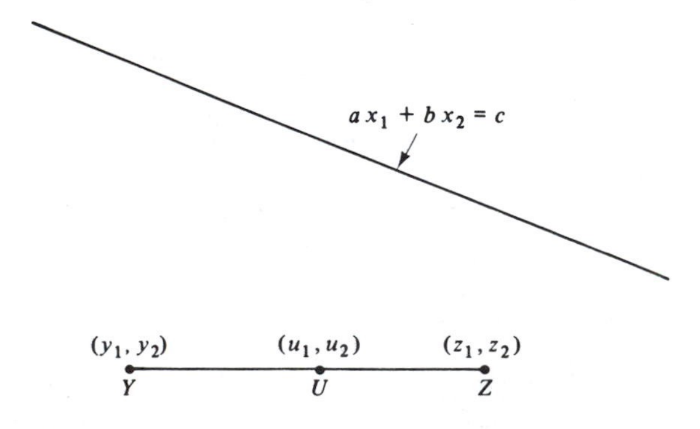

Figure 6-1. A closed half plane is a convex region.

since α and $(1 - \alpha)$ are nonnegative and since

$$ay_1 + by_2 \leq c \qquad \text{and} \qquad az_1 + bz_2 \leq c$$

Then

$$\begin{aligned} au_1 + bu_2 &\leq (1 - \alpha)c + \alpha c \\ &\leq c \end{aligned}$$

Thus, since the point U satisfies $au_1 + bu_2 \leq c$, it must lie in the closed half-plane. Further, since points Y, Z are any points in the half-plane and since U is any point on the line segment between Y and Z, the closed half-plane is a convex region.

Theorem 6-2

The intersection of any number of convex regions is convex.

Proof

Consider any two convex sets, A and B, in Figure 6-2 and their intersection, the set C. Now consider the points Y and Z that belong to set C. However, since C consists of all points common to both A and B, points Y and Z belong to both A and B. Further, since A and B are convex, the line segment drawn between Y and Z must belong to both A and B, and thus YZ belongs to set C. Thus, set C is convex. The same conclusion obviously holds true regardless of the number of convex sets forming the intersection, C.

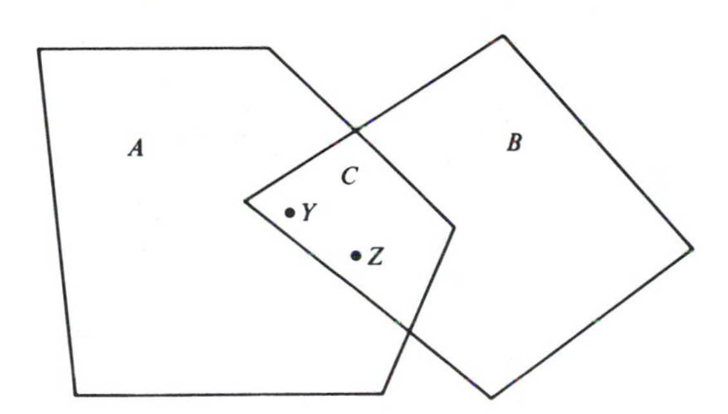

Figure 6-2. The intersection of any number of convex regions is convex.

By simply combining the results of Theorem 6-1 and 6-2, we may state Theorem 6-3.

Theorem 6-3

The intersection of any number of half-planes is a convex region.

These first three theorems show that the solution space formed by the constraints (i.e., the closed half-planes) of a linear programming problem is a convex set. We now go on to show that the optimal solution exists at the extreme points of this set.

AN EXTREME POINT (or VERTEX) A point, **x**, is termed an extreme point of a given convex set if and only if **x** does not lie on the line between two other distinct points of the convex set.

Theorem 6-4

Given a linear function, f:

a. If f has the same value at two distinct points, Y and Z, then f remains constant along the line YZ.
b. If, however, f has different values at Y and Z, then at each point on the open line segment YZ, f will have a value strictly between its values at Y and Z.

Proof

Let f have a value at Y of $f(y_1, y_2) = m$ and a value at Z of $f(z_1, z_2) = M$.

a. We first let $m = M$. Again, we consider any two points (Y and Z), the line segment between them (YZ), and any point on this line segment (U). We let

$$f(y_1, y_2) = m = ay_1 + by_2$$
$$f(z_1, z_2) = M = az_1 + bz_2$$

Thus, for point $U(u_1, u_2)$ on YZ, we have

$$\begin{aligned} f(u_1, u_2) &= a[(1 - \alpha)y_1 + \alpha z_1] + b[(1 - \alpha)y_2 + \alpha z_2] \\ &= (1 - \alpha)(ay_1 + by_2) + \alpha(az_1 + bz_2) \\ &= (1 - \alpha)f(y_1, y_2) + \alpha f(z_1, z_2) \\ &= (1 - \alpha)m + \alpha M \\ &= m + \alpha(M - m) \end{aligned}$$

but since $m = M$, $f(u_1, u_2) = m = M$, and thus $f(u_1, u_2)$ is *constant* if $m = M$.

b. Now, consider the case in which $m < M$. Using the same approach as in (a), we note that

$$f(u_1, u_2) = m + \alpha(M - m)$$

Thus, when $\alpha = 0, f(u_1, u_2) = m$ and when $\alpha = 1, f(u_1, u_2) = M$.

Further, for any two values of α, say $\alpha_1 < \alpha_2$, then

$$m + \alpha_1(M - m) < m + \alpha_2(M - m)$$

or

$$f(u_1, u_2)|_{\alpha_1} < f(u_1, u_2)|_{\alpha_2}$$

Thus, it may be seen that, if $m \neq M$, f has a value strictly between its value at Y and Z.

Theorem 6-4 may be stated more concisely as: A linear function defined over a closed line segment will assume its maximum and minimum values at the *end points* of that line segment. This allows us to proceed to Theorem 6-5.

Theorem 6-5

The maximum and minimum values of a linear function, f, defined over a given polygonal convex region, A, exist and are to be found at the extreme points of A.

Proof

An arbitrary convex region has been drawn in Figure 6-3. Observe first the line segment drawn through points P_1 and P_2. This is an *arbitrary* line segment

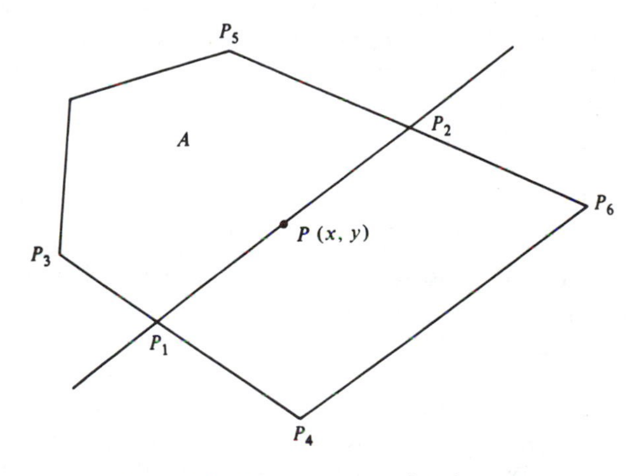

Figure 6-3. The extreme point theorem.

with two points, P_1 and P_2, on the boundaries of the convex region A. All points between P_1 and P_2 are interior points. Further, any points to the right of P_1 or left of P_1 are infeasible and thus are not of interest.

Point P is any point between P_1 and P_2 on the line segment. Thus, it represents any interior point of A. We shall now assume that

$$f(x_1, y_1) \leq f(x_2, y_2)$$

Thus, by Theorem 6-4,

$$f(x_1, y_1) \leq f(x, y) \leq f(x_2, y_2)$$

Now, consider point P_1 and assume that $f(x_3, y_3) \leq f(x_4, y_4)$. Applying Theorem 6-4 again, we have

$$f(x_3, y_3) \leq f(x_1, y_1) \leq f(x_4, y_4)$$

since P_1 is any point on the closed line segment between P_3 and P_4.

Combining the first two results, we note that

$$f(x_3, y_3) \leq f(x, y)$$

Returning to P_2 and under the assumption that $f(x_5, y_5) \leq f(x_6, y_6)$, we may state, by means of Theorem 6-4, that

$$f(x_5, y_5) \leq f(x_2, y_2) \leq f(x_6, y_6)$$

Combining all of the foregoing results, we have

$$f(x_3, y_3) \leq f(x, y) \leq f(x_6, y_6)$$

In words, f at any interior point P is bounded by the values of f on the convex set boundary and these, in turn, are bounded by the values of f at the extreme points. Thus, the extreme points of the convex set do represent the maximum and minimum values of f over that set.

With Theorem 6-5, we may now state the extreme point (or simplex filter) theorem.

Theorem 6-6

If the maximum or minimum value of a linear function defined over a polygonal convex region exists, it is found at the boundary or at an extreme point of the region.

This concludes our derivation of the extreme point theorem filter. We now proceed to a discussion of the manner in which the simplex method provides us with an indication of unboundedness, optimality, alternative optimal solutions, and infeasibility.

The Simplex Indicators

With the completion of the derivation of the extreme point theorem, we can now be assured that we need examine only the extreme points (i.e., basic feasible solutions) of the convex solution space. Earlier, we have found, through the use of slack, surplus, and artificial variables, that it is quite easy to identify an initial basic feasible solution—such a solution simply consists of the slacks and/or artificial variables used to convert the constraint set.

Unless the problem is trivial, the initial basic feasible solution formed by the slacks and/or artificials will have an associated objective function value, z, that may be improved upon by moving to an adjacent basic feasible solution. We now examine the procedure by which such a movement is made [1].

Improving the Basic Feasible Solution

Given any basis, $\mathbf{B}$, we may move to an adjacent extreme point (i.e., basic feasible solution) of the solution space by exchanging one of the columns in *the basis* ($\mathbf{b}_i$) for a column of $\mathbf{A}(\mathbf{a}_j)$ that is *not in the basis*. However, in making this one-for-one exchange, we must always maintain a feasible solution (all $x_j \geq 0$) and the objective function value (z) should increase (or, at least, not decrease).

There are essentially two rules to be followed in the one-for-one exchange. First, we must determine which vector in $\mathbf{A}(\mathbf{a}_j)$ is to be brought into the basis so that the solution will improve. Second, one of the vectors in $\mathbf{B}(\mathbf{b}_i)$ must leave and we select the departing vector so that the new basic solution is still feasible.

Under the assumption of a *maximizing* objective, we may now state the theorem that leads to the rule for the selection of the entering column (and associated variable).

Theorem 6-7

Given a basic feasible solution, $\mathbf{x}_B = \mathbf{B}^{-1}\mathbf{b}$, and associated objective function value, $z = \mathbf{c}_B\mathbf{x}_B$. If, for any column in $\mathbf{A}(\mathbf{a}_j)$ not in the basis, the condition $z_j - c_j < 0$ holds *and* at least one $y_{i,j} > 0$, then the replacement of one column in $\mathbf{B}$ by $\mathbf{a}_j$ will provide a new basic feasible solution in which $\hat{z} \geq z$.

Notice that our new objective function is denoted by $\hat{z}$. Our new basic feasible solution will be termed $\hat{\mathbf{x}}$, the new basis is $\hat{\mathbf{B}}$, and so on.

Theorem 6-7 shows that there is an indication of improvement available when employing simplex. The rule we employ to carry out this improvement process is termed the entering variable (or vector) rule. It should be stressed that it is not an exact rule but, rather, a very effective heuristic.

> **ENTERING VECTOR (VARIABLE) RULE** Given the conditions for improvement as stated in Theorem 6-7 (i.e., some $z_j - c_j < 0$ for which at least one $y_{i,j} > 0$), the vector to enter the basis will be that one associated with the most negative value of $z_j - c_j$. (Ties are broken arbitrarily.)

The vector that departs from $\mathbf{B}$, which we term $\mathbf{b}_r$, must be selected so that the new basis, $\mathbf{x}_B$ is feasible. The rule for the departing vector (or variable) is:

DEPARTING VECTOR (VARIABLE) RULE The column of **B**, $\mathbf{b}_r$, to be replaced by a_j is the one that satisfies

$$\frac{x_{B,r}}{y_{r,j}} = \min_i \left\{\frac{x_{B,i}}{y_{i,j}},\quad y_{i,j} > 0\right\} \tag{6.35}$$

The ratio $x_{B,r}/y_{i,j}$ is often denoted as θ and (6.35) is commonly referred to as the θ rule. (Ties are broken arbitrarily.)

Optimality

Optimality is reached when the following theorem holds.

Theorem 6-8

Given a basic feasible solution $\mathbf{x}_B = \mathbf{B}^{-1}\mathbf{b}$ with $z = \mathbf{c}_B\mathbf{x}_B$, z is maximized when $z_j - c_j \geq 0$ for every column $\mathbf{a}_j$ in **A**.

In practice, we normally check only the $\mathbf{a}_j$ columns corresponding to those vectors not in the basis, since $z_j - c_j$ for a vector in the basis is zero. The following example should clarify the preceding discussion.

Example 6-9

We use the following model to illustrate the entering and departing vector (variable) rules. In doing so, we are actually solving a linear programming problem by the simplex method.

$$\text{Minimize } z = -5x_1 + 2x_2 - 3x_3$$

$$\text{subject to}$$

$$\begin{aligned} 3x_1 + x_2 + 2x_3 &\leq 7 \\ x_1 + x_2 \qquad\quad &\leq 3 \\ \mathbf{x} &\geq \mathbf{0} \end{aligned}$$

First, we convert the constraints by adding a slack (x_4) to the first and adding a slack (x_5) to the second. The objective is then converted to the maximizing form and the new variables included. For convenience, we assume that the costs of the slack and surplus variables are zero. The resultant model is

$$\text{maximize } z' = 5x_1 - 2x_2 + 3x_3 + 0x_4 + 0x_5$$

$$\text{subject to}$$

$$\begin{aligned} 3x_1 + x_2 + 2x_3 + x_4 \qquad &= 7 \\ x_1 + x_2 \qquad\qquad + x_5 &= 3 \\ \mathbf{x} &\geq \mathbf{0} \end{aligned}$$

Thus,

$$\mathbf{c} = (5 \quad -2 \quad 3 \quad 0 \quad 0)$$

$$\mathbf{x} = \begin{pmatrix} x_1 \\ x_2 \\ x_3 \\ x_4 \\ x_5 \end{pmatrix}$$

$$\mathbf{A} = \begin{pmatrix} 3 & 1 & 2 & 1 & 0 \\ 1 & 1 & 0 & 0 & 1 \end{pmatrix}$$

$$\mathbf{b} = \begin{pmatrix} 7 \\ 3 \end{pmatrix}$$

Selecting the immediately obvious basis consisting of columns 4 and 5 (i.e., the slack variables), we form a basis **B** (which is an identity matrix). Thus,

$$\mathbf{x}_B = \mathbf{B}^{-1}\mathbf{b} = \begin{pmatrix} 1 & 0 \\ 0 & 1 \end{pmatrix}\begin{pmatrix} 7 \\ 3 \end{pmatrix} = \begin{pmatrix} 7 \\ 3 \end{pmatrix}$$

or

$$x_{B,1} = x_4 = 7$$

$$x_{B,2} = x_5 = 3$$

$$z' = \mathbf{c}_B\mathbf{x}_B = (0 \quad 0)\begin{pmatrix} 7 \\ 3 \end{pmatrix} = 0$$

Our initial basic feasible solution has $x_4 = 7$, $x_5 = 3$ for a value of z' of zero. We next determine whether improvement on this solution is possible and, if so, which variable enters and which departs. Since $\mathbf{a}_1$, $\mathbf{a}_2$, and $\mathbf{a}_3$ are not in the basis, we must determine $\mathbf{y}_1$, $\mathbf{y}_2$, $\mathbf{y}_3$, $z_1 - c_1$, $z_2 - c_2$, and $z_3 - c_3$.

Now,

$$\mathbf{y}_j = \mathbf{B}^{-1}\mathbf{a}_j$$

Thus,

$$\mathbf{y}_1 = \begin{pmatrix} 1 & 0 \\ 0 & 1 \end{pmatrix}\begin{pmatrix} 3 \\ 1 \end{pmatrix} = \begin{pmatrix} 3 \\ 1 \end{pmatrix}$$

$$\mathbf{y}_2 = \begin{pmatrix} 1 & 0 \\ 0 & 1 \end{pmatrix}\begin{pmatrix} 1 \\ 1 \end{pmatrix} = \begin{pmatrix} 1 \\ 1 \end{pmatrix}$$

$$\mathbf{y}_3 = \begin{pmatrix} 1 & 0 \\ 0 & 1 \end{pmatrix}\begin{pmatrix} 2 \\ 0 \end{pmatrix} = \begin{pmatrix} 2 \\ 0 \end{pmatrix}$$

$$z_j = \mathbf{c}_B\mathbf{y}_j$$

so

$$z_1 = (0 \quad 0)\begin{pmatrix} 3 \\ 1 \end{pmatrix} = 0$$

$$z_2 = 0$$

$$z_3 = 0$$

and

$$z_1 - c_1 = 0 - 5 = -5$$
$$z_2 - c_2 = 0 - (-2) = 2$$
$$z_3 - c_3 = 0 - 3 = -3$$

Notice that improvement on z' is possible since both $z_1 - c_1$ *and* $z_3 - c_3$ are negative (while at least one element in $\mathbf{y}_1$ and $\mathbf{y}_3$ is positive). The entering vector (variable) rule then suggests that $\mathbf{a}_1$ (the variable x_1 becomes basic) enters the basis (since $z_1 - c_1$ is more negative than $z_3 - c_3$).

To determine the vector (variable) to depart, we use (6.35) wherein the subscript j corresponds to the subscript on the entering vector (i.e., $\mathbf{a}_1$) or $j = 1$.

$$\frac{x_{B,r}}{y_{r,1}} = \min_i \left\{\frac{x_{B,1}}{y_{1,1}}, \frac{x_{B,2}}{y_{2,1}}\right\} \qquad \text{where } y_{i,j} > 0$$
$$= \min_i \left\{\frac{7}{3}, \frac{3}{1}\right\} = \frac{7}{3}$$

Thus, $x_{B,r}/y_{r,1}$ corresponds to $x_{B,1}/y_{1,1}$, which indicates that $\mathbf{b}_1$ must depart (i.e., $r = 1$). The new basis becomes

$$\hat{\mathbf{B}} = (\mathbf{a}_1 \quad \mathbf{a}_5) = (\mathbf{b}_1 \quad \mathbf{b}_2) = \begin{pmatrix} 3 & 0 \\ 1 & 1 \end{pmatrix}$$

and

$$\hat{\mathbf{x}}_B = \hat{\mathbf{B}}^{-1}\mathbf{b} = \begin{pmatrix} \frac{1}{3} & 0 \\ -\frac{1}{3} & 1 \end{pmatrix}\begin{pmatrix} 7 \\ 3 \end{pmatrix} = \begin{pmatrix} \frac{7}{3} \\ \frac{2}{3} \end{pmatrix}$$

or

$$\hat{x}_{B,1} = x_1 = \tfrac{7}{3}$$
$$\hat{x}_{B,2} = x_5 = \tfrac{2}{3}$$

and

$$\hat{z}' = \hat{\mathbf{c}}_B\hat{\mathbf{x}}_B = (5 \quad 0)\begin{pmatrix} \frac{7}{3} \\ \frac{2}{3} \end{pmatrix} = \frac{35}{3}$$

To determine whether or not improvement is now possible, we must check $\hat{z}_2 - c_2$, $\hat{z}_3 - c_3$, and $\hat{z}_4 - c_4$.

$$\hat{\mathbf{y}}_2 = \hat{\mathbf{B}}^{-1}\mathbf{a}_2 = \begin{pmatrix} \frac{1}{3} & 0 \\ -\frac{1}{3} & 1 \end{pmatrix}\begin{pmatrix} 1 \\ 1 \end{pmatrix} = \begin{pmatrix} \frac{1}{3} \\ \frac{2}{3} \end{pmatrix}$$
$$\hat{\mathbf{y}}_3 = \hat{\mathbf{B}}^{-1}\mathbf{a}_3 = \begin{pmatrix} \frac{1}{3} & 0 \\ -\frac{1}{3} & 1 \end{pmatrix}\begin{pmatrix} 2 \\ 0 \end{pmatrix} = \begin{pmatrix} \frac{2}{3} \\ -\frac{2}{3} \end{pmatrix}$$
$$\hat{\mathbf{y}}_4 = \hat{\mathbf{B}}^{-1}\mathbf{a}_4 = \begin{pmatrix} \frac{1}{3} & 0 \\ -\frac{1}{3} & 1 \end{pmatrix}\begin{pmatrix} 1 \\ 0 \end{pmatrix} = \begin{pmatrix} \frac{1}{3} \\ -\frac{1}{3} \end{pmatrix}$$

$$\hat{z}_2 = \hat{\mathbf{c}}_B \hat{\mathbf{y}}_2 = (5 \quad 0)\begin{pmatrix} \frac{1}{3} \\ \frac{2}{3} \end{pmatrix} = \frac{5}{3}$$

$$\hat{z}_3 = \hat{\mathbf{c}}_B \hat{\mathbf{y}}_3 = (5 \quad 0)\begin{pmatrix} \frac{2}{3} \\ -\frac{2}{3} \end{pmatrix} = \frac{10}{3}$$

$$\hat{z}_4 = \hat{\mathbf{c}}_B \hat{\mathbf{y}}_4 = (5 \quad 0)\begin{pmatrix} \frac{1}{3} \\ -\frac{1}{3} \end{pmatrix} = \frac{5}{3}$$

and

$$\begin{aligned} \hat{z}_2 - c_2 &= \tfrac{5}{3} - (-2) = \tfrac{11}{3} \\ \hat{z}_3 - c_3 &= \tfrac{10}{3} - 3 = \tfrac{1}{3} \\ \hat{z}_4 - c_4 &= \tfrac{5}{3} - 0 = \tfrac{5}{3} \end{aligned}$$

Since all $z_j - c_j$ for those $\mathbf{a}_j$ not in $\mathbf{B}$ are nonnegative, no improvement is possible and our optimal solution is thus

$$\begin{aligned} \mathbf{x}^* &= (\tfrac{7}{3} \quad 0 \quad 0 \quad 0 \quad \tfrac{2}{3}) \\ z^* &= -z'^* = -\tfrac{35}{3} \end{aligned}$$

We have now used the simplex algorithm to actually solve a linear programming problem. In Chapter 7 we see how this process may be *considerably* simplified through the use of a tabular format and a more streamlined algorithm. However, regardless of the refinements and/or form used, the simplex method follows the same general procedure illustrated by the foregoing example. That is, we move from one basis to another using our rules for the entering and departing vectors. However, in the general case there are other indications that are provided by simplex and of which one should be aware. We next discuss one of these events, the indication of an unbounded solution.

Unbounded Solutions

An unbounded problem exists when the conditions listed in Theorem 6-9 hold. Again, this theorem is based on the maximizing problem.

Theorem 6-9

Given any basic feasible solution, $\mathbf{x}_B = \mathbf{B}^{-1}\mathbf{b}$, to a linear programming problem. If, for some column $\mathbf{a}_j$ not in the basis, both

$$z_j - c_j < 0 \quad \text{and} \quad y_{i,j} \le 0 \qquad \text{for all } i$$

then the associated linear programming problem has an objective that may be made arbitrarily large. That is, the problem is unbounded.

From a practical point of view, we implement Theorem 6-9 by observing, at every iteration (i.e., change in basis) of the simplex method, the column vector $\mathbf{y}_j$ associated with every negative $z_j - c_j$ value. If *none* of the elements of $\mathbf{y}_j$

are positive, the problem is labeled as being unbounded and we may terminate our procedure.

Example 6-10

To illustrate the detection of an unbounded problem, consider the following example.

$$\text{Maximize } z = 3x_1 + x_2$$
$$\text{subject to}$$
$$x_1 - x_2 \leq 8$$
$$2x_1 - 2x_2 \leq 20$$
$$x_1, x_2 \geq 0$$

Adding slacks to both constraints and processing the objective (where slack cost is assumed zero), we have

$$\text{maximize } z = 3x_1 + x_2 + 0x_3 + 0x_4$$
$$\text{subject to}$$
$$x_1 - x_2 + x_3 = 8$$
$$2x_1 - 2x_2 + x_4 = 20$$
$$\mathbf{x} \geq \mathbf{0}$$

The initial basic feasible solution is just the slack variables x_3, x_4, so the initial basis contains $\mathbf{a}_3$ and $\mathbf{a}_4$ (in that order)

$$\mathbf{B} = \begin{pmatrix} 1 & 0 \\ 0 & 1 \end{pmatrix} = (\mathbf{a}_3 \quad \mathbf{a}_4) = (\mathbf{b}_1 \quad \mathbf{b}_2)$$

$$\mathbf{x}_B = \mathbf{B}^{-1}\mathbf{b} = \begin{pmatrix} 1 & 0 \\ 0 & 1 \end{pmatrix} \begin{pmatrix} 8 \\ 20 \end{pmatrix} = \begin{pmatrix} 8 \\ 20 \end{pmatrix}$$

$$x_{B,1} = x_3 = 8$$
$$x_{B,2} = x_4 = 20$$

$$z = \mathbf{c}_B\mathbf{x}_B = (0 \quad 0)\begin{pmatrix} 8 \\ 20 \end{pmatrix} = 0$$

Checking to see if this solution may be improved:

$$\mathbf{y}_1 = \mathbf{B}^{-1}\mathbf{a}_1 = \begin{pmatrix} 1 & 0 \\ 0 & 1 \end{pmatrix} \begin{pmatrix} 1 \\ 2 \end{pmatrix} = \begin{pmatrix} 1 \\ 2 \end{pmatrix}$$

$$\mathbf{y}_2 = \mathbf{B}^{-1}\mathbf{a}_2 = \begin{pmatrix} 1 & 0 \\ 0 & 1 \end{pmatrix} \begin{pmatrix} -1 \\ -2 \end{pmatrix} = \begin{pmatrix} -1 \\ -2 \end{pmatrix}$$

$$z_1 = \mathbf{c}_B\mathbf{y}_1 = (0 \quad 0)\begin{pmatrix}1\\2\end{pmatrix} = 0$$

$$z_2 = \mathbf{c}_B\mathbf{y}_2 = (0 \quad 0)\begin{pmatrix}-1\\-2\end{pmatrix} = 0$$

$$z_1 - c_1 = 0 - 3 = -3$$

$$z_2 - c_2 = 0 - 1 = -1$$

Now, looking at $z_1 - c_1$, we see that improvement appears possible since at least one element of $\mathbf{y}_1$ is positive. However, note that $z_2 - c_2$ is negative while $\mathbf{y}_2$ contains *no positive* elements. From Theorem 6-9 we see that our problem is unbounded and we need proceed no further. The reader is advised to solve this problem graphically also.

INFEASIBLE PROBLEMS

As we recall from our experience with the graphical technique, an infeasible problem is one in which there is no common intersection of all problem constraints. The solution space is empty and thus no point can satisfy all constraints. The simplex method detects the occurrence of an infeasible problem at the *conclusion* of its iterations. When artificial variables are employed to convert type II inequalities (this need not be done as we shall see later), the detection of an infeasible problem is accomplished by means of Theorem 6-10.

Theorem 6-10

Given a basic feasible solution, $\mathbf{x}_B = \mathbf{B}^{-1}\mathbf{b}$ to a linear programming problem and given that all $z_j - c_j \geq 0$ for every column $\mathbf{a}_j$ in $\mathbf{A}$, the problem is infeasible if any of the variables in $\mathbf{x}_B$ are artificial and have a positive value.

To employ Theorem 6-10 we merely check the final basic solution. If any $q > 0$ (where q is an artificial variable), we know that there is no mathematically feasible solution to the problem. We demonstrate such a case in Example 6-11.

Example 6-11

Let us simply assume that we have followed the steps of the simplex method (i.e., primarily the entering and departing vector rules) and find ourselves with an apparently optimal solution. That is, all $z_j - c_j \geq 0$ for each a_j and $\mathbf{x}_B = \mathbf{B}^{-1}\mathbf{b}$ is the final basic feasible solution. Examining $\mathbf{x}_B$, we may note that one (or more) of these basic variables correspond to an artificial variable. For example, assume that

$$x_{B,3} = q_1 = 7$$

Since $q_1 > 0$ in the final basis, the problem is infeasible. However, if $q_1 = 0$ (i.e., a degenerate solution), the problem is feasible as long as no other artificials appear in the basis at a positive value. We examine this case in more detail in Chapter 7.

ALTERNATIVE OPTIMAL SOLUTIONS

The final indicator that we discuss in this chapter is the signal for the existence of alternative optimal solutions. As with infeasible problems, the detection of alternative optimal solutions must wait until the iterations of the simplex algorithm are completed. We then employ Theorem 6-11.

Theorem 6-11

Given an optimal basic feasible solution, $\mathbf{x}_B = \mathbf{B}^{-1}\mathbf{b}$, to a linear programming problem (i.e., all $z_j - c_j \geq 0$ for every column $\mathbf{a}_j$ in $\mathbf{A}$ with no positive valued artificials in the basis). If, for some $\mathbf{a}_j$ not in the basis, $z_j - c_j = 0$, the problem has alternative optimal solutions.[a]

The check for alternative optimal solutions is made by simply examining every $z_j - c_j$ value for nonbasic $\mathbf{a}_j$ columns. If any of these $z_j - c_j$ values are zero, then alternative optimal solutions exist.

Example 6-12

We shall assume that we have reached an optimal, basic feasible solution (with no artificials in the basis) and that vectors $\mathbf{a}_1$ and $\mathbf{a}_4$ are in the basis while vectors $\mathbf{a}_2$, $\mathbf{a}_3$, and $\mathbf{a}_5$ are nonbasic. The corresponding values for $z_j - c_j$ are

$$z_2 - c_2 = 3$$
$$z_3 - c_3 = 0$$
$$z_5 - c_5 = 7$$

Since $z_3 - c_3 = 0$, there are alternative optimal solutions. As we recall from earlier material, these alternative optimal solutions lie at two (or more) adjacent extreme points *and* the edges joining these points. Thus, to find another alternative optimal solution, we may move to an adjacent extreme point with the same objective value. This may be accomplished, in our example, by bringing in $\mathbf{a}_3$ and removing either $\mathbf{a}_1$ or $\mathbf{a}_4$ (according to the rule for departing variables). We examine such problems in more detail in succeeding chapters.

Summary

A considerable amount of material has been covered, or at least presented, in this chapter. The diligent reader may, at this time, feel that simplex is not an easy technique to master. Nothing could be further from the truth. The discussion of the simplex method is somewhat complex; however, its actual employment

[a]If, for some $\mathbf{a}_j$ not in the basis, $z_j - c_j = 0$ *and* none of the elements in the associated $\mathbf{y}_j$ are positive, we have what is known as an alternative optimal ray.

is extremely straightforward. Simplex is one of those mathematical techniques that are far easier to use than to discuss. However, as we mentioned earlier, this discussion is necessary to lend understanding to the process.

We now summarize the highlights of Chapter 6.

The Simplex Method

The simplex method, in its most elementary form, consists of a systematic, directed process that begins at one basic feasible solution and moves, if possible, to an adjacent basic feasible solution at which the objective improves (or, at least, remains the same).

The extreme point theorem is the filter that permits us to examine only basic feasible solutions. If an optimal solution exists, it will exist at an extreme point. The movement from one extreme point to another is made, geometrically, by moving along the edge of the convex set that separates the two points. Algebraically, it is accomplished by two rules: the entering variable rule and the departing variable rule.

As an integral part of the simplex process we are given indications of optimality, improvement, unboundedness, infeasibility, and alternative optimal solutions. Since the simplex search process considers only the finite number of extreme points, and since it never moves to an extreme point that would degrade the solution, convergence in a finite number of steps occurs in practice.[b]

Model Preprocessing

The single-objective linear programming model should be preprocessed prior to the implementation of the simplex method. Such preprocessing includes:

1. Change a minimizing objective to a maximizing objective (multiply the objective by -1).
2. Convert all of the constraints (other than the nonnegativity conditions on the decision variables) as follows:

    a. $g_i(\mathbf{x}) \leq b_i$
    Add a slack variable.
    b. $g_i(\mathbf{x}) \geq b_i$
    Subtract a surplus variable and add an artificial variable.
    c. $g_i(\mathbf{x}) = b_i$
    Add an artificial variable.
    d. $g_i(\mathbf{x}) \leq -b_i$ or $g_i(\mathbf{x}) \geq -b_i$ or $g_i(\mathbf{x}) = -b_i$
    All b_i must be positive, so multiply through by -1 (this reverses the direction of any inequality).

---

[b]In theory, the method could cycle continuously under certain conditions. We shall briefly discuss this phenomenon later. However, we note that there are ways to prevent such cycling.

3. Modify the objective function by incorporating the slack, surplus, and artificial variables used to convert the constraints, where (a) for each slack or surplus, multiply the slack or surplus by its respective cost or profit (recall that, typically, a zero value is assumed); and (b) for each artificial, multiply by $-M$ (where M is some very large number).

Important Definitions and Notation

The preprocessed single-objective linear programming problem may be described, in concise matrix form, as

$$\text{maximize} \quad z = \mathbf{cx} \tag{6.36}$$

subject to

$$\mathbf{Ax} = \mathbf{b} \tag{6.37}$$

$$\mathbf{x} \geq \mathbf{0} \tag{6.38}$$

where $\mathbf{c}$ = $(1 \times n)$ row vector containing the coefficients of each x_j (which includes slacks, surplus, or artificial variables)

$\mathbf{A}$ = $(m \times n)$ matrix of the coefficients of the (converted) constraints

$\mathbf{b}$ = right-hand-side $(m \times 1)$ column vector

$\mathbf{x}$ = $(n \times 1)$ column vector of all decision, slack, surplus, and artificial variables

The basis matrix, denoted as **B**, is composed of m linearly independent columns from **A**. **B** is thus a $m \times m$ nonsingular matrix. Normally, the initial columns in **B** are those associated with the slacks and/or artificials in the converted constraints.

$$\mathbf{B} = (\mathbf{b}_1 \quad \mathbf{b}_2 \quad \cdots \quad \mathbf{b}_m) \tag{6.39}$$

where all $\mathbf{b}_i$ are linearly independent.

A basic solution designated as $\mathbf{x}_B$ is given by

$$\mathbf{x}_B = \mathbf{B}^{-1}\mathbf{b} \tag{6.40}$$

where

$$\mathbf{x}_B = \begin{pmatrix} x_{B,1} \\ x_{B,2} \\ \cdot \\ \cdot \\ \cdot \\ x_{B,m} \end{pmatrix}$$

If all $x_{B,i} \geq 0$, then $\mathbf{x}_B$ is a basic feasible solution.

Given a basic feasible solution, $\mathbf{x}_B$, the value of the objective function (z) is given as

$$z = \mathbf{c}_B \mathbf{x}_B \tag{6.41}$$

where

$$c_B = (c_{B,1} \quad c_{B,2} \quad \cdots \quad c_{B,m})$$

We shall define $\mathbf{y}_j$ as

$$\mathbf{y}_j = \mathbf{B}^{-1}\mathbf{a}_j \tag{6.42}$$

where

$$\mathbf{y}_j = \begin{pmatrix} y_{1,j} \\ \cdot \\ \cdot \\ \cdot \\ y_{m,j} \end{pmatrix}$$

and

$$z_j = \mathbf{c}_B \mathbf{y}_j \tag{6.43}$$

The Steps of the Simplex Method

Step 1. We begin our search with a basic feasible solution, $\mathbf{x}_B = \mathbf{B}^{-1}\mathbf{b}$ (where $\mathbf{x}_B$ is normally composed of slacks and/or artificials).

Step 2. Examine $z_j - c_j$ for all $\mathbf{a}_j$ not in the basis. If all $z_j - c_j \geq 0$, go to step 6. Otherwise, go to step 3.

Step 3. If, for any $\mathbf{a}_j$ for which $z_j - c_j$ is negative, there are no positive elements in $\mathbf{y}_j$, the problem is unbounded and we stop. Otherwise, we select the vector (and variable) associated with the most negative $z_j - c_j$ to enter the basis.

Step 4. Use (6.35) to determine the departing vector (and variable) from the basis.

Step 5. Establish the new basis, and find the new basic feasible solution and the new objective function value. Return to step 2.

Step 6. If any variable in the basis is both an artificial and has a positive value, the problem is infeasible. Otherwise, we have found the optimal solution. Note that if any $z_j - c_j$ equals zero for an $\mathbf{a}_j$, not in the basis, alternate optimal solutions exist.

REFERENCES

1. Hadley, G. *Linear Programming*. Reading, Mass.: Addison-Wesley, 1963.
2. Wolfe, C. S. *Linear Programming with FORTRAN*. Glenview, Ill.: Scott, Foresman, 1973.

PROBLEMS

6.1. Discuss the *physical* meaning of slack, surplus, and artificial variables. Provide illustrations.

6.2. Why are slacks, surplus, and artificial variables used in the simplex process?

6.3. Formulate a linear programming problem for which it would be unnecessary to employ either any slack, surplus, or artificial variables.

Process the following three problems so as to be in the proper form to begin the simplex procedure.

6.4. Maximize $z = 4x_1 - 7x_2 + x_3$ subject to

$$\begin{aligned} x_1 + 2x_2 - x_3 &= 12 \\ 3x_1 + x_2 \quad &\leq 18 \\ x_3 &\geq 2 \\ x_1, x_2, x_3 &\geq 0 \end{aligned}$$

6.5. Minimize $z = 2x_1 + x_2 - x_3$ subject to

$$\begin{aligned} x_1 + 3x_2 \quad &\leq 20 \\ 2x_1 + x_2 + x_3 &\geq 15 \\ x_1, x_2, x_3 &\geq 0 \end{aligned}$$

6.6. Maximize $z = x_1 - 4x_2 + x_3$ subject to

$$\begin{aligned} 2x_1 + x_2 + 3x_3 &= 10 \\ - 3x_2 + x_3 &= -4 \\ x_1 \quad + x_3 &= 3 \end{aligned}$$

Using the procedure described in this chapter and illustrated in Example 6-9, solve Problems 6.7 through 6.12.

6.7. Maximize $z = 10x_1 + 20x_2$ subject to

$$\begin{aligned} 5x_1 + 8x_2 &\leq 40 \\ 5x_1 + 3x_2 &\leq 30 \\ x_1, x_2 &\geq 0 \end{aligned}$$

6.8. Minimize $z = 3x_1 + 2x_2$ subject to

$$\begin{aligned} 5x_1 + x_2 &\geq 10 \\ 2x_1 + 2x_2 &\geq 12 \\ x_1 + 4x_2 &\geq 12 \\ x_1, x_2 &\geq 0 \end{aligned}$$

6.9. Maximize $z = 6x_1 + 8x_2$ subject to

$$\begin{aligned} 4x_1 + x_2 &\leq 20 \\ x_1 + 4x_2 &\leq 40 \\ x_1, x_2 &\geq 0 \end{aligned}$$

6.10. Maximize $z = 3x_1 + x_2$ subject to

$$\begin{aligned} 2x_1 + x_2 &\geq 4 \\ x_2 &\geq 2 \\ x_1, x_2 &\geq 0 \end{aligned}$$

6.11. Maximize $z = 6x_1 + 4x_2$ subject to

$$\begin{aligned} x_1 + x_2 &\leq 3 \\ 3x_1 + x_2 &\geq 10 \\ x_1, x_2 &\geq 0 \end{aligned}$$

6.12. Maximize $z = 30x_1 + 50x_2$ subject to

$$\begin{aligned} 2x_1 + x_2 &\leq 16 \\ x_1 + 2x_2 &\leq 11 \\ x_1 + 3x_2 &\leq 15 \\ x_1, x_2 &\geq 0 \end{aligned}$$

6.13. Once a variable departs from the basis (during the simplex procedure), can it ever return (i.e., become, once again, basic)? Why or why not?

6.14. For Problems 6.7 through 6.12, depict the results of the simplex procedure graphically and satisfy yourself that all moves are to adjacent vertices.

THE SIMPLEX METHOD: TABLEAUX AND COMPUTATION

Introduction

In Chapter 6 we solved several problems via the simplex method, utilizing the fundamental relationships summarized at the conclusion of that chapter. However, when solving problems by hand, it is generally far more convenient to employ one of the many tabular formats that have been developed. These tabular formats allow us to utilize exactly the same concepts cited earlier, but to do so in a more straightforward, systematic manner which, in addition, minimizes the amount of "bookkeeping" required.

In this chapter we present two types of simplex tableaux. The first is the more traditional: the *extended tableau.* A version of the simplex algorithm, fitted specifically to this tableau, will then be presented and its use illustrated by means of several examples.

The incorporation of artificial variables, as used with type II and equality constraints, presents a number of computational burdens. One of the earliest means to alleviate these difficulties is denoted as the *two-phase method.* We present this method for two reasons. First, it is of some historical interest. Second, and more important to our effort, is the fact that refinements to this two-phase method allow us, in Part Four, to handle multiple-objective linear programming problems.

Following the two-phase method, we present our preferred form of the simplex tableau: the *condensed tableau.* This tableau, which relieves some of the

7

drudgery of problem solving, is the one that is employed throughout this text. Concluding the chapter is a discussion of some of the practical complications that arise in problem formulation and solving, together with means to avoid them or, at least, minimize their impact.

At the risk of sounding old-fashioned, the author strongly suggests that the student gain as much practice in problem solving (particularly with the condensed tableau) as possible. It is usually far better to solve a large number of small problems than to expend a great amount of time and energy on a few large problems. It has been the author's observation that, not only does such practice sharpen mechanical skills, but it also increases one's understanding of simplex and linear programming.

The Extended Tableau

A tableau is simply a convenient place to store the information regarding the problem under investigation. Recall from Chapter 6 that such information consists primarily of:

1. $z = \mathbf{c}_B\mathbf{x}_B$, the objective value.
2. $\mathbf{x}_B = \mathbf{B}^{-1}\mathbf{b}$, the basic (feasible) solution.
3. $\mathbf{B} = (\mathbf{b}_1 \quad \mathbf{b}_2 \quad \cdots \quad \mathbf{b}_m)$, the basis matrix.

4. $\mathbf{y}_j = \mathbf{B}^{-1}\mathbf{a}_j$ and the marginal utility of variable j is $z_j - c_j$ (where $z_j = \mathbf{c}_B\mathbf{y}_j$), where these parameters indicate such conditions as improvement, unboundness, and optimality.

All of this information, and more, is conveniently summarized in the extended tableau—for every iteration of the simplex method. Table 7-1(a) depicts the general case, extended tableau (the rationale behind its name will be clear when we discuss the condensed tableau). In this chapter we use a slightly abbreviated version of the extended tableau, as depicted in Table 7-1(b).

Table 7-1. THE GENERAL CASE EXTENDED TABLEAU

Coefficients of the Basic Variables	*Basic Variable Labels*	*Variable Labels* x_1	x_2	$\cdots$	x_n	*Basic Feasible Solution Values*
$c_{B,1}$	$x_{B,1}$	$y_{1,1}$	$y_{1,2}$	$\cdots$	$y_{1,n}$	$x_{B,1}$
$\vdots$	$\vdots$	$\vdots$			$\vdots$	$\vdots$
$c_{B,m}$	$x_{B,m}$	$y_{m,1}$	$y_{m,2}$	$\cdots$	$y_{m,n}$	$x_{B,m}$
Indicator Row		$z_1 - c_1$	$z_2 - c_2$	$\cdots$	$z_n - c_n$	z

(a) Extended Tableau

	$c_j \longrightarrow$	c_1	c_2	$\cdots$	c_n	
$\mathbf{c}_B$	BV	x_1	x_2	$\cdots$	x_n	$\mathbf{x}_B$
$c_{B,1}$	$x_{B,1}$	$y_{1,1}$	$y_{1,2}$	$\cdots$	$y_{1,n}$	$x_{B,1}$
$\vdots$	$\vdots$	$\vdots$			$\vdots$	$\vdots$
$c_{B,m}$	$x_{B,m}$	$y_{m,1}$	$y_{m,2}$	$\cdots$	$y_{m,n}$	$x_{B,m}$
		$z_1 - c_1$	$z_2 - c_2$	$\cdots$	$z_n - c_n$	z

(b) Abbreviated Version

The extended tableau appears considerably more formidable than it really is, as we shall soon discover. Directing our attention to the tableau of Table 7-1(b), notice first that the second column (from the left) is *not* the same thing as the far right column. The column labeled BV contains the basic variable *labels*. Thus, if the variables in the basis were $x_{B,1} = x_3$, $x_{B,2} = x_1$, and $x_{B,3} = x_4$, this column would simply list the labels x_3, x_1, and x_4. On the other hand, the column to the far right (labeled as $\mathbf{x}_B$) gives the *values* for the basic variables. Thus, it would give, in the example above, the numerical values for x_3, x_1, and x_4. The first column (to the left) lists the numerical values of the objective function coefficients for each associated basic variable listed in column 2. Thus, it is, as labeled, simply the $\mathbf{c}_B$ vector (actually, the transpose of $\mathbf{c}_B$). The interior columns are the $\mathbf{y}_j$ vectors as associated with each variable x_j. The value of the objective function at any iteration, z, is given on the right-hand side of the bottom (indicator) row. This row also gives the values of $z_j - c_j$ for each variable. The basis matrix is not found explicitly in the tableau but may easily be constructed by simply observing which variables (and thus which vectors from **A**) are in the basis—as given in the second column under BV. Thus, if x_3, x_1, and x_4 are in the basis (in that order), then

$$\mathbf{B} = (\mathbf{b}_1 \quad \mathbf{b}_2 \quad \mathbf{b}_3) = (\mathbf{a}_3 \quad \mathbf{a}_1 \quad \mathbf{a}_4)$$

Getting to specifics, let us examine the construction of the extended tableau for an example problem. We deal initially with just the setup of the initial tableau. Also, as discussed in Chapter 6, we always deal with a maximizing form of the objective function.

Example 7-1

Given the following problem, set up the initial extended tableau:

$$\text{maximize} \quad z = 6x_1 + 8x_2$$

$$\text{subject to}$$

$$4x_1 + x_2 \leq 20$$

$$x_1 + 4x_2 \leq 40$$

$$x_1, x_2 \geq 0$$

Preprocessing the example[a] results in:

$$\text{maximize} \quad z = 6x_1 + 8x_2 + 0x_3 + 0x_4$$

$$\text{subject to}$$

$$4x_1 + x_2 + x_3 \qquad = 20$$

$$x_1 + 4x_2 \qquad + x_4 = 40$$

$$x_1, x_2 \geq 0$$

Since our simplex method always results in a nonnegative solution, the nonnegativity constraints are not included in the tableau. We fill in the tableau as follows.

1. The basic structure of the tableau is laid out as shown in Table 7-2.

Table 7-2. INITIAL STRUCTURE

	$c_j \longrightarrow$	6	8	0	0	
$\mathbf{c}_B$	BV	x_1	x_2	x_3	x_4	$\mathbf{x}_B$
0	x_3					
0	x_4					

[a] We assume zero coefficients for slacks unless otherwise noted.

2. Notice that our basic feasible solution consists solely of slack variables and thus the initial basis is

$$\mathbf{B} = \begin{pmatrix} 1 & 0 \\ 0 & 1 \end{pmatrix} = \mathbf{I}$$

In all cases, **B** will initially be an identity matrix and thus

$$\mathbf{y}_j = \mathbf{B}^{-1}\mathbf{a}_j = \mathbf{I}^{-1}\mathbf{a}_j = \mathbf{I}\mathbf{a}_j = \mathbf{a}_j \qquad \text{for all } j$$

This implies that our initial tableau has $\mathbf{y}_j$ columns that may simply be extracted from the **A** matrix. This is done in Table 7-3.

Table 7-3

	$c_j \longrightarrow$	6	8	0	0	
$\mathbf{c}_B$	BV	x_1	x_2	x_3	x_4	$\mathbf{x}_B$
0	x_3	4	1	1	0	
0	x_4	1	4	0	1	

3. We now compute the far right-hand-side column, the initial basic feasible solution:

$$\mathbf{x}_B = \mathbf{B}^{-1}\mathbf{b} = \mathbf{I}^{-1}\mathbf{b} = \mathbf{I}\mathbf{b} = \mathbf{b}$$

That is, since the initial basis is the identity matrix, the basic feasible solution vector is just the **b** vector. This is shown in Table 7-4.

Table 7-4

	$c_j \longrightarrow$	6	8	0	0	
$\mathbf{c}_B$	BV	x_1	x_2	x_3	x_4	$\mathbf{x}_B$
0	x_3	4	1	1	0	20
0	x_4	1	4	0	1	40

4. We next compute the values in the indicator row.

$$z = \mathbf{c}_B\mathbf{x}_B = (0 \quad 0)\begin{pmatrix} 20 \\ 40 \end{pmatrix} = 0$$

$$z_j - c_j = \mathbf{c}_B\mathbf{y}_j - c_j$$

Thus,

$$z_1 - c_1 = 0 - 6 = -6$$

$$z_2 - c_2 = 0 - 8 = -8$$

$$z_3 - c_3 = 0 - 0 = 0$$

$$z_4 - c_4 = 0 - 0 = 0$$

Placing those results into the tableau gives Table 7-5, the completed initial tableau.

Table 7-5

	$c_j \longrightarrow$	6	8	0	0	
$\mathbf{c}_B$	BV	x_1	x_2	x_3	x_4	$\mathbf{x}_B$
0	x_3	4	1	1	0	20
0	x_4	1	4	0	1	40
		−6	−8	0	0	0

Table 7-5 summarizes everything we need to know about the problem at this point, including:

$$\mathbf{x}_B = \begin{pmatrix} 20 \\ 40 \end{pmatrix}$$

$$z = 0$$

and since $z_1 - c_1$ (and $z_2 - c_2$) is negative and both $\mathbf{y}_1$ and $\mathbf{y}_2$ have at least one positive element, the solution can be improved.

As may be seen, the development of this initial tableau is quite easy due, in particular, to the fact that the initial basis is an identity matrix. In fact, the only calculations actually necessary are for finding the values in the indicator row—and they are not particularly difficult. We can summarize the setup of the initial tableau as given below.

Setup of Initial Extended Tableau

Step 1. Construct an extended tableau with the proper number of columns and rows. Label each interior ($\mathbf{y}_j$) column with its associated decision variable.

Step 2. The initial basic variables are those associated with the slack and/or artificial variables. List those variable labels in the second column from the left and list their associated objective function coefficients in the first column on the left.

Step 3. Under each decision variable label, place the associated $\mathbf{y}_j$ column. Since the initial basis is an identity matrix, these are just the $\mathbf{a}_j$ columns from $\mathbf{A}$.

Step 4. The far right-hand-side column gives the value of each associated basic variable in the second column (from the left). Since the initial basis is an identity matrix, this column is simply the $\mathbf{b}$ vector.

Step 5. Compute the indicator row, where

$$z = \mathbf{c}_B \mathbf{x}_B \tag{7.1a}$$

or

$$z = \sum_{i=1}^{m} c_{B,i} x_{B,i} \tag{7.1b}$$

and

$$z_j - c_j = \mathbf{c}_B \mathbf{y}_j - c_j \tag{7.2a}$$

or

$$z_j - c_j = \sum_{i=1}^{m} y_{i,j} c_{B,i} - c_j \tag{7.2b}$$

We are now ready to perform the steps of the associated simplex algorithm. The specific form of the algorithm that will be used with the extended tableau is given next.

THE SIMPLEX ALGORITHM/EXTENDED TABLEAU

Step 1. *Check for possible improvement.* Examine the $z_j - c_j$ values in the indicator row. If these are all nonnegative, go to step 2. If, however, any $z_j - c_j$ is negative, we go to step 3.

Step 2. *Check for optimality or infeasibility.* If all $z_j - c_j \geq 0$ and no artificial variable is in the basis at a positive value, the solution is optimal. Otherwise (if an artificial is in the basis at a positive value), the problem is (mathematically) infeasible. In either case, we are finished.

Step 3. *Check for unboundedness.* If, for any $z_j - c_j < 0$ there is no positive element in the associated $\mathbf{y}_j$ vector (the column directly above $z_j - c_j$ in the tableau), the problem is unbounded. Otherwise, improvement is possible and we go to step 4.

Step 4. *Determining the entering variable.* Select, as the entering variable, the (nonbasic) variable with the most negative $z_j - c_j$ value. Designate this variable as $x_{j'}$ and its corresponding column as j'.[b] Ties in the selection of j' may be broken arbitrarily. Go to step 5.

Step 5. *Determining the departing variable.* We use the relationship of (6-35) to determine the departing variable (vector). This is accomplished in the tableau by taking the ratio

$$\frac{x_{B,i}}{y_{i,j'}} \qquad (y_{i,j'} > 0) \tag{7.3}$$

for each row. Designate the row having the minimum ratio of $x_{B,i}/y_{i,j'}$ as row i'.[b] The basic variable associated with row i' is the departing variable.

Step 6. *Establishment of the new tableau.*

(a) Set up a new tableau with all $\mathbf{y}_j$, $z_j - c_j$, z and basic feasible solution ($\mathbf{x}_B$) values empty. Replace the departing basic variable row heading ($x_{B,i'}$) with the entering variable label ($x_{j'}$). Replace $c_{B,i'}$ with $c_{j'}$.

---

[b]Column j' is also known as the *pivot column*, while row i' is designated as the *pivot row*. The element at column j', row i' is called the *pivot number*. Again, ties are broken arbitrarily.

(b) Row i' of the new tableau is obtained by dividing row i' of the preceding tableau by $y_{i',j'}$ (the element at the intersection of the entering variable column and departing variable row).

(c) Column j' of the new tableau consists of all zero elements except for a 1 at $y_{i',j'}$.

(d) The remaining elements of the tableau are computed as follows. Let $\hat{x}_{B,i}$, $\hat{z}$, $\hat{z}_j - \hat{c}_j$, and $\hat{y}_{i,j}$ represent the *new* set of elements to be computed and let $x_{B,i}$, z, $z_j - c_j$, and $y_{i,j}$ represent the values for these elements from the *preceding* tableau. Then, for those elements *not* in row i' or column j':

$$\hat{y}_{i,j} = y_{i,j} - \frac{(y_{i',j})(y_{i,j'})}{y_{i',j'}} \tag{7.4}$$

$$\hat{x}_{B,i} = x_{B,i} - \frac{(x_{B,i'})(y_{i,j'})}{y_{i',j'}} \tag{7.5}$$

$$\hat{z}_j - \hat{c}_j = (z_j - c_j) - \frac{(z_{j'} - c_{j'})(y_{i',j})}{y_{i',j'}} \tag{7.6}$$

$$\hat{z} = z - \frac{(z_{j'} - c_{j'})(x_{B,i'})}{y_{i',j'}} \tag{7.7}$$

(e) Return to step 1.

The understanding of any algorithm is enhanced through the use of examples and, particularly, by practice. To illustrate the algorithm stated above, we solve the following example problems.

Example 7-2

We first solve the problem given in Example 7-1. The initial tableau is given in Table 7-5. The steps of the algorithm, as applied to this problem, are listed below.

Step 1. The initial tableau appears in Table 7-5. Since there are negative $z_j - c_j$ (both $z_1 - c_1$ and $z_2 - c_2$), we go to step 3.

Step 3. There are no $\mathbf{y}_j$ above a negative $z_j - c_j$ with all negative or zero elements. Thus, improvement is possible and we go to step 4.

Step 4. The most negative $z_j - c_j$ is -8 (for $z_2 - c_2$). Thus, $j' = 2$ and x_2 is the entering variable. Go to step 5.

Step 5. We now examine the ratios $x_{B,i}/y_{i,2}$ $(y_{i,2} > 0)$:

$$\frac{x_{B,1}}{y_{1,2}} = \frac{20}{1} = 20$$

$$\frac{x_{B,2}}{y_{2,2}} = \frac{40}{4} = 10 \qquad \text{(minimum value)}$$

Thus, $i' = 2$ and x_4 $(x_{B,2})$ is the departing variable.

Step 6. (a) Since x_4 is the departing variable and x_2 is the entering variable, x_2 replaces x_4 in the BV column. We also replace $c_{B,2} = 0$ with $c_2 = 8$ in the $\mathbf{c}_B$ column. (See Table 7-6)

Table 7-6

	$c_j \longrightarrow$	6	8	0	0	
$\mathbf{c}_B$	BV	x_1	x_2	x_3	x_4	$\mathbf{x}_B$
0	x_3					
8	x_2					

(b) Row 2 ($i' = 2$) of the new tableau is obtained by dividing row 2 ($i' = 2$) of the preceding tableau by $y_{i',j'} = y_{2,2} = 4$. (See Table 7-7.)

Table 7-7

	$c_j \longrightarrow$	6	8	0	0	
$\mathbf{c}_B$	BV	x_1	x_2	x_3	x_4	$\mathbf{x}_B$
0	x_3					
8	x_2	$\frac{1}{4}$	1	0	$\frac{1}{4}$	10

(c) Column 2 ($j' = 2$) of the new tableau has zeros as elements except for a one at $y_{i',j'} = y_{2,2}$. (See Table 7-8.)

Table 7-8

	$c_j \longrightarrow$	6	8	0	0	
$\mathbf{c}_B$	BV	x_1	x_2	x_3	x_4	$\mathbf{x}_B$
0	x_3		0			
8	x_2	$\frac{1}{4}$	1	0	$\frac{1}{4}$	10
			0			

(d) The remaining elements of the tableau are given by (7.4) through (7.7) as follows:

$$\hat{y}_{1,1} = 4 - \frac{(1)(1)}{4} = \frac{15}{4}$$

$$\hat{y}_{1,3} = 1 - \frac{0(1)}{4} = 1$$

$$\hat{y}_{1,4} = 0 - \frac{(1)(1)}{4} = -\frac{1}{4}$$

$$\hat{x}_{B,1} = 20 - \frac{(40)(1)}{4} = 10$$

$$\hat{z}_1 - \hat{c}_1 = -6 - \frac{(1)(-8)}{4} = -4$$

$$\hat{z}_3 - \hat{c}_3 = 0 - \frac{(0)(-8)}{4} = 0$$

$$\hat{z}_4 - \hat{c}_4 = 0 - \frac{(1)(-8)}{4} = 2$$

$$\hat{z} = 0 - \frac{(40)(-8)}{4} = 80$$

The completed, second tableau, is shown in Table 7-9.

Table 7-9

	$c_j \longrightarrow$	6	8	0	0	
$\mathbf{c}_B$	BV	x_1	x_2	x_3	x_4	$\mathbf{x}_B$
0	x_3	$\frac{15}{4}$	0	1	$-\frac{1}{4}$	10
8	x_2	$\frac{1}{4}$	1	0	$\frac{1}{4}$	10
		-4	0	0	2	80

(e) Return to step 1.

Step 1. Since $z_1 - c_1$ is negative, we go to step 3.

Step 3. There are positive elements in $\mathbf{y}_1$ ($\frac{15}{4}$ and $\frac{1}{4}$), so improvement is possible and we go to step 4.

Step 4. The most negative (and only negative) $z_j - c_j$ is associated with $z_1 - c_1$. Thus, x_1 enters and $j' = 1$. Go to step 5.

Step 5. The ratios $x_{B,i}/y_{i,2}$ are

$$\frac{x_{B,1}}{y_{1,2}} = \frac{10}{\frac{15}{4}} = \frac{40}{15} \qquad \text{(minimum)}$$

$$\frac{x_{B,2}}{y_{2,2}} = \frac{10}{\frac{1}{4}} = 40$$

Thus, $i' = 1$ and $x_{B,1}$ (x_3) is the departing variable.

Step 6. (a) Since x_3 is the departing variable and x_1 is the entering variable, x_1 replaces x_3 in the BV column. We also replace $c_{B,1} = 0$ with $c_1 = 6$ in the $\mathbf{c}_B$ column. (See Table 7-10.)

Table 7-10

	$c_j \longrightarrow$	6	8	0	0	
$\mathbf{c}_B$	BV	x_1	x_2	x_3	x_4	$\mathbf{x}_B$
6	x_1					
8	x_2					

(b) Row 1 ($i' = 1$) of the new tableau is obtained by dividing row 1 ($i' = 1$) of the preceding tableau by $y_{i',j'} = y_{1,1} = \frac{15}{4}$.

(c) Column 1 ($j' = 1$) of the new tableau has zeros as elements except for a one at $y_{i',j'} = y_{1,1}$. (See Table 7-11.)

Table 7-11

	$c_j \longrightarrow$	6	8	0	0	
$\mathbf{c}_B$	BV	x_1	x_2	x_3	x_4	$\mathbf{x}_B$
6	x_1	1	0	$\frac{4}{15}$	$-\frac{1}{15}$	$\frac{8}{3}$
8	x_2	0				
		0				

(d) The remaining elements of the tableau are given by (7.4) through (7.7) as follows:

$$\hat{y}_{2,2} = 1 - \frac{0(\frac{1}{4})}{\frac{15}{4}} = 1$$

$$\hat{y}_{2,3} = 0 - \frac{(1)(\frac{1}{4})}{\frac{15}{4}} = -\frac{1}{15}$$

$$\hat{y}_{2,4} = \frac{1}{4} - \frac{(-\frac{1}{4})(\frac{1}{4})}{\frac{15}{4}} = \frac{4}{15}$$

$$\hat{x}_{B,2} = 10 - \frac{(10)(\frac{1}{4})}{\frac{15}{4}} = \frac{28}{3}$$

$$\hat{z}_2 - \hat{c}_2 = 0 - \frac{(0)(-4)}{\frac{15}{4}} = 0$$

$$\hat{z}_3 - \hat{c}_3 = 0 - \frac{(1)(-4)}{\frac{15}{4}} = \frac{16}{15}$$

$$\hat{z}_4 - \hat{c}_4 = 2 - \frac{(-\frac{1}{4})(-4)}{\frac{15}{4}} = \frac{26}{15}$$

$$\hat{z} = 80 - \frac{(10)(-4)}{\frac{15}{4}} = \frac{272}{3}$$

Table 7-12 gives the completed, third tableau.

Table 7-12

	$c_j \longrightarrow$	6	8	0	0	
$\mathbf{c}_B$	BV	x_1	x_2	x_3	x_4	$\mathbf{x}_B$
6	x_1	1	0	$\frac{4}{15}$	$-\frac{1}{15}$	$\frac{8}{3}$
8	x_2	0	1	$-\frac{1}{15}$	$\frac{4}{15}$	$\frac{28}{3}$
		0	0	$\frac{16}{15}$	$\frac{26}{15}$	$\frac{272}{3}$

(e) We return to step 1.

Step 1. All $z_j - c_j > 0$, so we go to step 2.

Step 2. Since there are no artificial variables in the basis (or, actually, in the problem) at a positive value, the solution given in Table 7-12 is optimal. Thus,

$$x_1^* = \frac{8}{3}$$

$$x_2^* = \frac{28}{3}$$

$$x_3^* = 0$$

$$x_4^* = 0$$

$$z^* = \frac{272}{3}$$

Notice that x_1^* and x_2^* are read from the $\mathbf{x}_B$ column in Table 7-12. The values of x_3 and x_4 must be zero since they are nonbasic. The fact that x_3 and x_4 are zero implies that there is no slack in the two constraints. That is, all the resources are consumed.

It actually takes considerably longer to *describe* the process of the simplex algorithm than it does to *apply* it. Example 7-2 illustrates this, as it should take the average reader no more than a few minutes to solve a problem of this size once he or she has had a bit of practice. Let us now take a look at an example problem that includes artificial variables.

Example 7-3

$$\text{Maximize} \quad z = 6x_1 + 4x_2$$

subject to

$$x_1 + x_2 \leq 10$$

$$2x_1 + x_2 \geq 4$$

$$x_1, x_2 \geq 0$$

Preprocessing results in:

$$\text{maximize} \quad z = 6x_1 + 4x_2 + 0x_3 + 0x_4 - Mx_5$$

subject to

$$x_1 + x_2 + x_3 = 10$$

$$2x_1 + x_2 - x_4 + x_5 = 4$$

$$\mathbf{x} \geq 0$$

Note that x_3 is a slack variable, x_4 is a surplus variable, and x_5 is an artificial.

The initial simplex tableau is shown in Table 7-13. It should be stressed that we use $-M$ as a penalty when performing computations by hand. When employing the computer, some large negative number (say -100 or $-1{,}000$) in relation to the other coefficients in the problem must be used.

Table 7-13

	$c_j \longrightarrow$	6	4	0	0	$-M$	
$\mathbf{c}_B$	BV	x_1	x_2	x_3	x_4	x_5	$\mathbf{x}_B$
0	x_3	1	1	1	0	0	10
$-M$	x_5	2	1	0	-1	1	4
		$-2M$ -6	$-M$ -4	0	M	0	$-4M$

We then apply the simplex algorithm to the problem. Only the finished tableaux for each iteration are shown (Tables 7-13 through 7-15). The column (j') associated with the entering variables and the row (i') associated with the departing variables are circled in each tableau. The resulting solution is

$$x_1^* = 10$$
$$x_2^* = 0$$
$$x_3^* = 0$$
$$x_4^* = 16$$
$$x_5^* = 0$$

and

$$z^* = 60$$

The physical interpretation of $x_4 = 16$ is that there is 16 units of "surplus" in the second constraint. That is, the left side of this constraint, at $\mathbf{x}^*$, exceeds the right by 16 units. Since $x_3 = 0$, the first constraint has no slack (i.e., it is satisfied as an equality at $\mathbf{x}^*$).

Table 7-14

	$c_j \longrightarrow$	6	4	0	0	$-M$	
$\mathbf{c}_B$	BV	x_1	x_2	x_3	x_4	x_5	$\mathbf{x}_B$
0	x_3	0	$\frac{1}{2}$	1	$\frac{1}{2}$	$-\frac{1}{2}$	8
6	x_1	1	$\frac{1}{2}$	0	$-\frac{1}{2}$	$\frac{1}{2}$	2
		0	-1	0	-3	M $+3$	12

Table 7-15

	$c_j \longrightarrow$	6	4	0	0	$-M$	
$\mathbf{c}_B$	BV	x_1	x_2	x_3	x_4	x_5	$\mathbf{x}_B$
0	x_4	0	1	2	1	-1	16
6	x_1	1	1	1	0	0	10
		0	2	6	0	M	60

The Two-Phase Method

The use of $-M$ as the objective function coefficient for artificial variables serves its purpose (i.e., to drive these variables out of the basis, if possible), but only at the cost of tedious hand calculations. It also presents problems when using the computer. On the computer, M must be assigned some numerical value that must be considerably larger than any of the other objective coefficients. If M is too small, we might obtain a solution with an artificial in the basis at a positive value (signaling an infeasible problem) when actually the problem is feasible. However, if M is too large, it may tend to dominate the $z_j - c_j$ values. Round-off errors (inherent in any digital computer) could well result and impact our final solution.

The two-phase method was developed to avoid, or at least alleviate, such difficulties. However, it has an additional feature that is also of interest to us. With some modification, the two-phase method may be transformed into a *multiphase method* suitable for solving multiple-objective linear programming problems. It, in fact, forms the basis for the modified simplex method of goal programming that will be presented in Part Four.

The general approach of the two-phase method can be described as follows. We solve the linear programming problem in two (surprise!) phases. In phase I, an *artificial objective* function is used and all *artificial variables* are driven to a value of zero (if they cannot be, the problem is, of course, infeasible).

Phase II consists of replacing the artificial objective function by the original objective function and using the basic feasible solution of phase I as our starting point. If no artificial variables were in the basis at the end of phase I, we simply perform the simplex algorithm until an optimal solution is reached. If, however, an artificial variable was in the basis (at a zero value) at the conclusion of phase I, we must modify (slightly) our *departing variable rule*. The specific steps of the two-phase algorithm are given below and illustrated via an example problem.

Two-Phase Algorithm

Step 1. Establish the problem formulation in a form suitable for the implementation of the simplex algorithm (i.e., convert the objective function to a maximization form and convert all the constraints by adding the slack, surplus, or artificial variables that are required).

Step 2. The artificial objective function of phase I is constructed by changing all the coefficients of the variables in the original objective as follows:
(a) The coefficients of any artificial variables will be -1's.
(b) The coefficients of all other variables in the objective will be zeros.

Step 3. *Phase I.* Employ the simplex algorithm of the previous section on the problem constructed in steps 1 and 2. However, note that we may terminate the process (and phase I) as soon as the value of z (the value of the artificial objective) is *zero*. If the simplex process ends with either $z = 0$ or all $z_j - c_j \geq 0$ and there are no artificials in the basis at a positive value, we go to step 4 (phase II). Otherwise, the problem is (mathematically) infeasible and we stop.

Phase II:

Step 4. Assign the actual objective function coefficient (the original c_j's) to each variable except for artificial variables. Any artificial variables in the basis at a zero level are given a c_j value of 0 in phase II. Any artificials not in the basis may be dropped from consideration by striking out their entire associated column in the tableau.

Step 5. The first tableau of phase II is the final tableau of phase I except for the objective coefficients and the indicator row values. Using (7.1) and (7.2), we recompute the indicator row ($z_j - c_j$ and z) values.

Step 6. If no artificial variables were in the basis (at zero values) at the end of phase I, we now simply use the simplex algorithm and proceed as usual. If, however, there are artificials in the basis, go to step 7.

Step 7. We must make sure that the artificial variables in the basis do not ever become positive in phase II. This is accomplished by modifying the departing variable rule of the simplex algorithm as follows:

(a) Determine the entering variable and its associated column (j') in the usual manner.

(b) Examine the $y_{i,j'}$ values for each artificial variable. If any of these are negative, let an artificial with a negative $y_{i,j'}$ depart. Otherwise, employ the usual departing variable rule.

Example 7-4

We solve the following problem using the two-phase method.

$$\text{Minimize} \quad z = 3x_1 + 2x_2 + 4x_3$$

$$\text{subject to}$$

$$x_1 + x_2 + x_3 \geq 12$$

$$4x_1 - x_2 \geq 6$$

$$x_1, x_2, x_3 \geq 0$$

The preprocessed problem is then:

$$\text{maximize} \quad z' = -3x_1 - 2x_2 - 4x_3 + 0x_4 - Mx_5 + 0x_6 - Mx_7$$

$$\text{subject to}$$

$$x_1 + x_2 + x_3 - x_4 + x_5 = 12$$

$$4x_1 - x_2 - x_6 + x_7 = 6$$

$$\mathbf{x} \geq 0$$

Note that x_4 and x_6 are surplus variables and x_5 and x_7 are artificials.

The artificial objective function, for use during phase I, is:

$$\text{maximize } Z = 0x_1 + 0x_2 + 0x_3 + 0x_4 - x_5 + 0x_6 - x_7$$

We now proceed to phase I and attempt to drive all artificials to zero (and thus, $Z = 0$). Tables 7-16 through 7-18 summarize these results.

Table 7-16

c_B	BV	x_1	x_2	x_3	x_4	x_5	x_6	x_7	$\mathbf{x}_B$
	$c'_j \longrightarrow$	0	0	0	0	−1	0	−1	
−1	x_5	1	1	1	−1	1	0	0	12
−1	x_7	4	−1	0	0	0	−1	1	6
		−5	0	−1	1	0	1	0	−18

Table 7-17

c_B	BV	x_1	x_2	x_3	x_4	x_5	x_6	x_7	$\mathbf{x}_B$
	$c'_j \longrightarrow$	0	0	0	0	−1	0	−1	
−1	x_5	0	$\frac{5}{4}$	1	−1	1	$\frac{1}{4}$	$-\frac{1}{4}$	$\frac{21}{2}$
0	x_1	1	$-\frac{1}{4}$	0	0	0	$-\frac{1}{4}$	$\frac{1}{4}$	$\frac{3}{2}$
		0	$-\frac{5}{4}$	−1	1	0	$-\frac{1}{4}$	$\frac{5}{4}$	$-\frac{21}{2}$

Table 7-18

c_B	BV	x_1	x_2	x_3	x_4	x_5	x_6	x_7	$\mathbf{x}_B$
	$c'_j \longrightarrow$	0	0	0	0	−1	0	−1	
0	x_2	0	1	$\frac{4}{5}$	$-\frac{4}{5}$	$\frac{4}{5}$	$\frac{1}{5}$	$-\frac{1}{5}$	$\frac{42}{5}$
0	x_1	1	0	$\frac{1}{5}$	$-\frac{1}{5}$	$\frac{1}{5}$	$-\frac{1}{5}$	$\frac{1}{5}$	$\frac{18}{5}$
		0	0	0	0	1	0	1	0

Notice in particular that we would stop phase I at the end of the second iteration (Table 7-18) even if there were a negative-valued $z_j - c_j$ indicator. This is, again, because of the fact that all we wish to do in phase I is to drive the artificials to zero, which is accomplished as soon as $Z = 0$.

Having driven the artificial variables x_5 and x_7 from the basis, their columns may be dropped from the phase II tableaux. Reinserting the objective function coefficients from z', we may begin phase II with the tableau of Table 7-19. This is the same tableau as shown in Table 7-18 except for the column under $\mathbf{c}_B$ and the indicator row.

Table 7-19

c_B	BV	x_1	x_2	x_3	x_4	x_6	$\mathbf{x}_B$
	$c_j \longrightarrow$	−3	−2	−4	0	0	
−2	x_2	0	1	$\frac{4}{5}$	$-\frac{4}{5}$	$\frac{1}{5}$	$\frac{42}{5}$
−3	x_1	1	0	$\frac{1}{5}$	$-\frac{1}{5}$	$-\frac{1}{5}$	$\frac{18}{5}$
		0	0	$\frac{9}{5}$	$\frac{11}{5}$	$\frac{1}{5}$	$-\frac{138}{5}$

Since all $z_j - c_j \geq 0$ for this tableau, we can stop as we have found the optimal solution wherein

$$x_1^* = \tfrac{18}{5}$$
$$x_2^* = \tfrac{42}{5}$$
$$x_3^* = 0$$
$$x_4^* = 0$$
$$x_5^* = 0$$
$$x_6^* = 0$$
$$x_7^* = 0$$

and

$$z = -z' = \tfrac{138}{5}$$

Example 7-5

Let us now use the two-phase method on:

$$\text{maximize} \quad z = 2x_1 + 6x_2$$
$$\text{subject to}$$
$$x_1 + x_2 = 8$$
$$2x_1 + x_2 \leq 6$$
$$x_1, x_2 \geq 0$$

The preprocessed model is:

$$\text{maximize} \quad z = 2x_1 + 6x_2 - Mx_3 + 0x_4$$
$$\text{subject to}$$
$$x_1 + x_2 + x_3 \qquad = 8$$
$$2x_1 + x_2 \qquad + x_4 = 6$$
$$\mathbf{x} \geq 0$$

Variable x_3 is an artificial and x_4 is a slack. The artificial objective function is given as

$$Z = 0x_1 + 0x_2 - x_3 + 0x_4$$

The first phase of the two-phase method is presented in Tables 7-20 and 7-21. (Notice how and *why* the tie between x_1 and x_2, as entering variables, was broken in the first iteration.) The final tableau (Table 7-21) indicates optimality (all $z_j - c_j \geq 0$) but Z is *not* zero (since an artificial is in the basis at a positive value). Consequently, this problem, as modeled, is mathematically infeasible and we may terminate our procedure (and try to decide what can be done to find a workable solution to the actual problem).

Table 7-20

	$c'_j \longrightarrow$	0	0	−1	0	
$\mathbf{c}_B$	BV	x_1	x_2	x_3	x_4	$\mathbf{x}_B$
−1	x_3	1	1	1	0	8
0	x_4	2	1	0	1	6
		−1	−1	0	0	−8

Table 7-21

	$c'_j \longrightarrow$	0	0	−1	0	
$\mathbf{c}_B$	BV	x_1	x_2	x_3	x_4	$\mathbf{x}_B$
−1	x_3	−1	0	1	−1	2
0	x_2	2	1	0	1	6
		1	0	0	1	−2

Example 7-6

Our final illustration of the two-phase method involves a problem in which artificials remain in the basis, at zero values, at the end of phase I. This is often caused by *redundant* constraints in the model formulation.

Table 7-22 is the first tableau, for a particular problem, of phase II. Variables x_6 and x_7 are *artificial variables* and are both in the basis at zero values. Under the normal simplex rules, variable x_2 would enter the basis and x_4 would leave—resulting in x_6 *and* x_7 both going positive. However, using the modified departing variable rule of the two-phase method, we should select either x_6 or x_7 to leave the basis. The reader is invited to complete the operations of phase II.

Table 7-22

	$c_j \longrightarrow$	2	1	3	0	0	0	0	
$\mathbf{c}_B$	BV	x_1	x_2	x_3	x_4	x_5	x_6	x_7	$\mathbf{x}_B$
2	x_1	1	−1	3	0	1	0	0	3
0	x_6	0	−2	1	0	3	1	0	0
0	x_4	0	3	−1	1	−1	0	0	9
0	x_7	0	−1	0	0	2	0	1	0
		0	−3	3	0	2	0	0	6

The Condensed Tableau

Hand computation of linear programming problems is not a particularly enjoyable task. Although the mathematical operations themselves are simple, the repetition of the iterative simplex process becomes tedious. However, it is still important that the student of linear programming solve, by hand, a variety of problems.

In this section we introduce the condensed tableau [3] and its associated simplex algorithm. The tableau does *not* change the number of iterations required for convergence, but it does relieve a bit of the computational burden by reducing the number of columns that we must carry in each tableau.

The reader should now pause to reexamine the previous tableaux of this chapter. Notice that in every (complete) tableau, there are exactly m (m = number of constraints) columns that contain a single 1 element and all the rest of the elements are zeros. These columns are, of course, associated with the variables in the basis. The position of the single 1 element in the column corresponds to the position of the associated basic variable in the BV (basic variable) column (or, also, to the position of the associated $\mathbf{a}_j$ vector in the basis matrix). Since this must always hold true, the inclusion of the columns under the basic variables is redundant and wasteful.

The general format of the condensed tableau is shown in Table 7-23. Notice that:

Table 7-23

V	1	$\cdots$	s	$\cdots$	S	$\mathbf{x}_B$
1	$y_{1,1}$		$y_{1,s}$		$y_{1,S}$	$x_{B,1}$
$\vdots$			$\vdots$			
i			$y_{i,s}$			$x_{B,i}$
$\vdots$			$\vdots$			
m	$y_{m,1}$		$y_{m,s}$		$y_{m,S}$	$x_{B,m}$
	$z_1 - c_1$	$\cdots$	$z_s - c_s$	$\cdots$	$z_S - c_S$	z

1. Only the nonbasic column vectors (S in number) are shown.
2. The row labels (under V for "variables") are the *subscripts* of the respective basic variables (m in number).
3. The column labels (to the right of V) are the *subscripts* of the respective nonbasic variables (S in number).
4. We have omitted the c_j listing (that appears on the top of the extended tableau) and the $\mathbf{c}_B$ column (that appeared as the far left-hand side column of the extended tableau). This information is available from the original problem formulation and is thus redundant.
5. The remaining information contained in the tableau is similar to that given in the extended tableau.
6. As the basis is changed (during the simplex iterations), the associated row and column labels must also be interchanged.

Example 7-7

Place the following problem into the condensed tableau format.

$$\text{Minimize} \quad z = 4x_1 - 3x_2 + 6x_3$$

$$\text{subject to}$$

$$2x_1 + x_2 + x_3 \geq 20$$

$$x_2 \leq 7$$

$$\mathbf{x} \geq \mathbf{0}$$

The preprocessed model is:

$$\text{maximize} \quad z' = -4x_1 + 3x_2 - 6x_3 + 0x_4 - Mx_5 + 0x_6$$

$$\text{subject to}$$

$$2x_1 + x_2 + x_3 - x_4 + x_5 = 20$$

$$x_2 + x_6 = 7$$

$$\mathbf{x} \geq \mathbf{0}$$

Variable x_4 is a surplus, x_5 is an artificial, and x_6 is a slack. The resultant initial condensed tableau is shown in Table 7-24.

Table 7-24

V	1	2	3	4	$\mathbf{x}_B$
5	2	1	1	-1	20
6	0	1	0	0	7
	$-2M$ $+4$	$-M$ -3	$-M$ $+6$	$-M$	$-20M$

Some readers may wish to continue to include column $\mathbf{c}_B$ and row $\mathbf{c}_j$ (actually, $\mathbf{c}_s$), particularly in the initial tableau. We leave this choice to individual preference.

The associated simplex algorithm for the condensed tableau performs the exact same functions as always but, because of the modified tableau format, some minor changes in implementation are necessary.

The Simplex Algorithm/Condensed Tableau

Steps 1 through 5. Steps 1 through 5 are the same as listed previously for the extended tableau except that we replace the subscript j by s (e.g., the entering column is the s' column).

Step 6. *Establishment of the New (Condensed) Tableau*

(a) Set up a new tableau with all $\mathbf{y}_s$, $z_s - c_s$, z, and basic feasible solution ($\mathbf{x}_B$) values empty. Interchange the departing variable row heading (the subscript i') and the entering variable column heading (the subscript s').

(b) Row i' of the new tableau, except for $\hat{y}_{i',s'}$, is obtained by dividing row i' of the preceding tableau by $y_{i',s'}$.
(c) Column s' of the new tableau, except for $\hat{y}_{i',s'}$, is obtained by dividing column s' by $-y_{i',s'}$.
(d) The new element at position $\hat{y}_{i',s'}$ is the reciprocal of $y_{i',s'}$. The remaining elements of the new tableau are computed as follows:

$$\hat{y}_{i,s} = y_{i,s} - \frac{(y_{i',s})(y_{i,s'})}{y_{i',s'}} \tag{7.8}$$

$$\hat{x}_{B,i} = x_{B,i} - \frac{(x_{B,i'})(y_{i,s'})}{y_{i',s'}} \tag{7.9}$$

$$\hat{z}_s - \hat{c}_s = (z_s - c_s) - \frac{(z_{s'} - c_{s'})(y_{i',s})}{y_{i',s'}} \tag{7.10}$$

$$\hat{z} = z - \frac{(z_{s'} - c_{s'})(x_{B,i'})}{y_{i',s'}} \tag{7.11}$$

(e) Return to step 1.

We now use this algorithm to solve some example problems.

Example 7-8

So that the reader may examine in more detail the differences between the extended and condensed tableaux, we first resolve the problem presented earlier in Example 7-2. The tableaux (for each completed iteration) for this problem are given in Tables 7-25 through 7-27. They should be compared with Tables 7-5, 7-9, and 7-12.

As may be easily seen, the condensed tableau is considerably more compact, with even greater efficiency evidenced in larger problems. The reader should assure himself

Table 7-25

V	1	2	$\mathbf{x}_B$
3	4	1	20
4	1	4	40
	−6	−8	0

Table 7-26

V	1	4	$\mathbf{x}_B$
3	$\frac{15}{4}$	$-\frac{1}{4}$	10
2	$\frac{1}{4}$	$\frac{1}{4}$	10
	−4	2	80

Table 7-27

V	3	4	$\mathbf{x}_B$
1	$\frac{4}{15}$	$-\frac{1}{15}$	$\frac{8}{3}$
2	$-\frac{1}{15}$	$\frac{4}{15}$	$\frac{28}{3}$
	$\frac{16}{15}$	$\frac{26}{15}$	$\frac{272}{3}$

or herself that he or she may form the equivalent extended tableau representation for each of the tables.

Example 7-9

Find $\mathbf{x}$ so as to

$$\text{maximize} \quad z = 15x_1 + 45x_2$$

$$\text{subject to}$$

$$x_1 + 1.6x_2 \leq 240$$

$$0.5x_1 + 2x_2 \leq 162$$

$$x_2 \leq 50$$

$$x_1, x_2 \geq 0$$

Preprocessing, we obtain:

$$\text{maximize} \quad z = 15x_1 + 45x_2 + 0x_3 + 0x_4 + 0x_5$$

$$\text{subject to}$$

$$x_1 + 1.6x_2 + x_3 = 240$$

$$0.5x_1 + 2x_2 + x_4 = 162$$

$$x_2 + x_5 = 50$$

$$\mathbf{x} \geq \mathbf{0}$$

The tableaux for this problem are given in Tables 7-28 through 7-31.

Table 7-28

V	1	2	$\mathbf{x}_B$
3	1	1.6	240
4	0.5	2	162
5	0	1	50
	−15	−45	0

Table 7-29

V	1	5	$\mathbf{x}_B$
3	1	−1.6	160
4	0.5	−2	62
2	0	1	50
	−15	45	2,250

Table 7-30

V	4	5	$\mathbf{x}_B$
3	−2	2.4	36
1	2	−4	124
2	0	1	50
	30	−15	4,110

Table 7-31

V	4	3	$\mathbf{x}_B$
5	$-\frac{5}{6}$	$\frac{5}{12}$	15
1	$-\frac{4}{3}$	$\frac{5}{3}$	184
2	$\frac{5}{6}$	$-\frac{5}{12}$	35
	17.5	6.25	4,335

The resultant optimal solution is given by

$$x_1^* = 184$$
$$x_2^* = 35$$
$$x_3^* = 0$$
$$x_4^* = 0$$
$$x_5^* = 15$$

and

$$z^* = 4,335$$

Since x_5 is the only nonzero slack, constraint 3 is the only constraint having idle (or slack) resources.

Example 7-10

Our final example in this section follows.

$$\text{Maximize} \quad 10x_1 + 20x_2$$
$$\text{subject to}$$
$$x_1 + x_2 = 150$$
$$x_1 \geq 20$$
$$x_2 \leq 40$$
$$\mathbf{x} \geq \mathbf{0}$$

The preprocessed problem is:

$$\text{maximize} \quad z = 10x_1 + 20x_2 - Mx_3 + 0x_4 - Mx_5 + 0x_6$$
$$\text{subject to}$$
$$x_1 + x_2 + x_3 = 150$$
$$x_1 - x_4 + x_5 = 20$$
$$x_2 + x_6 = 40$$
$$\mathbf{x} \geq 0$$

The solution process is given in Tables 7-32 through 7-35.

Table 7-32

V	1	2	4	$\mathbf{x}_B$
3	1	1	0	150
5	1	0	-1	20
6	0	1	0	40
	-10 $-2M$	-20 $-1M$	$+M$	$-170M$

Table 7-33

V	5	2	4	$\mathbf{x}_B$
3	-1	1	1	130
1	1	0	-1	20
6	0	1	0	40
	10 $+2M$	-20 $-M$	-10 $-M$	200 $-130M$

Table 7-34

V	5	6	4	$\mathbf{x}_B$
3	−1	−1	1	90
1	1	0	−1	20
2	0	1	0	40
	10 +2M	20 +M	−10 −M	1,000 −90M

Table 7-35

V	5	6	3	$\mathbf{x}_B$
4	−1	−1	1	90
1	0	−1	1	110
2	0	1	0	40
	0 +M	10	10 +M	1,900

The resultant solution is

$$x_1^* = 110$$
$$x_2^* = 40$$
$$x_3^* = 0$$
$$x_4^* = 90$$
$$x_5^* = 0$$
$$x_6^* = 0$$

and

$$z = 1{,}900$$

Since x_4 is a surplus variable, we see that the second constraint has 90 units of surplus.

Computational Aspects: Some Practical Considerations

There is a wide variety of somewhat unusual occurrences that will be encountered in the course of solving a large number, and types, of linear programming problems. This is why some understanding of the simplex method (as presented in Chapter 6) and experience through both hand computation and computer-assisted problem solving is so important. In this section we briefly touch on some of the more typical problems that will be encountered. Further details are provided in Chapter 11.

Unrestricted Variables

All of the material so far presented has stressed that all variables must be restricted to nonnegative values when employing the simplex algorithm. This restriction is one that may be easily circumvented by simply substituting, for every unrestricted variable, the difference between two restricted variables. For example, if x_3 is unrestricted, we may simply let

$$x_3 = x_3^+ - x_3^-$$

where

$$-\infty < x_3 < +\infty$$
$$x_3^+ \geq 0$$
$$x_3^- \geq 0$$

Wherever x_3 appears in the problem, we substitute $x_3^+ - x_3^-$. This increases, of course, the size of the problem we shall have to solve. Since x_3^+ and x_3^- are dependent, only one (at most) can appear in the basis at a positive value. If x_3^+ is in the basis, then $x_3 \geq 0$, whereas if x_3^- is in the basis, then $x_3 \leq 0$.

Redundancy

It is quite possible, particularly with large problems, to incorporate a number of redundant constraints in the model. Now, there are ways to check this [2], but in general, they take more time and effort than they are worth.

If an artificial or artificials appear in the basis, at the final iteration, at zero levels, then it is *possible* that redundancy may exist within the constraints. Normally, this is not a factor of considerable interest to the problem solver.

Degeneracy

Degeneracy is evident whenever one or more basic variables have a value of zero. This result does not infer that anything is wrong with the solution. However, degeneracy could possibly create two related problems:

1. The objective function (z) may not improve when we move from one basis to another.
2. We might, in fact, cycle forever (repeating a sequence of bases) and not ever reach the optimal solution.

The second phenomenon may be avoided via several approaches, including revision of the pivoting rules. Consequently, if the reader is interested in this problem, we direct him or her to the references [1, 2].

Round-Off Errors

Normally, when solving problems by hand, we employ fractions rather than decimals. Round-off is then not a problem. However, computers, despite their highly acclaimed accuracy, do deal with decimals and do have a limit to the number of decimal places that may be carried. The two-phase method was one of the earliest approaches to reducing the round-off errors due to the large objective function coefficients of artificial variables. However, despite the best efforts of many analysts, round-off errors persist and can cause some surprising results. Problems with a feasible solution may have indications of infeasibility. Problems with a finite optimal solution may be designated as unbounded. In essence, the solution you obtain may be in error. This seems to amaze many trusting souls who credit the digital computer with an infallibility that just does not exist.

One way to partially check a solution is to insert it into the constraints. Are they really satisfied? Do the slack or surplus variable values check with what the computer printed out? Be careful of the degree of faith you place in the computer.

A related problem occurs because of errors (the computer scientist seems to prefer the term "bugs") in the way the simplex algorithm has been coded. Many of these "bugs" occur only during the processing of very large problems and, as a result, are seldom caught. This author has used a wide variety of linear programming (simplex) computer codes over the past two decades and has yet to find one that did *not* have at least a few bugs in it. This includes some of the better known and more costly computer programs. Again, this fact seems to often surprise students.

Miscellaneous Aspects

Invariably, when teaching the linear programming course, the author has a number of students wander into his office in regard to difficulties with homework assignments. Some of these difficulties exist because the student has really not put his or her full effort into the problem. On other occasions, there is a great deal of effort but little use of logic. The problems that seem to be most often encountered are:

1. A "simple" mathematical error. The student multiplies $\frac{3}{5}$ by $\frac{4}{3}$ and gets, say, $\frac{12}{20}$.
2. The student selects the wrong departing variable. As a result, one or more of the basic variables will go negative. However, ignoring this, he or she continues on several more iterations and then wonders what happened.
3. Instead of using the minimum $x_{B,i}/y_{i,j'}$, ratio (where $y_{i,j'} > 0$) to determine the departing variable, the student takes the ratio $x_{B,i}/y_{i,j'}$ where $y_{i,j'}$ is *negative* and, to compound the error, selects this ratio as the "minimum" ratio.

4. The objective function is not converted to a maximizing form or, if it was converted, the student forgets to multiply the resulting final objective (z) value by -1.
5. The student (not using the two-phase method) uses -1 as the objective function coefficient for artificial variables. The result may be a problem that appears to be infeasible, although it actually has a solution.

The only advice that the author can give is to obtain a good deal of practice, *think* about *what* you are doing, *why* you are doing it, and what the results *mean*—from a *physical* as well as mathematical view. When some difficulty is experienced, try to approach the problem in a systematic, logical manner. Don't go to the instructor without first making this effort. Having someone else always solve your problems does nothing to build up either your own skills or confidence. In fact, the reverse is generally true.

Some Additional Examples

So as to provide the reader with some acquaintance with the detection of unboundedness, infeasibility, and alternative optimal solutions, we now present examples of each. (It might also be wise to solve the first two examples graphically.)

Example 7-11

Consider the following problem.

$$\text{Maximize} \quad z = x_1 + 4x_2$$

$$\text{subject to}$$

$$-8x_1 + 10x_2 \geq 80$$

$$2x_1 + x_2 \geq 16$$

$$x_1 x_2 \geq 0$$

Preprocessing, we have:

$$\text{maximize} \quad z = x_1 + 4x_2 + 0x_3 - Mx_4 + 0x_5 - Mx_6$$

$$\text{subject to}$$

$$-8x_1 + 10x_2 - x_3 + x_4 = 80$$

$$2x_1 + x_2 - x_5 + x_6 = 16$$

$$\mathbf{x} \geq 0$$

The variables x_3 and x_5 are surplus, while x_4 and x_6 are artificials.

Table 7-36 presents the third tableau (at the end of the second iteration). The

Table 7-36

V	6	4	3	5	$\mathbf{x}_B$
2	$\frac{2}{7}$	$\frac{1}{14}$	$-\frac{1}{14}$	$-\frac{2}{7}$	$\frac{72}{7}$
1	$\frac{5}{14}$	$-\frac{1}{28}$	$\frac{1}{28}$	$-\frac{5}{14}$	$\frac{20}{7}$
	$\frac{3}{2}$ $+M$	$\frac{1}{4}$ $+M$	$-\frac{1}{4}$	$-\frac{3}{2}$	44

problem can be designated as (mathematically) unbounded at this point since $z_j - c_j$ for x_5 is negative and there are no positive elements in the associated $\mathbf{y}_j$ vector.

For a real-life problem, the next step would be to review the model. Has an error been made in the constraint formulation, or (more likely) have we overlooked a limited resource or other restrictions that, when included in the formulation, will bound the objective value?

Example 7-12

We now solve the following problem.

$$\text{Maximize} \quad z = 5x_1 + 3x_2$$
$$\text{subject to}$$
$$x_1 + 2x_2 \leq 8$$
$$-x_1 + x_2 \geq 7$$
$$x_1, x_2 \geq \mathbf{0}$$

Preprocessing:

$$\text{maximize} \quad z = 5x_1 + 3x_2 + 0x_3 + 0x_4 - Mx_5$$
$$\text{subject to}$$
$$x_1 + 2x_2 + x_3 = 8$$
$$-x_1 + x_2 - x_4 + x_5 = 7$$
$$\mathbf{x} \geq \mathbf{0}$$

Variable x_3 is a slack, x_4' is a surplus, and x_5 is an artificial. Table 7-37 presents the final tableau and, as may be seen, the problem is infeasible since x_5 is in the basis at a

Table 7-37

V	1	3	4	$\mathbf{x}_B$
2	$\frac{1}{2}$	$\frac{1}{2}$	0	4
5	$-\frac{3}{2}$	$-\frac{1}{2}$	-1	3
	$-\frac{7}{2}$ $+3M$	$\frac{3}{2}$ $+M/2$	0 $+M$	12 $-3M$

positive value. Note also that the value of z also reflects the presence of the artificial. Again, in a real-life situation we must further analyze this problem. There are two possibilities: (1) the model has been constructed incorrectly, or (2) we need to "relax" one or more of the rigid constraints. In Part Four we learn how this dilemma can often be completely avoided using the multiple-objective approach.

Example 7-13

We illustrate two things in this example problem. First, we see that it is not always necessary to include artificial variables when dealing with equality constraints. Second, the tabular indication of alternative optimal solutions will be presented.

$$\text{Maximize} \quad z = 4x_1 - 2x_2 + 6x_3$$

$$\text{subject to}$$

$$\begin{aligned} 2x_1 + x_2 + 3x_3 &= 36 \\ 5x_1 \qquad - x_3 &\leq 2 \\ \mathbf{x} &\geq \mathbf{0} \end{aligned}$$

Normally, we would add an artificial variable to the first constraint—to assure an initial basic feasible solution. However, note that x_2 appears, with a coefficient of 1, in only a *single* constraint. Thus, x_2, in conjunction with the slack variable (x_4) for the second constraint, gives us an initial basic feasible solution—without the need to employ an artificial variable. The preprocessed model is then:

$$\text{maximize} \quad z = 4x_1 - 2x_2 + 6x_3 + 0x_4$$

$$\text{subject to}$$

$$\begin{aligned} 2x_1 + x_2 + 3x_3 \qquad &= 36 \\ 5x_1 \qquad - x_3 + x_4 &= 2 \\ \mathbf{x} &\geq \mathbf{0} \end{aligned}$$

Tables 7-38 and 7-39 summarize the simplex operations. Since there is a zero element, in the indicator row, under a nonbasic variable, there are alternative optimal solutions (one of which is obtained by letting x_1 enter the basis).

Table 7-38

V	1	3	$\mathbf{x}_B$
2	2	3	36
4	5	−1	2
	−8	−12	−72

Table 7-39

V	1	2	$\mathbf{x}_B$
3	$\frac{2}{3}$	$\frac{1}{3}$	12
4	$\frac{17}{3}$	$\frac{1}{3}$	14
	0	4	72

Summary

In this chapter we have presented two tableau types: the extended and the condensed tableau, together with the two-phase method for dealing with artificial variables. Although some practice with the extended tableau and two-phase method is encouraged, the thrust of one's effort should center about the more convenient condensed tableau. In Chapter 9 we couple the condensed tableau with a technique that allows us to avoid, entirely, the use of either artificial or surplus variables, making computation considerably easier and faster (reducing, in fact, the usual number of iterations required).

REFERENCES

1. HADLEY, G. *Linear Programming*. Reading, Mass.: Addison-Wesley, 1963.
2. LLEWELLYN, R. W. *Linear Programming*. New York: Holt, Rinehart and Winston, 1964.
3. WOLFE, C. S. *Linear Programming with FORTRAN*. Glenview, Ill.: Scott, Foresman, 1973.

PROBLEMS

7.1. Solve the following problem using the simplex algorithm and the *extended* tableau.

$$\text{Maximize} \quad z = 20x_1 + 24x_2$$

subject to

$$\begin{aligned} 2x_1 + x_2 &\leq 24 \\ 2x_1 + 3x_2 &\leq 48 \\ x_1 + x_2 &\leq 20 \\ -2x_1 + 3x_2 &\geq 0 \\ x_1, x_2 &\geq 0 \end{aligned}$$

7.2. Solve Problem 7.1 with the simplex algorithm and the extended tableau but ignore the fourth constraint ($-2x_1 + 3x_2 \geq 0$). What can you conclude from your results?

7.3. Solve the following problem using the simplex algorithm and the extended tableau.

$$\text{Maximize} \quad z = 45x_1 + 15x_2$$

subject to

$$\begin{aligned} 4x_1 + \quad x_2 &\leq 364 \\ x_2 &\leq 50 \\ 0.8x_1 + 0.5x_2 &\leq 120 \\ x_1, x_2 &\geq 0 \end{aligned}$$

7.4. Repeat Problem 7.1 but use the two-phase method to solve the problem. Is the two-phase method really necessary here?

7.5. Solve Example 7-3 by the two-phase method.

7.6. Finish Example 7-6 by the two-phase method.

7.7. Solve Problem 7.1 by the simplex method and the *condensed* tableau. When finished, transform the final condensed tableau into the equivalent extended tableau.

Solve Problems 7.8 through 7.15 via the simplex method and the condensed tableau.

7.8. Maximize $z = 10x_1 + 8x_2$ subject to

$$\begin{aligned} 2x_1 + x_2 &\leq 4 \\ 3x_1 + x_2 &\geq 10 \\ x_1, x_2 &\geq 0 \end{aligned}$$

7.9. Maximize $z = 2x_1 + 4x_2$ subject to

$$\begin{aligned} x_1 - x_2 &\leq 1 \\ -x_1 + x_2 &\leq 1 \\ x_1, x_2 &\geq 0 \end{aligned}$$

7.10. Maximize $z = 5x_1 + 4x_2 + 2x_3$ subject to

$$\begin{aligned} 2x_1 + \ x_2 + \ x_3 &\leq 4 \\ 4x_1 + 4x_2 + 2x_3 &\leq 12 \\ x_1, x_2, x_3 &\geq 0 \end{aligned}$$

7.11. Minimize $z = x_1 - 2x_2$ subject to

$$\begin{aligned} x_1 - \ x_2 &\leq 5 \\ x_1 + 2x_2 &\leq 20 \\ -x_1 + \ x_2 &\leq 4 \\ x_1, x_2 &\geq 0 \end{aligned}$$

7.12. Maximize $z = 8x_1 + 10x_2$ subject to

$$\begin{aligned} x_1 \qquad\quad &= 2 \\ x_1 + \quad x_2 &\leq 9 \\ x_1 + 0.5x_2 &\geq 3 \\ x_1, x_2 &\geq 0 \end{aligned}$$

7.13. Maximize $z = 4x_1 + 6x_2$ subject to

$$\begin{aligned} x_1 + 1.5x_2 &\leq 12 \\ 2x_1 + \quad x_2 &\leq 16 \\ x_1, x_2 &\geq 0 \end{aligned}$$

7.14. Minimize $z = 2x_1 - x_2$ subject to

$$\begin{aligned} x_1 + x_2 &\geq 10 \\ 3x_1 - x_2 &\geq 6 \\ x_1 \geq 0, x_2 &\text{ unrestricted} \end{aligned}$$

7.15. Maximize $z = 4x_1 + 3x_2 + x_3$ subject to

$$\begin{aligned} 2x_1 - 3x_2 + x_3 &\geq 22 \\ x_1 + 2x_2 + 3x_3 &= 30 \\ -x_1 + 5x_2 + 2x_3 &= 42 \\ x_1, x_2, x_3 &\geq 0 \end{aligned}$$

7.16. The Red Tide Shell Company dredges the bay for oyster shells and produces various products from the shells (e.g., a base for paints, chicken feed, etc.). The firm is extremely proud of the *number* of different products that they make from the shells and intends to offer a $10,000 reward to any employee who comes up with a new, marketable use for the shells. As a consultant to the firm, you note that their manufacturing process may be modeled as a linear program with exactly 22 constraints. The single objective of the firm is to maximize profits. Presently, they manufacture 31 different products from the shells. What advice would you offer the firm?

7.17. Fred's Foods operates a "catering" service. A van is sent to a large manufacturing plant at lunch time and prepared, cold sandwiches are sold to the employees. The selling price and costs of the bread and ingredients are known and Fred has contracted with a grocery to deliver a fixed amount of the ingredients each day. Given that Fred can sell as many sandwiches as he can make, what is the program that will maximize profit? (Assume, for convenience, that fractions of sandwiches can be sold.)

Ingredient	*Cost*
Buns	$1.00 per dozen
White bread	$0.10 per slice
Rye bread	$0.12 per slice
Turkey loaf	$3.00 per pound
Salami	$4.00 per pound

Ingredient	*Sandwich Type*				
	Fred's Best	*The Deluxe*	*Supreme*	*Big Bird*	*Amount Ordered Each Day*
Buns	1	0	0	0	120
White bread	0	2 slices	0	2 slices	100 slices
Rye bread	0	0	2 slices	0	100 slices
Turkey loaf	0.3 pound	0.3 pound	0.1 pound	0.2 pound	100 pounds
Salami	0.1 pound	0	0.2 pound	0	100 pounds
Selling price	\$2.50	\$2.00	\$2.00	\$1.50	

7.18. What advice would you give Fred (in Problem 7.17) besides the optimal program?

7.19. Trivia Products Ltd. makes cast-iron paperweights in three sizes: 6, 12, and 18 ounces. All weights must be cast, ground, and polished prior to shipment. The per unit returns are:

6 ounces: \$1.00
12 ounces: \$1.75
18 ounces: \$2.30

The firm wishes to determine the product mix that maximizes profit given the requirements listed in the table. Formulate and solve.

Paper Weight (Ounces)	*Time Required (Minutes)*		
	Casting	*Grinding*	*Polishing*
6	10	10	15
12	18	16	28
18	22	30	32
Total hours available per day	10	20	24

DUALITY

Introduction

One of the most important, if not *the* most important, of the concepts associated with linear programming is that of *duality*. We define duality as:

DUALITY Given any linear programming problem

$$\left.\begin{aligned} \text{maximize} \quad z &= \mathbf{cx} \\ \text{subject to} & \\ \mathbf{Ax} &\leq \mathbf{b} \\ \mathbf{x} &\geq \mathbf{0} \end{aligned}\right\} \tag{8.1}$$

there is a related linear programming problem

$$\left.\begin{aligned} \text{minimize} \quad Z &= \mathbf{b}^T\mathbf{v} \\ \text{subject to} & \\ \mathbf{A}^T\mathbf{v} &\geq \mathbf{c}^T \\ \mathbf{v} &\geq \mathbf{0} \end{aligned}\right\} \tag{8.2}$$

If we term the problem in (8.1) as the *primal*, the problem formulated in (8.2) is termed the *dual*. The two problems are related in such a manner that the optimal solution of one also provides all the information needed to determine the optimal solution of the other. That is, if we have solved one problem, we have solved *both*.

8

This concept of duality is not unique to linear programming; it also appears in mathematics, physics, statistics, and engineering. Those who have been introduced to electrical engineering, for example, may recall that for every electrical network, there is an equivalent dual network. Babinets' eloquent principle of duality plays a major role in the analysis of antennas. Duality is also exhibited in two-person game theory wherein one person's problem is the "dual" of the opponent's problem.

Duality is far more than just a curious relationship. In linear programming we use duality in a wide variety of both theoretical and practical ways. Included among these are:

1. It may, in some cases, be easier (less iterations, etc.) to solve the dual than the primal.
2. The dual variables (v_i) provide an important economic interpretation of the results obtained when solving a linear programming problem.
3. Duality is used as an aid when investigating changes in the coefficients or formulation of a given linear programming problem (i.e., in sensitivity analysis).
4. Duality will be utilized to allow us to employ simplex to solve problems in which the initial basis is *infeasible* (the technique itself is known as *dual simplex*).
5. Duality will be utilized to allow us to completely avoid the necessity for

surplus or artificial variables—and to solve the resulting problem in generally fewer iterations (the technique is known as the primal–dual algorithm).

6. Duality is used to develop a number of important theorems in linear programming. We list the results of this development in a succeeding section.

In this chapter we demonstrate the formulation of the dual from, first, the *canonical* form of the linear programming problem and then from the *general* form of the linear programming problem. We then list and discuss some of the major primal–dual relationships. Finally, we discuss the use of the dual variables in the economic interpretation of a linear programming problem. In Chapter 9 we employ the duality relationship so as to develop a simpler, more efficient technique for problem solution, the primal–dual algorithm.

Formulation of the Linear Programming Dual

The formulation of the dual is a mechanical, straightforward process. However, the mechanics differ somewhat according to the form of the primal. We shall then first discuss three common forms of the (primal) linear programming problem.

The Canonical Form

The definition employed in the introduction to this chapter is actually one that employs the canonical form. A canonical form of a linear programming problem is one in which (1) the objective is of maximizing form, (2) all constraints are of type I form ($\leq$), and (3) all variables are restricted to nonnegative values. Thus, the primal, in canonical form, is:

$$\left.\begin{aligned} \text{maximize} \quad z &= \mathbf{cx} \\ \text{subject to}& \\ \mathbf{Ax} &\leq \mathbf{b} \\ \mathbf{x} &\geq \mathbf{0} \end{aligned}\right\} \tag{8.3}$$

Notice in particular that, in the canonical form, elements of the right-hand side vector (**b**) may be *negative*. Consequently, *any* linear programming problem may be placed into the canonical form.

If (8.3) is specified to be the primal, its dual is given by (8.4) as:

$$\left.\begin{aligned} \text{minimize} \quad Z &= \mathbf{b}^T\mathbf{v} \\ \text{subject to}& \\ \mathbf{A}^T\mathbf{v} &\geq \mathbf{c}^T \\ \mathbf{v} &\geq \mathbf{0} \end{aligned}\right\} \tag{8.4}$$

Note that the designation of "primal" and "dual" is strictly arbitrary. That is, if (8.3) is the primal, (8.4) is its dual. However, we could just as well designate (8.4) as the primal, and then (8.3) is its dual.

The rules for constructing the dual from the primal (or primal from the dual) when using the canonical form are:

1. If the objective of one problem is to be maximized, the objective of the other is to be minimized.
2. The maximization problem must have all type I ($\leq$) constraints and the minimization problem has all type II ($\geq$) constraints.
3. All primal and dual variables ($\mathbf{x}$ and $\mathbf{v}$, respectively) must be nonnegative.
4. Each *constraint* in one problem corresponds to a *variable* (and vice versa) in the other. For example, given m primal constraints, there are m dual variables and, given n primal variables, there are n dual constraints. Consequently, if one problem is of order $m \times n$, the other is of order $n \times m$.
5. The elements of the right-hand side of the constraints in one problem are the respective coefficients of the objective function in the other problem.
6. The matrix of constraint coefficients for one problem is the transpose of the matrix of constraint coefficients for the other problem.

Rule 4 provides a clue as to (just) *one* practical use of the dual. Note that a $m \times n$ primal is transformed into a $n \times m$ dual. Now, generally speaking, the number of simplex iterations (pivots) drive the computation time. The fewer the number of constraints, the fewer the iterations that are normally required for convergence. Consequently, we would usually prefer to solve a problem with as few constraints as possible. Given a primal with 100 constraints and 10 variables, the corresponding dual has 10 constraints and 100 variables. Normally, the dual will be solved much faster than the primal in such a case.

Example 8-1

Find the dual of the following problem:

$$\text{maximize} \quad z = 4x_1 + 2x_2 + 5x_3$$

$$\text{subject to}$$

$$x_1 + x_2 \geq 2$$

$$x_1 + 2x_2 + 3x_3 \leq 15$$

$$\mathbf{x} \geq \mathbf{0}$$

If we choose to use the maximizing form of the objective (the choice is strictly arbitrary and, because of symmetry, provides equivalent results), then all the primal constraints must be of type I form. Thus, the proper, canonical form for the example is:

$$\text{maximize} \quad z = 4x_1 + 2x_2 + 5x_3$$
$$\text{subject to}$$
$$-x_1 - x_2 \qquad \leq -2$$
$$x_1 + 2x_2 + 3x_3 \leq 15$$
$$\mathbf{x} \geq \mathbf{0}$$

Since there are two primal constraints, there must be two dual variables and, since there are three primal variables, there must be three dual constraints. Further, the coefficients of the primal objective $\mathbf{c} = (4 \quad 2 \quad 5)$ become (after being transposed to a column vector) the right-hand side values of the dual. The right-hand side values of the primal $(-2 \quad 15)$ become (after being transposed) the coefficients in the dual objective function. Finally, note that the dual must have a minimizing objective and all type II constraints. The resultant dual is:

$$\text{minimize} \quad Z = -2v_1 + 15v_2$$
$$\text{subject to}$$
$$-v_1 + v_2 \geq 4$$
$$-v_1 + 2v_2 \geq 2$$
$$3v_2 \geq 5$$
$$\mathbf{v} \geq \mathbf{0}$$

Example 8-2

We now find the dual for:

$$\text{minimize} \quad z = 2x_1 + 3x_2 + x_3$$
$$\text{subject to}$$
$$x_1 + x_2 + x_3 \leq 40$$
$$2x_1 - x_2 + x_3 \geq 17$$
$$\mathbf{x} \geq \mathbf{0}$$

Again, we must decide whether to use the maximizing or minimizing form of the objective function. Arbitrarily, we select the minimizing form, which requires all constraints to be in type II form. (The reader is advised to try the same problem—but use the maximizing form of the objective and, of course, all type I constraints.) The proper canonical form is then:

$$\text{minimize} \quad z = 2x_1 + 3x_2 + x_3$$
$$\text{subject to}$$
$$-x_1 - x_2 - x_3 \geq -40$$
$$2x_1 - x_2 + x_3 \geq 17$$
$$\mathbf{x} \geq \mathbf{0}$$

and the dual is:

$$\text{maximize} \quad Z = -40v_1 + 17v_2$$

$$\text{subject to}$$

$$-v_1 + 2v_2 \leq 2$$

$$-v_1 - v_2 \leq 3$$

$$-v_1 + v_2 \leq 1$$

$$\mathbf{v} \geq \mathbf{0}$$

Example 8-3

As a final example, for the canonical form, we find the dual of:

$$\text{maximize} \quad z = 4x_1 + x_2 + 7x_3$$

$$\text{subject to}$$

$$x_1 + x_2 + x_3 = 10$$

$$5x_1 - x_2 + x_3 \geq 12$$

$$x_1 + 7x_2 - 3x_3 \leq 4$$

$$\mathbf{x} \geq \mathbf{0}$$

We select, arbitrarily, the maximizing form of the objective, and thus all constraints must be of type I form. To accomplish this, we replace the first, equality constraint, by two inequality constraints. That is, given

$$g_1(\mathbf{x}) = b_1$$

the equivalent inequality form is

$$g_1(\mathbf{x}) \leq b_1$$

$$g_1(\mathbf{x}) \geq b_1$$

Thus,

$$x_1 + x_2 + x_3 = 10$$

is replaced by

$$x_1 + x_2 + x_3 \leq 10$$

$$x_1 + x_2 + x_3 \geq 10$$

Any values of x satisfying both of the two inequality constraints must also satisfy the original, equality form.

The resultant dual is then:

$$\text{minimize} \quad Z = 10v_1 - 10v_2 - 12v_3 + 4v_4$$

$$\text{subject to}$$

$$v_1 - v_2 - 5v_3 + v_4 \geq 4$$

$$v_1 - v_2 + v_3 + 7v_4 \geq 1$$

$$v_1 - v_2 - v_3 - 3v_4 \geq 7$$

$$\mathbf{v} \geq \mathbf{0}$$

It is not necessary to use the canonical form when formulating the dual. However, because of the symmetry that exists, some analysts do tend to prefer this approach. For those who might prefer another approach, let us now consider the formulation of the dual from the *general* form of the primal (and vice versa).

The General Form

There are few requirements as to the general form of a linear programming problem. The objective may be either of a maximizing or minimizing form, variables may be restricted or unrestricted, and the constraints may be of any form (type I, II, or equalities) and of any mixture of forms. As a result, we save one step in the dual formulation process: the preprocessing in terms of the canonical form. We gain this advantage at the cost of a few additional rules. We summarize the rules for finding the dual from the general form of the primal as follows:

1. If the objective of one problem is to be maximized, the objective of the other is to be minimized.
2. For a maximizing objective, all constraints must be of type I *or* equality form. For a minimizing objective, all constraints must be of type II *or* equality form.
3. Each *constraint* in one problem corresponds to a *variable* (and vice versa) in the other.
4. The elements of the right-hand side of the constraints in one problem are the respective coefficients of the objective function in the other problem.
5. The matrix of constraint coefficients for one problem is the transpose of the matrix of constraint coefficients for the other.
6. If a variable in one problem is *unrestricted*, its associated constraint in the other problem will be an *equality* (and vice versa).

To help illustrate the reasoning behind the foregoing rules, let us return to the dual finally realized in Example 8-3.

$$\text{Minimize} \quad Z = 10v_1 - 10v_2 - 12v_3 + 4v_4$$

subject to

$$\begin{aligned} v_1 - v_2 - 5v_3 + v_4 &\geq 4 \\ v_1 - v_2 + v_3 + 7v_4 &\geq 1 \\ v_1 - v_2 - v_3 - 3v_4 &\geq 7 \\ \mathbf{v} &\geq \mathbf{0} \end{aligned}$$

Notice that we may replace $v_1 - v_2$ by the *unrestricted* variable $v_{1,2}$ and the dual is then written as:

$$\text{minimize} \quad Z = 10v_{1,2} - 12v_3 + 4v_4$$
$$\text{subject to}$$
$$v_{1,2} - 5v_3 + v_4 \geq 4$$
$$v_{1,2} + v_3 + 7v_4 \geq 1$$
$$v_{1,2} - v_3 - 3v_4 \geq 7$$
$$v_3, v_4 \geq 0$$

The dual formulation given above is one of several proper and equivalent ways to write the dual of the original problem. The original problem, having an equality as its *first constraint*, results in a dual (when using the general form) with its *first variable* unrestricted ($v_{1,2}$). Consider the following examples.

Example 8-4

Find the dual of:

$$\text{maximize} \quad z = 5x_1 + 3x_2$$
$$\text{subject to}$$
$$x_1 - 6x_2 \geq 2$$
$$5x_1 + 7x_2 = -4$$
$$x_1 \leq 0$$
$$x_2 \geq 0$$

Changing the first constraint ($x_1 - 6x_2 \geq 2$) to type I, we may immediately write the dual as:

$$\text{minimize} \quad Z = -2v_1 - 4v_2 + 0v_3$$
$$\text{subject to}$$
$$-v_1 + 5v_2 + v_3 = 5$$
$$6v_1 + 7v_2 \geq 3$$
$$v_1, v_3 \geq 0$$
$$v_2 \text{ unrestricted}$$

The first dual constraint is an equality since the first primal variable is unrestricted. The second dual variable is unrestricted because the second primal constraint is an equality.

Example 8-5

Find the dual of:

$$\text{minimize} \quad z = 4x_1 + 2x_2 - x_3$$

$$\text{subject to}$$

$$\begin{aligned} x_1 + x_2 + x_3 &= 20 \\ 2x_1 - x_2 \quad &\geq 6 \\ x_3 &\leq 4 \\ x_1, x_2 &\geq 0 \end{aligned}$$

The dual, after converting the constraint $x_3 \leq 4$ to a type II, is then:

$$\text{maximize} \quad Z = 20v_1 + 6v_2 - 4v_3$$

$$\text{subject to}$$

$$\begin{aligned} v_1 + 2v_2 \quad &\leq 4 \\ v_1 - v_2 \quad &\leq 2 \\ v_1 \quad - v_3 &= -1 \\ v_2, v_3 &\geq 0 \end{aligned}$$

THE STANDARD FORM

There is one other form of the primal that is commonly used. This is the form in which the constraints of the primal are *all* equalities. A problem in this form is said to be in the *standard* form. Our rules for the general form may be applied to the standard form, and thus it simply represents a special case of the general form. Since all constraints in the primal are equalities, all dual variables will be unrestricted.

Relationships in Duality

There are a number of useful primal–dual relationships which one often needs to refer to in either problem solving or theoretical work. We have already mentioned several but, for completeness, we shall repeat these, together with some new relationships.

Relationship 1. The dual of the dual is the primal.

Relationship 2. A $m \times n$ primal gives an $n \times m$ dual (note that when the canonical form is used, and when *equations* are converted, the value of m must be adjusted for this relationship to hold true).

Relationship 3. For each primal constraint, there is a related dual variable, and vice versa.

Relationship 4. For each primal variable, there is a related dual constraint, and vice versa.

Relationship 5. In the general form, an unrestricted variable in one problem gives an associated equality constraint in the other, and vice versa.

Relationship 6. Given the canonical form of the dual with z the objective function value of the maximizing primal, Z the objective function value of the minimizing dual, $\mathbf{x}_0$ a feasible solution to the primal, and $\mathbf{v}_0$ a feasible solution to the dual:

(a) $z \leq Z$

(b) $z^* = Z^*$

(c) If $\mathbf{c}\mathbf{x}_0 = \mathbf{v}_0\mathbf{b}$, then $\mathbf{x}_0 = \mathbf{x}^*$ and $\mathbf{v}_0 = \mathbf{v}^*$.

Relationship 7. If one problem has an optimal solution, the other has an optimal solution.

Relationship 8. If the primal is unbounded, the dual is infeasible.

Relationship 9. If the primal is infeasible, the dual may be either unbounded *or* infeasible.

These are by no means all of the primal–dual relationships that we shall exploit. Others will be discussed later.

There is a very helpful diagram that often clarifies some of the foregoing relationships, particularly those of relationship 6. Figure 8-1 depicts the primal–dual space for the canonical form. In Figure 8-1 we note that we initiate our simplex procedure on the primal at z_0 and, since we are maximizing, z increases

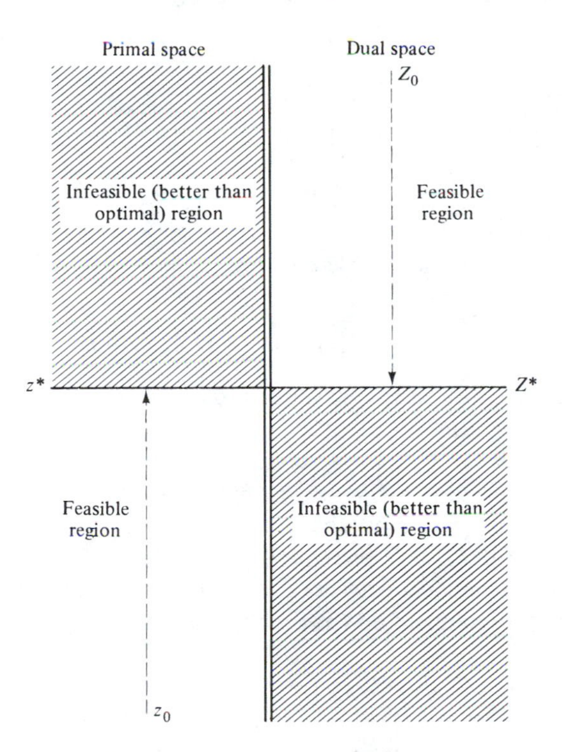

Figure 8-1. The primal-dual space.

until the value z^* is reached. At the same time that this is occurring, we are actually also solving the dual. When $z = z_0$, the dual objective is at Z_0. When $z = z^*$, the dual objective has been minimized at a value of Z^*. From the figure, we see that $z^* = Z^*$.

Notice the infeasible regions for the primal and dual. A value of z in this infeasible region is "better than optimal" since it is greater than z^*. However, the constraints of the primal are violated. The same thing holds true for the dual except that we refer to minimization rather than maximization. Finally, note that the *only* time *both* problems are simultaneously feasible is when *both* are optimal.

Primal–Dual Tableau Relationships

As mentioned earlier and as implied by the relationships cited in the preceding section, if we solve one problem (i.e., the primal or the dual), we have solved the other. Finding a solution to one problem from the other's tableau is relatively straightforward.[a]

Let $z_k - c_k$ be the indicator row value associated with the *original* basic primal variable in the kth primal constraint. The value of the kth dual *decision* variable (i.e., *not* a slack, surplus, or artificial) is given by[b]

$$v_k = |(z_k - c_k) + c_k| \tag{8.5}$$

Thus, for example, if the indicator row value under an original basic variable for the third primal constraint is equal to $-10 + M$, and the basic variable is an artificial variable (thus $c_k = -M$), then:

$$v_3 = |(-10 + M) - M| = 10$$

The same approach may be used to read the primal decision variables from the dual tableau. More precisely,

$$x_k = |(Z_k - b_k) + b_k| \tag{8.6}$$

where $Z_k - b_k$ is the indicator row value under the original basic dual variable for the kth dual constraint.

Typically, our interest centers about the values of the decision variables (i.e., either dual or primal). However, we may also read the values of the dual slack variables from the primal tableau or vice versa. To accomplish this, we utilize either (8.5) or (8.6), as appropriate, and note that: (a) the value of the kth dual slack variable is associated, via (8.5), with the indicator row value of the kth primal *decision* variable, and (b) the value of the kth primal slack variable

---

[a] And it will be made even easier when we employ the primal–dual algorithm of Chapter 9.

[b] Strictly speaking, (8.5) and (8.6) are valid only if v_k and x_k are restricted nonnegative. Otherwise, these values could be negative.

is associated, via (8.6), with the indicator row value of the kth dual *decision* variable.

Example 8-6

Consider the following problem, which we shall arbitrarily term as the primal. Its final tableau is given in Table 8-1.

$$\text{Maximize} \quad z = 8x_1 + 4x_2$$
$$\text{subject to}$$
$$x_1 + x_2 \leq 10$$
$$5x_1 + x_2 \leq 15$$
$$\mathbf{x} \geq \mathbf{0}$$

Table 8-1

V	3	4	$\mathbf{x}_B$
2	$\frac{5}{4}$	$-\frac{1}{4}$	$\frac{35}{4}$
1	$-\frac{1}{4}$	$\frac{1}{4}$	$\frac{5}{4}$
	3	1	45

The slack variable x_3 has been added to the first constraint (with $c_3 = 0$) and slack variable x_4 was added to the second constraint (with $c_4 = 0$).

The tableau of Table 8-1 indicates that the optimal solution is given by

$$x_1^* = \frac{5}{4}$$
$$x_2^* = \frac{35}{4}$$
$$z^* = 45$$

The solution to the dual must have a value of $Z^* = z^*$. Thus, immediately we know that $Z^* = 45$. Next, we determine the values of the dual decision variables (v_1 and v_2 since there are two constraints in the primal). Thus,

$$\begin{aligned} v_1 &= |(z_3 - c_3) + c_3| \\ &= |(3 - 0) + 0| = 3 \\ v_2 &= |(z_4 - c_4) + c_4| \\ &= |(1 - 0) + 0| = 1 \end{aligned}$$

The optimal solution to the dual is then

$$v_1^* = 3$$
$$v_2^* = 1$$
$$Z^* = 45$$

Example 8-7

Now let us consider the dual formulation for the problem in Example 8-6. The dual and its preprocessed form are listed below. Variables v_3 and v_5 are surplus (with $b_3 = b_5 = 0$) and v_4 and v_6 are artificials (with $b_4 = b_6 = -M$).

$$\text{Minimize} \quad Z = 10v_1 + 15v_2$$

$$\text{subject to}$$

$$v_1 + 5v_2 \geq 8$$

$$v_1 + v_2 \geq 4$$

$$\mathbf{v} \geq \mathbf{0}$$

or

$$\text{maximize} \quad Z' = -10v_1 - 15v_2 + 0v_3 - Mv_4 + 0v_5 - Mv_6$$

$$\text{subject to}$$

$$v_1 + 5v_2 - v_3 + v_4 = 8$$

$$v_1 + v_2 - v_5 + v_6 = 6$$

$$\mathbf{v} \geq \mathbf{0}$$

Table 8-2 gives the final tableau for this problem. Notice that $v_1^* = 3$ and $v_2^* = 1$, as already determined from the primal tableau of Example 8-6. Also, $Z = -Z' = -(-45) = 45$, which also was determined previously.

Table 8-2

V	4	6	3	5	$\mathbf{v}_B$
2	$\frac{1}{4}$	$-\frac{1}{4}$	$-\frac{1}{4}$	$\frac{1}{4}$	1
1	$-\frac{1}{4}$	$\frac{5}{4}$	$\frac{1}{4}$	$-\frac{5}{4}$	3
	$-\frac{5}{4}$ $+M$	$-\frac{35}{4}$ $+M$	$\frac{5}{4}$	$\frac{35}{4}$	-45

To read the *primal* solution from the tableau in Table 8-2, we implement (8.6). The values of x_1 and x_2 are then given by looking under variables v_4 and v_6—the artificial variables (since they were the original set of basic variables). Thus,

$$\begin{aligned} x_1^* &= |(Z_4 - b_4) + b_4| \\ &= |(-\tfrac{5}{4} + M) - M| \\ &= \tfrac{5}{4} \\ x_2^* &= |(Z_6 - b_6) + b_6| \\ &= |(-\tfrac{35}{4} + M) - M| \\ &= \tfrac{35}{4} \end{aligned}$$

and

$$z^* = Z^* = 45$$

and we see that we find the primal solution in the dual tableau.

Example 8-8

As our last example in this section, consider the problem (which we shall designate as the primal) shown below. We shall solve the primal and read the dual solution from the final primal tableau. The reader is encouraged to attempt the same thing, but with the dual tableau.

$$\text{Minimize} \quad z = 2x_1 + 4x_2 + x_3$$

$$\text{subject to}$$

$$\begin{aligned} x_1 + 2x_2 - x_3 &\leq 5 \\ 2x_1 - x_2 + 2x_3 &= 2 \\ -x_1 + 2x_2 + 2x_3 &\geq 1 \\ \mathbf{x} &\geq \mathbf{0} \end{aligned}$$

The preprocessed model is then:

$$\text{maximize} \quad z' = -2x_1 - 4x_2 - x_3 + 0x_4 - Mx_5 + 0x_6 - Mx_7$$

$$\text{subject to}$$

$$\begin{aligned} x_1 + 2x_2 - x_3 + x_4 \qquad\qquad &= 5 \\ 2x_1 - x_2 + 2x_3 \qquad + x_5 \qquad &= 2 \\ -x_1 + 2x_2 + 2x_3 \qquad\qquad - x_6 + x_7 &= 1 \\ \mathbf{x} &\geq \mathbf{0} \end{aligned}$$

Variable x_4 is a slack, x_5 and x_7 are artificials, and x_6 is a surplus. We let $c_4 = c_6 = 0$.

The final tableau for this problem is shown in Table 8-3. We read the primal solution as

$$\begin{aligned} x_1^* &= 0 \\ x_2^* &= 0 \\ x_3^* &= 1 \\ z^* &= -z' = -(-1) = 1 \end{aligned}$$

Table 8-3

V	5	2	7	1	$\mathbf{x}_B$
4	$\frac{1}{2}$	$\frac{3}{2}$	0	2	6
6	1	-3	-1	3	1
3	$\frac{1}{2}$	$-\frac{1}{2}$	0	1	1
	$-\frac{1}{2}$ $+M$	$\frac{9}{2}$	M	1	-1

The solution to the dual is given by (8.5). Thus,

$$
\begin{aligned}
v_1^* &= |(z_4 - c_4) + c_4| \\
&= |0 + 0| \\
&= 0 \\
v_2^* &= |(z_5 - c_5) + c_5| \\
&= |(-\tfrac{1}{2} + M) - M| \\
&= \tfrac{1}{2} \\
v_3^* &= |(z_7 - c_7) + c_7| \\
&= |M - M| \\
&= 0 \\
Z^* &= z^* = 1
\end{aligned}
$$

Economic Interpretation

Consider the canonical form of the primal and its dual in summation form:
Primal:

$$\text{maximize} \quad z = \sum_{j=1}^{n} c_j x_j$$

subject to

$$\sum_{j=1}^{n} a_{i,j} x_j \leq b_i \qquad (i = 1, 2, \ldots, m)$$

$$x_j \geq 0 \qquad (j = 1, 2, \ldots, n)$$

Dual:

$$\text{minimize} \quad Z = \sum_{i=1}^{m} b_i v_i$$

subject to

$$\sum_{i=1}^{m} a_{i,j} v_i \geq c_j \qquad (j = 1, 2, \ldots, n)$$

$$v_i \geq 0 \qquad (i = 1, 2, \ldots, m)$$

Notice that, in the primal, we may define the dimensions of each parameter as follows:

$$
\begin{aligned}
z &= \text{return} \\
c_j &= \text{return/(units of variable } j) \\
x_j &= \text{units of variable } j \\
a_{i,j} &= \text{(units of resource } i)/(\text{units of variable } j)
\end{aligned}
$$

The only new parameters introduced in the dual formulation are Z and v_j. It is obvious that, since $z^* = Z^*$, the dimensions of Z are in terms of "return." The question remaining is: What are the dimensions associated with the dual variables, v_j?

Let us first rewrite the primal in terms of its dimensions:

Primal:

$$\text{maximize } z = \sum_{j=1}^{n} \left(\frac{\text{return}}{\text{units of variable } j}\right)(\text{units of variable } j) = \text{return}$$

subject to

$$\sum_{j=1}^{n} \left(\frac{\text{units of resource } i}{\text{units of variable } j}\right)(\text{units of variable } j) \leq$$

$$\text{total amount of resource } i \qquad \text{for } i = 1, 2, \ldots, m$$

$$(\text{units of variable } j) \geq 0$$

The dual may be expressed in a similar manner as:

$$\text{minimize } Z = \sum_{i=1}^{m} (\text{units of resource } i) v_i = \text{return}$$

subject to

$$\sum_{i=1}^{m} \left(\frac{\text{units of resource } i}{\text{units of variable } j}\right) v_i \geq \left(\frac{\text{return}}{\text{units of variable } j}\right) \qquad \text{for } j = 1, 2, \ldots, n$$

$$v_i \geq 0$$

Examining this formulation, the dimensions of the dual variable are easily resolved. For the dimensions to be consistent, the dual variable (v_i) must be expressed in terms of *return per unit of resource i.*

There is an important implication to this result. Given an optimal solution to a linear programming problem, the dual variable then indicates the per unit contribution of the ith resource toward the increase in the (presently) optimal value of the objective. For example, if $v_3 = 9$, we interpret this to mean that, for every additional unit (*up to a limit* to be discussed later) of resource 3 (the resource associated with constraint 3 in the primal), the objective value (z) will increase by 9 units.

Not only is this a useful result, in practice it is often of far greater importance and interest than is the optimal solution. The reason is that most companies wish to improve on the status quo. The optimal *solution* (z^* and $\mathbf{x}^*$) to a linear programming problem tells them only how to best allocate their resources for their *present* state. The dual variables, on the other hand, provide the company with the information needed to expand and to increase profit—if one knows how to interpret them. Those who do not take advantage of this information are overlooking one of the most important factors of linear programming.

Before proceeding, we might mention that the dual variables have been assigned various names, including shadow prices, marginal input, and so forth. Regardless of the terminology used, they are interpreted as discussed above and as illustrated in the following examples.

Example 8-9

The author had some minor involvement in the example to be discussed herein. It was, if nothing else, an educational experience that comes to mind whenever the interpretation of dual variables is considered. For reasons to become obvious, the example has been changed somewhat. However, the pertinent factors remain.

A June graduate accepted a job offer with the relatively small firm of Odum and Odum. The job was somewhat of a combination purchasing agent, marketing man, and "troubleshooter." Odum and Odum performed the basic, straightforward (and monetarily rewarding) task of serving as middlemen for the farmers in a county within a state in the Deep South. What Odum and Odum did, in essence, was to store certain crops (tobacco, corn, and peanuts, for example) for the farmers and to serve as their agent in selling these crops. They made their money by taking a percentage of the amount received.

The recent graduate, whom we shall call Harry, had received a bachelor's degree in business. Included in his coursework was a course in linear programming. Harry built, on his own time, a mathematical model of the firm's operations. Basically, it was of a form similar to that shown below. Let

$$x_1 = \text{tons of corn stored}$$

$$x_2 = \text{tons of tobacco stored}$$

$$x_3 = \text{tons of peanuts stored}$$

$$\text{Profit} = z = 10x_1 + 18x_2 + 7x_3 \tag{8.7}$$

subject to

$$40x_1 + 50x_2 + 60x_3 \leq 10{,}000 \tag{8.8}$$

$$3x_1 + 6x_2 + 2x_3 \leq 600 \tag{8.9}$$

$$x_1 \leq 130 \tag{8.10}$$

$$x_2 \leq 80 \tag{8.11}$$

$$x_3 \leq 200 \tag{8.12}$$

$$\mathbf{x} \geq \mathbf{0} \tag{8.13}$$

Equation (8.7) is the profit relationship. Equation (8.8) reflects the limited floor space (10,000 square feet) of the Odum and Odum warehouse. Equation (8.9) is due to the limited hours for processing (600 person-hours total over the time period of interest) for each crop. Finally, (8.10) through (8.12) reflect the county's predicted output of each crop. Obviously, the coefficients in this problem are *all estimates*. For the moment, however, we treat this as a deterministic problem.

The final tableau for our problem is given in Table 8-4. Observe that

Table 8-4

V	5	6	4	$\mathbf{x}_B$
3	$-\frac{5}{26}$	$-\frac{9}{26}$	$\frac{3}{130}$	$\frac{915}{13}$
1	0	1	0	130
7	$-\frac{9}{39}$	$\frac{5}{13}$	$\frac{1}{130}$	$\frac{890}{13}$
2	$\frac{9}{39}$	$-\frac{5}{13}$	$-\frac{1}{130}$	$\frac{150}{13}$
8	$\frac{5}{26}$	$\frac{9}{26}$	$-\frac{3}{130}$	$\frac{[illegible]{,}685}{13}$
	$\frac{73}{26}$	$\frac{17}{26}$	$\frac{3}{130}$	$\frac{26{,}005}{13}$

$$x_1^* = 130$$

$$x_2^* = \frac{150}{13} = 11.54$$

$$x_3^* = \frac{915}{13} = 70.38$$

$$z^* = Z^* = \frac{26{,}005}{13} = \$2{,}000.38$$

and

$v_1^* = \frac{3}{130} = 0.023$ (per unit contribution of floor space)

$v_2^* = \frac{73}{26} = 2.81$ (per unit contribution of person-hours)

$v_3^* = \frac{17}{26} = 0.65$ (per unit contribution of corn)

$v_4^* = 0$ (per unit contribution of tobacco)

$v_5^* = 0$ (per unit contribution of peanuts)

Although the program designated above will maximize profit *for the conditions assumed*, an increase in profit is obtainable by noting the dual variables. First, v_4 and v_5 are both zero, and thus any additional amounts of tobacco or peanuts will not (for the moment) increase total profit. What will increase total profit is extra: (1) floor space, (2) labor, or (3) corn. To obtain extra corn, we must seek sources outside the county. To obtain extra floor space (on such short notice), we would have to rent. Finally, to obtain extra labor, we must either put the present force on overtime or hire additional labor.

We can also use the dual variables to determine the prices we should pay for any extra floor space, labor, or corn. This price should be less then the value of the corresponding dual variable value. For example, any amount paid for extra labor should be below $2.81 per hour. Unless floor space can be rented for less than $0.023 per square foot, we should not consider it. Finally, we must be able to obtain at least $0.65 per ton profit on the extra corn purchased.

There are a host of other factors that must also be considered:

1. How valid are the estimates of profit, market availability, and so forth in the deterministic model that is being used to approximate this stochastic system?
2. What is the most extra corn, floor space, and labor that can be purchased before the values of the respective dual variables change?
3. Is the model of (8.7) through (8.13) complete? That is, do we need to also consider market saturation (e.g., too much corn)?

Such questions will be answered when we proceed to Chapters 9 and 10.

To finish this particular tale—Harry proceeded with a complete analysis of the model and brought his findings to the owners. They were dubious. Odum and Odum had prospered for several decades as a small organization, run by "seat-of-the-pants" judgment, and had never once failed to earn a highly respectable profit (this same emotional argument, by the way, is constantly used by both giant corporations and Mom and Pop shops).

Harry made a dramatic gesture. He would forgo half of his salary for the next 2 years if the plan were implemented. The owners were suitably impressed. Harry was given permission to move outside the county for his purchases, to obtain extra floor space through short-term leasing, and to set up a (temporary) plan for overtime compensation.

Odum and Odum had, as a result, an increase in profit of well over $2 million (when compared to that which would have been obtained under the optimum program for the original conditions). Harry did not have to forgo half of his salary. His reward? Harry received a handshake and an award of $100 (the maximum under the Odum and Odum employee suggestion plan).

Example 8-10

A university professor is engaged in consulting on a part-time basis. Through such efforts, his combined income, although still less than a New York city sanitation worker, allows for a modest standard of living. The professor, whom we shall call Oliver, has learned the fundamental law of consulting compensation: "The worth of an expert is directly proportional to the distance between the firm and the expert." That is, he is invariably paid more for tasks involving distant organizations than those nearby. Unfortunately, the amount of travel required also increases. At present, he has more consulting opportunities available than time available and he would like to analyze the situation in more detail. Should he hire some help and, if so, how much should he pay?

We let

x_1 = number of hours per month devoted to consulting for firm *A* (at $10 profit per hour)

x_2 = number of hours per month devoted to consulting for firm *B* (at $12 profit per hour)

x_3 = number of hours per month devoted to consulting for firm *C* (at $16 profit per hour)

and assume further that Oliver has 40 hours per month of his own time available for consulting and wishes to keep his travel time, per month, to less than 24 hours.

Travel time per hour of compensated consulting is 0.1, 0.2, and 0.2 hour for firms *A*, *B*, and *C*. Oliver estimates a maximum level of consulting for each firm, per month, to be 80, 60, and 20 hours, respectively. The single-objective problem formulation is then

$$\text{compensation/month} = z = 10x_1 + 12x_2 + 16x_3$$

such that

$$\begin{aligned} x_1 + x_2 + x_3 &\leq 40 \quad &&\text{(hours available for consulting/month)} \\ 0.1x_1 + 0.2x_2 + 0.2x_3 &\leq 24 \quad &&\text{(hours available for travel/month)} \\ x_1 &\leq 80 \quad &&\text{maximum available consulting/month for } A \\ x_2 &\leq 60 \quad &&\text{maximum available consulting/month for } B \\ x_3 &\leq 20 \quad &&\text{maximum available consulting/month for } C \\ & x_1, x_2, x_3 \geq 0 \end{aligned}$$

Solving this problem, our final tableau is shown in Table 8-5. The optimal primal and dual programs and solution are:

Table 8-5

V	1	4	8	$\mathbf{x}_B$
2	1	1	−1	20
5	−0.1	−0.2	0	16
6	1	0	0	80
7	−1	−1	1	40
3	0	0	1	20
	2	12	4	560

$$\begin{aligned} x_1^* &= 0 \text{ hours/month} \\ x_2^* &= 20 \text{ hours/month} \\ x_3^* &= 20 \text{ hours/month} \\ z^* &= Z^* = \$560 \text{ compensation/month} \\ v_1^* &= \$12 \text{ per unit of resource 1 (hours/month)} \\ v_2^* &= 0 \text{ per unit of resource 2 (travel hours/month)} \\ v_3^* &= 0 \text{ per unit of resource 3 (company } A \text{ consulting)} \\ v_4^* &= 0 \text{ per unit of resource 4 (company } B \text{ consulting)} \\ v_5^* &= 4 \text{ per unit of resource 5 (company } C \text{ consulting)} \end{aligned}$$

Notice that we may increase Oliver's monthly compensation by \$12 per month for every additional hour (above 40 hours) that is put into consulting. This determines (to a point) the compensation to be offered for additional help—something less than \$12 per hour will increase the monthly receipts.

In addition, we see that the value of company C's consulting, over the maximum 20 hours per month now available, is \$4 per hour per month. Since it has been assumed that only 20 hours per month is available, we cannot take advantage of this result.

Now, it would appear that Oliver should pay someone some amount less than \$12 per hour to help him each month. For each additional hour that his part-time employee works, Oliver will receive (12 − employee's hourly salary) dollars extra per month. Obviously, there has to be an upper limit to this result; otherwise, the amount of additional compensation is unlimited. We shall see how to determine this in Chapter 10.

As may be seen, the dual variables may play an important part in problem analysis. However, until we pursue the aspect of sensitivity or postoptimality analysis (Chapter 10), the presentation provides only a hint of what can and should be done.

Summary

The concept of duality has been introduced in this chapter. We have seen that it may, on occasion, shorten problem solving and also give a clue as to the potential for increased revenue returns. In the next two chapters we exploit duality even further so as to give an essentially complete picture of its uses in practical applications.

PROBLEMS

8.1. Formulate the canonical form of the dual for Problem 7.8.

8.2. Formulate the canonical form of the dual for Problem 7.9.

8.3. Formulate the canonical form of the dual for Problem 7.10.

8.4. Formulate the canonical form of the dual for Problem 7.11.

8.5. Formulate the canonical form of the dual for Problem 7.12.

8.6. Formulate the canonical form of the dual for Problem 7.13.

8.7. Formulate the canonical form of the dual for Problem 7.14.

8.8. Formulate the canonical form of the dual for Problem 7.15.

8.9. Repeat Problem 8.1 for the general form of the dual.

8.10. Repeat Problem 8.2 for the general form of the dual.

8.11. Repeat Problem 8.3 for the general form of the dual.

8.12. Repeat Problem 8.4 for the general form of the dual.

8.13. Repeat Problem 8.5 for the general form of the dual.

8.14. Repeat Problem 8.6 for the general form of the dual.

8.15. Repeat Problem 8.7 for the general form of the dual.

8.16. Repeat Problem 8.8 for the general form of the dual.

8.17. Solve the dual formulation of Problem 8.1 using the condensed tableau. List the solutions to both the dual and the primal.

8.18. Solve the dual formulation of Problem 8.2 using the condensed tableau. List the solutions to both the dual and the primal.

8.19. Solve the dual formulation of Problem 8.3 using the condensed tableau. List the solutions to both the dual and the primal.

8.20. Solve the dual formulation of Problem 8.4 using the condensed tableau. List the solutions to both the dual and the primal.

8.21. Solve the dual formulation of Problem 8.5 using the condensed tableau. List the solutions to both the dual and the primal.

8.22. Solve the dual formulation of Problem 8.6 using the condensed tableau. List the solutions to both the dual and the primal.

8.23. Solve the dual formulation of Problem 8.7 using the condensed tableau. List the solutions to both the dual and the primal.

8.24. Solve the dual formulation of Problem 8.8 using the condensed tableau. List the solutions to both the dual and the primal.

8.25. Formulate the dual of Problem 7.1 and solve by simplex, using the condensed tableau, and also graphically.

8.26. Construct a two-variable problem for which *both* the primal and dual formulations are infeasible.

8.27. Construct the dual of Problem 2.13 wherein the single objective selected is that of profit maximization and the objective of maximizing the number of microcomputers sold is ignored. Solve the dual problem and discuss the value of any extra assembly-line time.

DUAL SIMPLEX AND THE PRIMAL–DUAL ALGORITHM

Introduction

Duality is a *property*. Arbitrarily designating a given linear programming problem as the "primal," we know that, through certain transformations, we may arrive at a related linear programming problem which we term the "dual." Since both problems are linear programming problems, both may be solved via the simplex algorithm; but, since the solution of one problem also gives the solution of the other, we usually select the problem (i.e., the primal or the dual) that appears easiest to solve. It is further noted that every iteration of one problem is associated with a corresponding iteration of the other. Therefore, when we are solving one problem, be it the primal or the dual, we may imagine that the other is also being simultaneously solved, in its own "dimension."

In this chapter we take advantage of such properties to develop two specific algorithms for solving linear programming problems. The first is the dual simplex algorithm developed initially by Lemke [1]. In its typical form, the dual simplex algorithm is limited to the solution of very specific types of linear programming problems. Even so, it is a useful tool, as we shall see, for dealing with sensitivity analysis in linear programming (and in aiding certain integer programming algorithms).

The second algorithm is termed the primal–dual algorithm[a] Unfortunately, there are a number of algorithms that bear this same name in linear programming. Our particular primal–dual algorithm is unrestricted in the type of

---

[a]Where the algorithm to be described is a *heuristic* approach.

9

linear programming problem that it may solve (where it is understood that we are referring to traditional single-objective linear programming problems with continuous variables). It has three very practical advantages over the simplex algorithm previously presented: (1) it does not require the addition of surplus or artificial variables, (2) it normally requires fewer iterations, and (3) it eases the reading of the dual variables from the primal tableau (or vice versa).

From this point on, we shall refer to the original simplex algorithm of Chapter 7 (for either the extended or condensed tableau) as the *primal simplex* method, to differentiate between algorithms. We now present the dual simplex algorithm.

The Dual Simplex Algorithm

For convenience in discussion, we normally refer to the problem that has been placed into the simplex tableau as the primal. The primal simplex method is then an algorithm that always deals with a basic feasible primal solution. We terminate the primal simplex algorithm as soon as we reach optimality ($z_s - c_s \geq 0$) or until the indication of unboundedness.

The dual simplex problem also addresses the primal and its tableau. However, although it is the primal problem that we see in the tableau before us, it is its dual that is actually being operated on. Such an approach is possible only if the problem is both *dual feasible and primal infeasible*. That is, if one or more $x_{B,i}$ are negative *and* all $z_s - c_s \geq 0$, the dual simplex algorithm may be applied. Figure 9-1 indicates such conditions on our mapping of the primal and dual space.

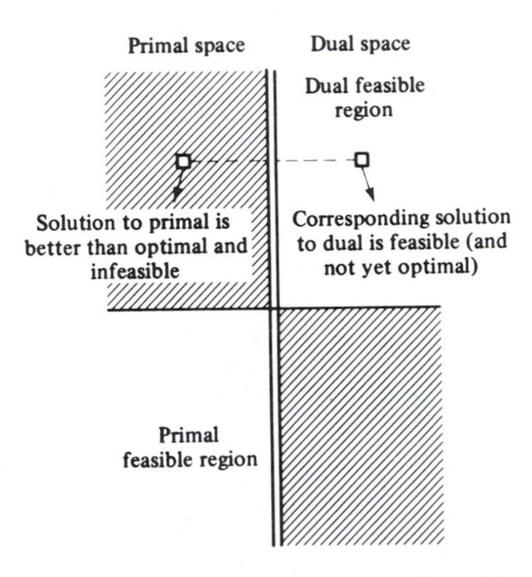

Figure 9-1. Conditions for the dual simplex algorithm.

The basic thrust of the dual simplex algorithm is quite simple. One always attempts to retain "optimality" (i.e., $z_s - c_s \geq 0$) while bringing the primal back to feasibility (i.e., $x_{B,i} \geq 0$ for all i). This may be accomplished by bringing the equivalent feasible dual solution to optimality—but all computations are performed with the *primal* tableau. The form of the dual simplex algorithm that we shall employ is given below.

Dual Simplex Algorithm

Step 1. To employ this algorithm, the problem must be dual feasible and primal infeasible. That is, all $z_s - c_s \geq 0$ and one or more $x_{B,i} < 0$. If these conditions are met, go to step 2.

Step 2. Select the row associated with the most negative $x_{B,i}$ element. The basic variable associated with this row is the departing variable. Denote this row as row i'.

Step 3. Determine the column ratios for only those columns having a *negative* element in row i' (i.e., $y_{i',s} < 0$). The column ratio is given by

$$\phi_s = \min_s \left\{ \left| \frac{z_s - c_s}{y_{i',s}} \right| \right\} \tag{9.1}$$

where $y_{i',s} < 0$ and $z_s - c_s \geq 0$. Designate the column associated with the minimum ϕ_s as column s'. The nonbasic variable associated with column s' is the new entering variable.

Step 4. Using the same procedure as with the original simplex algorithm of Chapter 7, exchange the departing variable for the entering variable and establish the new tableau.

Step 5. If all $x_{B,i}$ are now positive, we stop, having found the optimal, feasible solution. If not, return to step 2.

Example 9-1

The dual simplex algorithm is generally considered unequal to the task of performing as a general-purpose linear programming algorithm because of the difficulty in finding an initial basic solution that is both dual feasible and primal infeasible. Consequently, the following example has obviously been contrived so as to exploit the dual simplex properties. The reader should recognize the improbability of finding such problems in practice.

We shall assume that we wish to:

$$\text{minimize} \quad z = 5x_1 + 2x_2 + 3x_3$$

$$\text{subject to}$$

$$x_1 + 2x_2 - x_3 \geq 5$$

$$2x_1 + x_2 + x_3 \geq 4$$

$$\mathbf{x} \geq \mathbf{0}$$

Rather than preprocessing as usual, we shall, instead, change the objective to the maximizing form and multiply both constraints through by -1. Adding slack variable x_4 to the first constraint and x_5 to the second (assuming $c_4 = c_5 = 0$), we have:

$$\text{maximize} \quad z' = -5x_1 - 2x_2 - 3x_3 + 0x_4 + 0x_5$$

$$\text{subject to}$$

$$-x_1 - 2x_2 + x_3 + x_4 = -5$$

$$-2x_1 - x_2 - x_3 + x_5 = -4$$

$$\mathbf{x} \geq \mathbf{0}$$

The initial tableau for the resultant model is then given in Table 9-1. Notice that

Table 9-1

V	1	2	3	$\mathbf{x}_B$
4	-1	-2	1	-5
5	-2	-1	-1	-4
	5	2	3	0

$$\phi_1 = \left|\frac{5}{-1}\right| = 5$$

$$\phi_2 = \left|\frac{2}{-2}\right| = 1$$

the initial basis is primal infeasible ($x_4 = -5$ and $x_5 = -4$) and dual feasible (all $z_s - c_s \geq 0$). Thus, the conditions for the employment of the dual simplex algorithm are satisfied.

Our second and third steps of the dual simplex algorithm have been indicated in Table 9-1. We first select x_4 ($i' = 1$) as the departing variable because it is the most negative $x_{B,i}$. (This step actually corresponds to the determination of the *entering* variable in the associated dual problem.)

From step 3, we find the minimum ϕ ratio to be associated with the second column ($s' = 2$). Thus, the entering variable is x_2. (Again, this corresponds to selecting a *departing* variable in the associated dual.) The next tableau is given in Table 9-2.

Table 9-2

V	1	4	3	$\mathbf{x}_B$
2	$\frac{1}{2}$	$-\frac{1}{2}$	$-\frac{1}{2}$	$\frac{5}{2}$
5	$-\frac{3}{2}$	$-\frac{1}{2}$	$-\frac{3}{2}$	$-\frac{3}{2}$
	4	1	4	-5

$$\phi_1 = \left|\frac{4}{-\frac{3}{2}}\right| = \frac{8}{3}$$

$$\phi_2 = \left|\frac{1}{-\frac{1}{2}}\right| = 2$$

$$\phi_3 = \left|\frac{4}{-\frac{3}{2}}\right| = \frac{8}{3}$$

Notice that, in Table 9-2, the value of the objective (z') has *decreased*, the indicator row still indicates optimality, and we are not yet feasible. There is no choice as to the departing variable in the second tableau, as only x_5 is negative (thus, $i' = 2$). The entering variable is obtained by the ϕ rule (thus, $s' = 2$). The next and final tableau is given in Table 9-3.

We may read the primal and dual solutions from Table 9-3 as

$$\begin{aligned}
x_1^* &= 0 \\
x_2^* &= 4 \\
x_3^* &= 0 \\
x_4^* &= 3 \\
x_5^* &= 0 \\
z^* &= -(z') = 8 = Z^* \\
v_1^* &= 0 \\
v_2^* &= 2 \\
v_3^* &= 1 \\
v_4^* &= 0 \\
v_5^* &= 1
\end{aligned}$$

Table 9-3

V	1	5	3	x_B
2	2	−1	1	4
4	3	−2	3	3
	1	2	1	−8

Figure 9-2 summarizes the process, in both primal and dual space, for the dual simplex algorithm for our example.

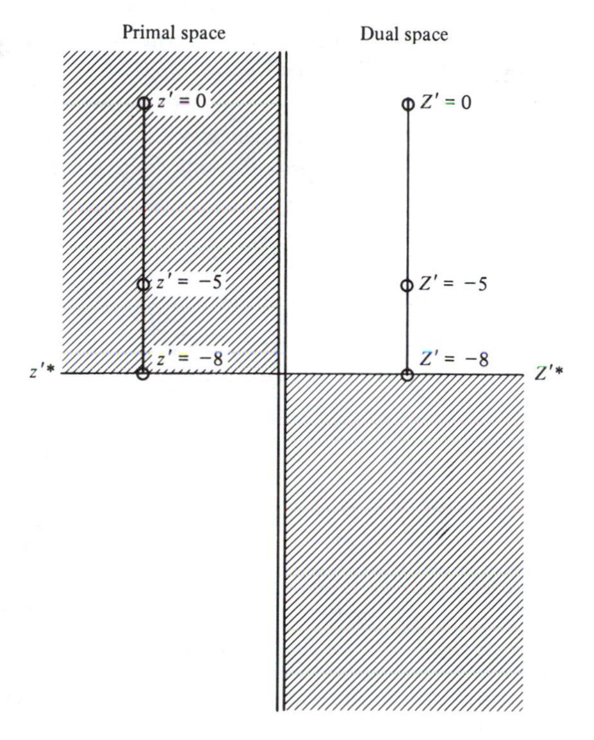

Figure 9-2

An Extended Dual Simplex Algorithm

The type of problem that we encountered in Example 9-1 is, as we warned, simply not typical of what we would expect in practice. As a result, effort has been made to extend the dual simplex algorithm into a form that is more robust. One such extension has been termed the "artificial constraint method," and we briefly illustrate its approach in this section.

The artificial constraint method is intended for a linear programming problem in which one or more $x_{B,i} < 0$ and one or more $z_s - c_s < 0$. Given these conditions, we add the artificial constraint

$$\sum_{s \in P} x_s \leq M \tag{9.2}$$

where

$$P = \{s \mid z_s - c_s < 0\} \tag{9.3}$$

M is a positive number that should be substantially larger than any other value that is expected to be encountered in the computations.

The artificial constraint of (9.2) is changed into an equality by adding a slack variable. The resulting preprocessed constraint is then included in the simplex tableau. We then select, as the *entering variable*, that variable associated with the most negative $z_s - c_s$ value, just as we did with primal simplex. The *departing variable*, however, must be the slack variable associated with the artificial constraint. This initialization process will always result in *all* $z_s - c_s$ becoming nonnegative, since if we let

$$a = \text{the most negative } z_s - c_s$$

$$b = \text{any other negative } z_s - c_s$$

$$c = \text{any nonnegative } z_s - c_s$$

then

$$\hat{a} = \frac{a}{-1} = -a > 0$$

$$\hat{b} = b - \frac{1(a)}{1} = b - a \geq 0$$

$$\hat{c} = c - \frac{0(a)}{1} = c \geq 0$$

where $\hat{a}$, $\hat{b}$, and $\hat{c}$ are the new values of the indicator row elements as obtained through our normal computational process. Once the indicator row is nonnegative, dual simplex may be used to obtain feasibility. The entire process may be explained graphically as seen in the following examples.

Example 9-2

$$\text{Maximize} \quad z = 3x_1 + 6x_2$$

subject to

$$x_1 + 2x_2 \geq 6$$

$$3x_1 + x_2 \geq 9$$

$$7x_1 + 5x_2 \leq 35$$

$$x_1, x_2 \geq 0$$

To preprocess, we change all constraints to type I and add slack variables x_3, x_4, and x_5 (where $c_3 = c_4 = c_5 = 0$) to constraints 1, 2, and 3, respectively. The graphical representation of the problem is given in Figure 9-3 and the initial tableau is shown in Table 9-4.

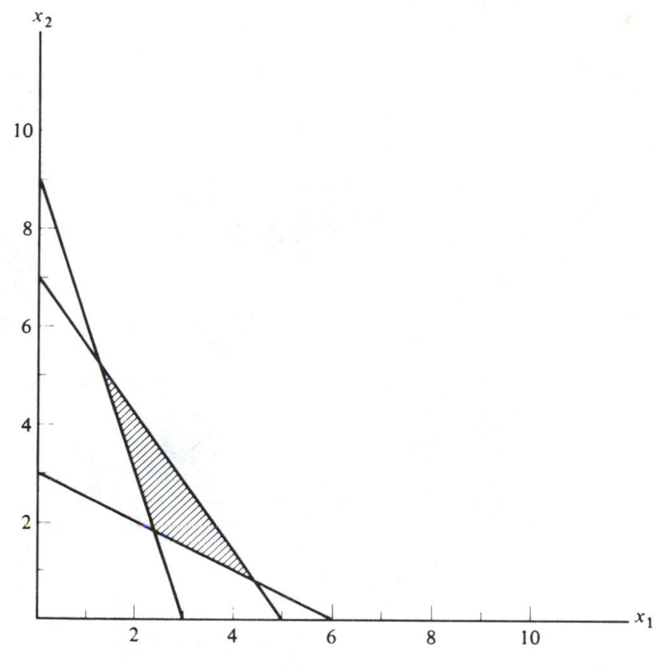

Figure 9-3

Table 9-4

V	1	2	$\mathbf{x}_B$
3	−1	−2	−6
4	−3	−1	−9
5	7	5	35
	−3	−6	0

Since the problem is both infeasible and nonoptimal, we add the artificial constraint $x_1 + x_2 \leq 10$ [note that we have let $M = 10$ in equation (9-2)]. The resultant graph and tableau are given in Figure 9-4 and Table 9-5, respectively.

Notice from Table 9-5 that the initial problem solution (i.e., the initial basis) is given by $x_1 = x_2 = 0$. This corresponds to the origin in Figure 9-4, which has been boxed. As may be seen from the graph, our initial solution is neither feasible nor is it near optimal. However, any point on the artificial constraint ($x_1 + x_2 \leq 10$), while not feasible, is certainly greater than optimal. Our first step consists, therefore, of moving from the origin to a point on the artificial constraint. This point ($x_1 = 0$,

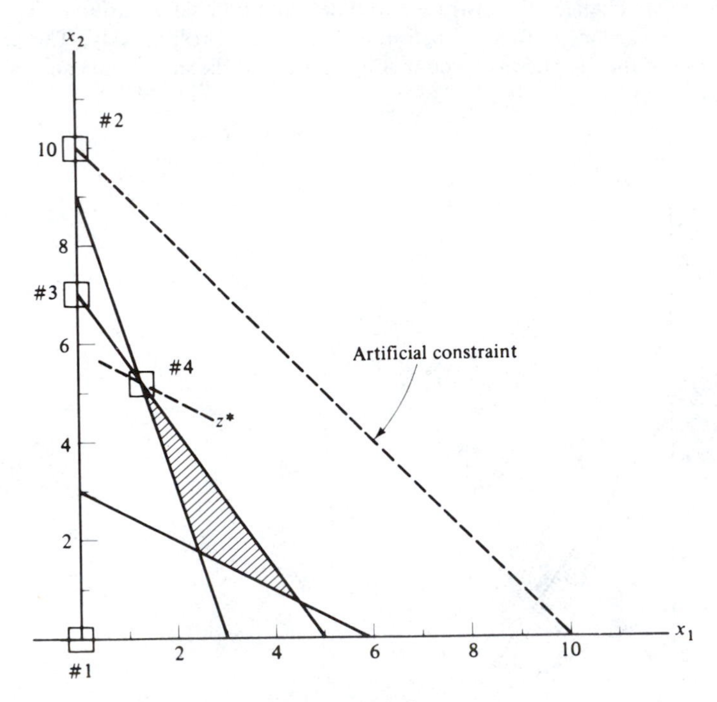

Figure 9-4

Table 9-5

V	1	2	$\mathbf{x}_B$
6	1	1	10
3	−1	−2	−6
4	−3	−1	−9
5	7	5	35
	−3	−6	0

$x_2 = 10$) has also been boxed and, since it is adjacent to the initial solution, we may reach the point in one iteration.

Returning to the tableau of Table 9-5, we see that x_2 enters and x_6 departs, leading to Table 9-6, which indicates that the solution is: $x_1 = 0$, $x_2 = 10$, and all $z_s - c_s$ have become nonnegative. The process is continued in Tables 9-7 and 9-8.

Tracing these results graphically, we see that our initial solution was at point 1 ($x_1 = x_2 = 0$), the second (point 2) was *on* the artificial constraint ($x_1 = 0$, $x_2 = 10$), the third (point 3) is at $x_1 = 0$, $x_2 = 7$, and the final (point 4) is at the optimal vertex of the feasible set. We later compare this result with that obtained by the primal–dual algorithm that is to be discussed in the next section.

Table 9-6

V	1	6	$\mathbf{x}_B$
2	1	1	10
3	1	2	14
4	-2	1	1
5	2	-5	-15
	3	6	60

Table 9-7

V	1	5	$\mathbf{x}_B$
2	$\frac{7}{5}$	$\frac{1}{5}$	7
3	$\frac{9}{5}$	$\frac{2}{5}$	8
4	$-\frac{8}{5}$	$\frac{1}{5}$	-2
6	$-\frac{2}{5}$	$-\frac{1}{5}$	3
	$\frac{27}{5}$	$\frac{6}{5}$	42

Table 9-8

V	4	5	$\mathbf{x}_B$
2	$\frac{7}{8}$	$\frac{3}{8}$	$\frac{21}{4}$
3	$\frac{9}{8}$	$\frac{5}{8}$	$\frac{23}{4}$
1	$-\frac{5}{8}$	$-\frac{1}{8}$	$\frac{5}{4}$
6	$-\frac{1}{4}$	$-\frac{1}{4}$	$\frac{7}{2}$
	$\frac{27}{8}$	$\frac{15}{8}$	$\frac{141}{4}$

When employing the artificial constraint method, there are three possible outcomes:

1. The associated artificial constraint slack variable is in the final basis at a *positive* value. In this case we have reached the optimum solution.
2. The associated artificial constraint slack variable is either nonbasic or in the final basis at a *zero* value. In this case either the value of M was too small or the problem may be unbounded.
3. One (or more) $x_{B,i}$ element is negative, but there is no possible pivot (for *any* of the negative $x_{B,i}$). That is, the corresponding $y_{i',s}$ values are all nonnegative. In this case the dual is unbounded and the primal is *infeasible*.

Outcome 1 has already been illustrated in Example 9-2. Notice that $x_6^* = \frac{7}{2}$ and x_6 is the artificial constraint slack variable.

The second outcome may indicate either a poor choice (too small) for the value of M, or an unbounded problem, or both. To illustrate, consider Example

9-2 except that the artificial constraint is given as $x_1 + x_2 \leq 5$. If the reader plots out this constraint, on Figure 9-3 for example, he or she will see that the artificial constraint serves to exclude a region (and, in this case, the optimal region) of the feasible solution space. The same result (i.e., the artificial constraint slack variable in the final basis at a zero value) can be made to occur by dropping the constraint $7x_1 + 5x_2 \leq 35$ from this example. We can see from Figure 9-4 that, if this is done, the problem is actually unbounded but, since an artificial constraint has been added, we will obtain an indication of optimality and feasibility with a "solution" at $x_1 = 0$ and $x_2 = 10$. As long as M is a finite value, the artificial constraint itself will bound the unbounded problem.

The third and final outcome occurs, as discussed, whenever the primal is infeasible. The reader can easily demonstrate this outcome for himself or herself by solving a simple, infeasible example such as:

$$\text{maximize} \quad z = 3x_1 + 2x_2$$

$$\text{subject to}$$

$$x_1 + x_2 \leq 5$$

$$x_1 + x_2 \geq 10$$

$$x_1, x_2 \geq 0$$

Example 9-3

Now consider the following problem:

$$\text{maximize} \quad z = 3x_1 - x_2 + 2x_3$$

$$\text{subject to}$$

$$x_1 - x_2 + 2x_3 \leq 10$$

$$-x_1 + x_2 \qquad \geq 3$$

$$\mathbf{x} \geq \mathbf{0}$$

To preprocess, we first shall change the second constraint to a type I (i.e., multiply by -1) and then add two slack variables, x_4 to the first constraint and x_5 to the second. Letting $c_4 = c_5 = 0$, we obtain the first tableau as indicated in Table 9-9. As may be seen, the problem is both infeasible ($x_5 = -3$) and nonoptimal ($z_1 - c_1 = -3$ and $z_3 - c_3 = -2$).

We now add the constraint $x_1 + x_3 \leq 100$ (we have let $M = 100$) to the problem with x_6 as the slack variable. The resultant tableau is given in Table 9-10.

Table 9-9

V	1	2	3	$\mathbf{x}_B$
4	1	−1	2	10
5	1	−1	0	−3
	−3	1	−2	0

Table 9-10

V	1	2	3	$\mathbf{x}_B$
6	1	0	1	100
4	1	−1	2	10
5	1	−1	0	−3
	−3	1	−2	0

At this point we select x_1 to enter and x_6 to depart. This drives all $z_j - c_j$ nonnegative, as seen in Table 9-11. The dual simplex may then be initiated. The reader is invited to complete the example.

Table 9-11

V	6	2	3	$\mathbf{x}_B$
1	1	0	1	100
4	−1	−1	1	−90
5	−1	−1	−1	−103
	3	1	1	300

The Primal–Dual Algorithm[b]

The original dual simplex algorithm employs, exclusively, the dual simplex process. When an artificial constraint is added, the resultant process consists of a single, initial primal simplex iteration followed by a series of strictly dual simplex steps. The primal–dual algorithm of this section is a natural extension of these concepts employing, based on certain conditions, either primal simplex or dual simplex at each iteration. It shares the advantages of the dual simplex or the artificial constraint method (e.g., no artificial or surplus variables are employed) while having a few additional advantages of its own. Because of its computational savings, it will be the algorithm used in Chapter 10 in our discussion of sensitivity analysis [2].

It should again be stressed that there are a number of algorithms that are called the primal–dual algorithm. Although they share the same name, their procedures and efficiencies often differ considerably. Our particular primal–dual algorithm tries to achieve, heuristically, the same thing that the artificial constraint/dual simplex method tries to achieve: that is, to drive an infeasible, nonoptimal problem (i.e., some $x_{B,i} < 0$ and some $z_s - c_s < 0$) to feasibility and optimality. However, the primal–dual method requires no additional (artificial) constraint (nor do we need to worry about its impact on the solution space) and, in general, requires fewer iterations to converge.

Our discussion of the primal–dual algorithm will first be directed at the

[b]While satisfactory for textbook illustration, it is emphasized that the convergence of this algorithm has not been proven.

preprocessing requirements and the algorithm. We then compare the primal–dual procedure with that of the artificial constraint method.

The primal–dual algorithm, including the initial preprocessing, is presented next.

Primal–Dual Algorithm

Step 1. *Problem form.* All constraints must be converted to type I ($\leq$) form and the objective must be of the maximization form.

Step 2. Add a slack variable to each constraint and establish the condensed simplex tableau for the problem. (Note that the initial basic solution will always consist of strictly slack variables.)

Step 3. Evaluate the *impact* (i.e., the numerical change in value) on the objective function by *both* primal simplex and dual simplex as follows:

(a) *Primal simplex impact.* If a primal simplex pivot is possible,[c] designate the associated pivot row and column as i' and s', respectively. The impact is then

$$\left| \frac{(z_{s'} - c_{s'})(x_{B,i'})}{y_{i',s'}} \right| \tag{9.4}$$

(b) *Dual simplex impact.* If a dual simplex pivot is possible,[d] designate the associated pivot row and column as i' and s', respectively. The impact is then given by (9.4).

Step 4. Select either the primal or dual simplex pivot according to which has the largest impact value in step 3. If neither a dual simplex or a primal simplex pivot is possible, we terminate the process. Otherwise, return to step 3.

Note carefully in step 3 that if a particular pivot is not possible for the most negative $z_s - c_s$ (for the primal impact) or the most negative $x_{B,i}$ (for the dual simplex impact), we evaluate the next-most-negative $z_s - c_s$ (for the primal) or $x_{B,i}$ (for the dual). We continue in this manner until either a pivot (primal *or* dual or both) is found or until we determine that no pivots are possible.

To illustrate the implementation of the primal–dual algorithm, we first solve the same problem as earlier approached by the artificial constraint method. The results may then be compared with those obtained in Example 9-2.

Example 9-4

$$\text{Maximize} \quad z = 3x_1 + 6x_2$$

subject to

$$\begin{aligned} x_1 + 2x_2 &\geq 6 \\ 3x_1 + x_2 &\geq 9 \\ 7x_1 + 5x_2 &\leq 35 \\ x_1, x_2 &\geq 0 \end{aligned}$$

---

[c] The conditions for a primal simplex pivot are $z_{s'} - c_{s'} \leq 0$, $x_{B,i'} \geq 0$, and $y_{i',s'} > 0$.

[d] The conditions for a dual simplex pivot are $z_{s'} - c_{s'} \geq 0$, $x_{B,i'} \leq 0$, and $y_{i',s'} < 0$.

Preprocessing according to steps 1 and 2 of the primal–dual algorithm, we have:

$$\text{maximize} \quad z = 3x_1 + 6x_2 + 0x_3 + 0x_4 + 0x_5$$

$$\text{subject to}$$

$$\begin{aligned}
-x_1 - 2x_2 + x_3 \qquad\qquad &= -6 \\
-3x_1 - x_2 \qquad + x_4 \qquad &= -9 \\
7x_1 + 5x_2 \qquad\qquad + x_5 &= 35 \\
\mathbf{x} \geq \mathbf{0}&
\end{aligned}$$

The initial tableau for the resultant model is shown in Table 9-12. As may be seen, only a primal simplex pivot (column 2, row 3) is possible.

Table 9-12

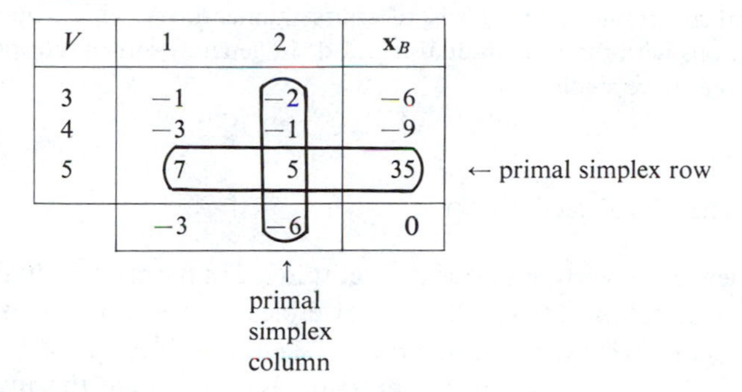

V	1	2	$\mathbf{x}_B$
3	-1	-2	-6
4	-3	-1	-9
5	7	5	35
	-3	-6	0

The primal simplex pivot impact is

$$\text{primal simplex impact} = \left| \frac{(35)(-6)}{5} \right| = 42$$

Since the primal simplex is the *only* pivot, we have no choice but to select it as a pivot to obtain our new tableau.

The next tableau is shown in Table 9-13. Since there is now no primal simplex pivot possible (i.e., all $z_s - c_s \geq 0$), we have no choice but to select dual simplex here.

The next and final tableau is shown in Table 9-14. Comparing our results with those of the artificial constraint method, we see that it only required two iterations (three tableaux) using the primal–dual algorithm, whereas it required three iterations

Table 9-13

V	1	5	$\mathbf{x}_B$
3	$\frac{9}{5}$	$\frac{2}{5}$	8
4	$-\frac{8}{5}$	$\frac{1}{5}$	-2
2	$\frac{7}{5}$	$\frac{1}{5}$	7
	$\frac{27}{5}$	$\frac{6}{5}$	42

(four tableaux) with the artificial constraint method (and each tableau was larger due to the added row).

Table 9-14

V	4	5	$\mathbf{x}_B$
3	$\frac{9}{8}$	$\frac{5}{8}$	$\frac{23}{4}$
1	$-\frac{5}{8}$	$-\frac{1}{8}$	$\frac{5}{4}$
2	$\frac{7}{8}$	$\frac{3}{8}$	$\frac{21}{4}$
	$\frac{27}{8}$	$\frac{15}{8}$	$\frac{141}{4}$

It is also interesting to compare the two procedures graphically. Examining Tables 9-12 through 9-14 and Figure 9-4, we see that the primal–dual algorithm began at point 1 (on Figure 9-4), moved to point 3, and then to the optimal point 4. Movement to the artificial constraint (point 2) was, of course, unnecessary. This is one reason why we save iterations with the primal–dual method. In general, with larger problems, the savings are even more significant.

Reading the Dual Solution

Since there are no artificial variables employed in the primal–dual algorithm, reading the dual solution from the primal tableau (or vice versa) is simplified. The original set of basic variables are *always* slack variables and, *if their objective function coefficients are zero*, all we need do is to examine the indicator row beneath the slack variables. The dual variable values appear there directly.

As an example, returning to Table 9-14, we see that x_3, x_4, and x_5 were the original basic variables. Thus, the dual variables are

$$v_1^* = 0$$
$$v_2^* = \frac{27}{8}$$
$$v_3^* = \frac{15}{8}$$

and, of course,

$$Z^* = \frac{141}{4}$$

Example 9-5

$$\text{Minimize} \quad z = -3x_1 + x_2 - 2x_3$$

subject to

$$x_1 - x_2 + 2x_3 \leq 10$$
$$-x_1 + x_2 \geq 3$$
$$\mathbf{x} \geq \mathbf{0}$$

Preprocessing for the primal–dual, we have:

$$\text{maximize} \quad z' = 3x_1 - x_2 + 2x_3 + 0x_4 + 0x_5$$

subject to

$$x_1 - x_2 + 2x_3 + x_4 \qquad = 10$$

$$x_1 - x_2 \qquad + x_5 = -3$$

$$\mathbf{x} \geq \mathbf{0}$$

The tableaux are given in Tables 9-15 and 9-16. We may stop at the second tableau because there is an indication of unboundedness (examine the column under x_2). This is actually the same problem that we attempted in Example 9-3. Comparing with the work in Example 9-3, we see that even after three tableaux with the artificial constraint

Table 9-15

V	1	2	3	$\mathbf{x}_B$
4	1	−1	2	10
5	1	−1	0	−3
	−3	1	−2	0

$$\text{dual simplex impact} = \left|\frac{(-3)(1)}{-1}\right| = 3$$

$$\text{primal simplex impact} = \left|\frac{(10)(-3)}{1}\right| = 30 \text{ (maximum)}$$

Table 9-16

V	4	2	3	$\mathbf{x}_B$
1	1	−1	2	10
5	−1	0	−2	−13
	3	−2	4	30

method, we had not (and will not until the final tableau) detected the fact that this problem was unbounded. The artificial constraint, in fact, disguises this facet, and yet another advantage of the primal–dual is realized.

We have seen that the indication of an unbounded problem is the same, with the primal–dual algorithm, as with primal simplex. Indications of alternative optimal solutions are, obviously, achieved in the same manner. But what about infeasibility? The primal simplex algorithm used the existence of an artificial variable in the basis (with a positive value) in the final tableau to indicate infeasibility. However, there are no artificials in the primal–dual approach. Detection of infeasibility is obtained, rather, by the inability to find a primal or dual simplex pivot even though the problem has not yet reached feasibility and/or optimality. We illustrate this in the next example.

Example 9-6

$$\text{Maximize} \quad z = 6x_1 + 3x_2$$
$$\text{subject to}$$
$$x_1 + x_2 \geq 10$$
$$2x_1 + 3x_2 \leq 12$$
$$\mathbf{x} \geq \mathbf{0}$$

The preprocessed problem is:

$$\text{maximize} \quad z = 6x_1 + 3x_2 + 0x_3 + 0x_4$$
$$\text{subject to}$$
$$-x_1 - x_2 + x_3 = -10$$
$$2x_1 + 3x_2 + x_4 = 12$$
$$\mathbf{x} \geq \mathbf{0}$$

The tableaux for this (obviously) infeasible problem are given in Tables 9-17 and 9-18. As may be seen, there is no pivot possible in the second tableau, even though it has not yet reached feasibility.

Table 9-17

V	1	2	$\mathbf{x}_B$
3	−1	−1	−10
4	2	3	12
	−6	−3	0

(no dual simplex pivot possible)

Table 9-18

V	4	2	$\mathbf{x}_B$
3	$\frac{1}{2}$	$\frac{1}{2}$	−4
1	$\frac{1}{2}$	$\frac{3}{2}$	6
	3	6	36

(no primal or dual simplex pivot possible)

Degeneracy may, and often does, occur naturally in a linear programming problem. However, when employing the primal–dual algorithm we often have a type of "forced" degeneracy. This occurs whenever the original problem has one or more *equality* constraints. Such constraints must be replaced by two type I (actually, a type I and converted type II constraint) constraints. For example, if the constraint

$$x_1 + 2x_2 - x_3 = 12$$

is to be included, we use

$$x_1 + 2x_2 - x_3 \leq 12$$
$$x_1 + 2x_2 - x_3 \geq 12$$

or

$$\begin{aligned} x_1 + 2x_2 - x_3 + x_4 \qquad &= \quad 12 \\ -x_1 - 2x_2 + x_3 \qquad + x_5 &= -12 \end{aligned}$$

where x_4 and x_5 are slack variables.

If the equality constraint is to be satisfied, then *both* x_4 and x_5 must be zero in the final tableau. One of these variables will appear in the basis, and thus an indication of "degeneracy" is given. This should be noted when working the problems at the end of the chapter.

Equality Constraints—A Simplification

In the event that there are a large number of equality constraints, the use of the primal–dual algorithm will result in a substantial increase in the number of constraints (since each equality constraint is replaced by two inequalities). However, there is a straightforward and easy way to avoid this difficulty.

Given a system of equality constraints such as

$$\begin{gathered} a_{1,1}x_1 + a_{1,2}x_2 + \cdots + a_{1,n}x_n = b_1 \\ \vdots \\ a_{m,1}x_1 + a_{m,2}x_2 + \cdots + a_{m,n}x_n = b_m \end{gathered}$$

we may replace these by the following *equivalent* system:

$$\begin{gathered} a_{1,1}x_1 + \cdots + a_{1,n}x_n \leq b_1 \\ \vdots \\ a_{m,1}x_1 + \cdots + a_{m,n}x_n \leq b_m \\ \sum_{i=1}^{m} (a_{i,1}x_1 + \cdots + a_{i,n}x_n) \geq \sum_{i=1}^{m} b_i \end{gathered}$$

and thus only one additional constraint is imposed.

Summary

The material in this chapter, particularly the dual simplex and primal–dual algorithms, is used extensively in the following, important chapter on sensitivity analysis. Consequently, it is strongly advised that the reader become familiar with both tools and attempt the problem set that follows.

REFERENCES

1. Lemke, C. E. "The Dual Method of Solving the Linear Programming Problem," *Naval Research Logistics Quarterly*, Vol. 1, 1954, pp. 48–54.
2. Wolfe, C. S. *Linear Programming with FORTRAN*. Glenview, Ill.: Scott, Foresman, 1973.

PROBLEMS

9.1. Solve the following problem via the dual simplex algorithm.

$$\text{Minimize } z = 8x_1 + 10x_2 + x_3$$
$$\text{subject to}$$
$$x_1 + x_2 + 3x_3 \geq 18$$
$$2x_1 - x_2 + x_3 \geq 10$$
$$x_1, x_2, x_3 \geq 0$$

9.2. Solve by the dual simplex algorithm.

$$\text{Maximize } z = -x_1 - 4x_2 - 3x_3$$
$$\text{subject to}$$
$$2x_1 + x_2 + 3x_3 \geq 4$$
$$x_1 + 2x_2 + 2x_3 \geq 3$$
$$x_1, x_2, x_3 \geq 0$$

9.3. List both the primal and dual solutions to Problems 9.1 and 9.2.

9.4. Use the extended dual simplex algorithm to solve the following problems. Plot the results graphically.

(a) Maximize $z = 6x_1 + 4x_2$ subject to

$$x_1 + x_2 \leq 10$$
$$2x_1 + x_2 \geq 4$$
$$x_1, x_2 \geq 0$$

(b) Maximize $z = 2x_1 + 6x_2$ subject to

$$\begin{aligned} x_1 + x_2 &= 8 \\ 2x_1 + x_2 &\leq 6 \\ x_1, x_2 &\geq 0 \end{aligned}$$

(c) Finish Example 9-3.

9.5. Use the primal–dual algorithm to solve Problem 9.4. Compare the results graphically.

9.6. List both the primal and dual solutions to Problem 9.5.

9.7. Solve Problem 5.3 by the primal–dual algorithm.

9.8. Solve Problem 5.4 by the primal–dual algorithm.

9.9. Solve Problem 5.5 by the primal–dual algorithm.

9.10. Solve Problem 6.4 by the primal–dual algorithm.

9.11. Solve Problem 6.5 by the primal–dual algorithm.

9.12. Solve Problem 6.6 by the primal–dual algorithm.

9.13. Solve Problem 6.7 by the primal–dual algorithm.

9.14. Solve Problem 6.8 by the primal–dual algorithm.

9.15. Solve Problem 6.9 by the primal–dual algorithm.

9.16. Solve Problem 6.10 by the primal–dual algorithm.

9.17. Solve Problem 6.11 by the primal–dual algorithm.

9.18. Solve Problem 6.12 by the primal–dual algorithm.

9.19. Solve Problem 7.1 by the primal–dual algorithm.

9.20. The "trim problem" is another classical application of linear programming. Typically, the problem is discussed in terms of cutting paper stock or dress patterns. However, one recent example encountered by the author provides yet another variation. The CRT display (i.e., television monitor) for a missile checkout and launch center is set up so that messages of up to 120 characters in width and 100 characters in length may be displayed. Messages are received in three standard widths (all are 100 characters in length): 64 characters (type *A*), 40 characters (type *B*), and 26 characters (type *C*). The number of expected type *A*, *B*, and *C* messages are 40, 70, and 35, respectively. If these messages are to be simultaneously received, what is the minimum number of CRT displays needed? Use both the regular primal simplex algorithm and the primal–dual algorithm to solve and compare the amount of effort required.

SENSITIVITY ANALYSIS IN LINEAR PROGRAMMING[a]

Introduction

Throughout Chapters 1 through 9, we have cited numerous instances in which simplifying assumptions have been made so as to ease both model development and analysis. It should be obvious that real problems will seldom, if ever, strictly satisfy all of these assumptions. Data used may be subject to error, costs and resource availabilities can change with time, and the system itself may be modified. Since management must deal with the future as well as the present, and deal with various price and commodity flunctuations, it should be apparent that an approach is needed to include such considerations within the linear programming technique. We term this approach *sensitivity analysis.*

It is then the purpose of this chapter to present an approach for determining the impact of changes in the linear programming model structure on the resulting problem solution. Such changes can be broken into two separate classes: discrete changes and continuous or range variations. The latter is also denoted as parametric changes.

To clarify, consider the sensitivity analysis in which one factor, say c_3 (i.e., the "cost" associated with each unit of control variable x_3), is to be analyzed. Now, if we wish to determine the impact of a change in the value of c_3 from its present value (c_3) to a specific new value (c'_3), we are dealing with a discrete

[a]All material in this chapter is based on the use of the condensed tableau in conjunction with the primal-dual algorithm of Chapter 9.

10

change. However, if what we want to examine is the impact over the *entire range* of values of c_3 from, say, c_3' to c_3'', then problem deals with a parametric change.

In this chapter we consider the following set of discrete changes:

—Changes in c_j
—Changes in b_i
—Changes in $a_{i,j}$
—Addition of a new constraint
—Addition of a new variable

Our parametric changes, however, will be limited to simply:

—Changes in c_j
—Changes in b_i

Before presenting the techniques of sensitivity analysis, it may be appropriate to review briefly the elements of the condensed simplex tableau for single-objective linear programming problems. We do this in the next section.

Tableau Interpretation

A general, condensed tableau is shown in Table 10-1. Each element in this tableau has both a meaning and a use. Sensitivity analysis involves the utilization of all these elements, as we see in the sections that follow.

Table 10-1. Illustrative Condensed Tableau

	$\mathbf{c}_N \longrightarrow$	$c_{N,1}$ ⋯	$c_{N,s}$ ⋯	$c_{N,S}$	
$\mathbf{c}_B$	V	$x_{N,1}$ ⋯	$x_{N,s}$ ⋯	$x_{N,S}$	$\mathbf{x}_B$
$c_{B,1}$	$x_{B,1}$	$y_{1,1}$ ⋯	$y_{1,s}$ ⋯	$y_{1,S}$	$x_{B,1}$
⋮	⋮	⋮	⋮	⋮	⋮
$c_{B,i}$	$x_{B,i}$	⋮	$y_{i,s}$	⋮	$x_{B,i}$
⋮	⋮	⋮	⋮	⋮	⋮
$c_{B,m}$	$x_{B,m}$	$y_{m,1}$ ⋯	$y_{m,s}$ ⋯	$y_{m,S}$	$x_{B,m}$
		$(z_1 - c_{N,1})$ ⋯	$(z_s - c_{N,s})$ ⋯	$(z_S - c_{N,S})$	z

Before discussing each individual element, note that we have used the subscripts B and N to refer to either basic or nonbasic variables (or associated parameters), respectively. Thus, $c_{N,3}$ is the "cost" associated with the third *nonbasic* variable, $x_{B,2}$ is the label (if under column V) or value (if under column $\mathbf{x}_B$) of the second *basic* variable, and so on.

Nonbasic Cost Coefficients, $c_{N,s}$

The elements on the top of the tableau, above the column heading labels, indicate the objective function coefficients associated with the respective nonbasic variables. For example, $c_{N,2}$ is the objective function coefficient for the second nonbasic variable. We denote these coefficients as $c_{N,s}$ ($s = 1, 2, \ldots, S$ and $S = n - m$) and recall that the subscript simply refers to the *ordering* of the nonbasic variable *in the condensed tableau.* Thus, $c_{N,4}$ could, for example, be the nonbasic variable above the fourth column in the tableau and could actually refer, for example, to the original coefficient, c_7.

The value of the $c_{N,s}$ elements remain the same throughout the iterations of the simplex method, but their *positions* can change with any change in the position (a move from nonbasic to basic, and vice versa) of the associated control variable. If, in a sensitivity analysis, we wish to investigate an actual change in the value of a $c_{N,s}$ element, it should be noted that such a change will have an impact only on the associated indicator row element, $z_s - c_{N,s}$, since

$$z_s - c_{N,s} = \mathbf{c}_B \mathbf{y}_s - c_{N,s}$$

No other elements in the tableau are affected (at least directly) by such a change, but note that since there is an impact on $z_s - c_{N,s}$, the problem solution can go from optimal to nonoptimal (i.e., $z_s - c_{N,s}$ may become negative as a result of the change in $c_{N,s}$).

Basic Cost Coefficients, $c_{B,i}$

The elements in the stub to the far left of the tableau, under $\mathbf{c}_B$, are the values of the associated basic variable coefficients from the original objective function. For example, $c_{B,3}$ is the objective function coefficient for the basic variable associated with the third row ($x_{B,3}$). Such coefficients are denoted in general as $c_{B,i}$ ($i = 1, 2, \ldots, m$) and, as with the nonbasic cost coefficients, the subscript i indicates the ordering of the nonbasic variable in the tableau rows, and not to the order in the original objective function.

Note also that a coefficient $c_{B,i}$ can be changed to a $c_{N,s}$ if the basic variable $x_{B,i}$ is exchanged for the nonbasic variable $x_{N,s}$. Thus, as discussed for nonbasic cost coefficients, although the values of these coefficients do not change with each simplex iteration, their position can.

A change in a nonbasic cost coefficient can affect the *entire* indicator row since $z_s - c_{N,s} = \mathbf{c}_B\mathbf{y}_s - c_{N,s}$ ($s = 1, 2, \ldots, S$) and $z = \mathbf{c}_B\mathbf{x}_B$.

Basic Variable Values, $\mathbf{x}_B$

The elements in the $\mathbf{x}_B$ column (to the far right) represent the values of the basic variables at any given iteration. All other variables, being nonbasic, are zero-valued.

In the *initial* simplex tableau, these $x_{B,i}$ elements also correspond to the original right-hand side values of each constraint. Thus, a change in the right-hand side values (b_i) will be reflected in the resultant $x_{B,i}$ values. Such changes occur when resource levels change, when legal restrictions are tightened or relaxed, and so forth. The impact of such changes is of particular importance in sensitivity analysis.

Indicator Row

The indicator row, the bottom row of the condensed tableau, consists of both the $z_s - c_{N,s}$ and z elements. The element z, of course, is simply the value of the objective function (given by $z = \mathbf{c}_B\mathbf{x}_B$) at the given iteration represented by the tableau. Its value indicates just how well we have been able to maximize (or minimize) the objective function measure. The value of z is not a parameter that is considered in the sensitivity analysis of a problem. However, z may change as a result of changes in *other* parameters within the model.

$z_s - c_{N,s}$ is the indicator row value associated with the nonbasic variable $x_{N,s}$ (the variable whose label heads the column above $z_s - c_{N,s}$). The value of $z_s - c_{N,s}$ is used to represent several things. First, if this value is negative, the problem has not yet been optimized. Thus, $z_s - c_{N,s}$ is used simply as an indication of the achievement or nonachievement of optimality. More important, from the point of view of analysis, the element $z_s - c_{N,s}$ reflects the *per unit contribution* of nonbasic variable $x_{N,s}$ toward the maximization (or minimization) of the objective function. In practice, $z_s - c_{N,s}$ is also denoted as the *shadow price* of the "product" represented by $x_{N,s}$. If $z_s - c_{N,s}$ is negative, z (the objective function value) will be *increased* $z_s - c_{N,s}$ units per every unit of $x_{N,s}$ introduced into the basis. On the other hand, if $z_s - c_{N,s}$ is positive, z will decrease exactly $z_s - c_{N,s}$ units per every unit of $x_{N,s}$ introduced into the basis. A third interpretation of $z_s - c_{N,s}$ is associated with its use as an indicator of the dual variable values (as discussed in Chapter 8). As with z, the values of $z_s - c_{N,s}$ are not, themselves, considered for change, although they too will vary as a function of changes in other model parameters.

INTERIOR ELEMENTS ($y_{i,s}$) AND THE $\mathbf{B}^{-1}$ MATRIX

The interior elements of the tableau are denoted as $y_{i,s}$, where $y_{i,s}$ is the element in the ith row (associated with the ith constraint or ith basic variable) and sth column (associated with the sth nonbasic variable). In the *initial* condensed tableau, these elements will also correspond to the original $a_{i,s}$ coefficients in each constraint. As would be expected, any change in the value of the original $a_{i,s}$ will be reflected in the resultant values of the $y_{i,s}$ elements.

One interpretation of the $a_{i,s}$ coefficient that helps clarify its physical meaning is that it is the per unit resource consumption rate of the ith resource by the sth variable. These coefficients are also often denoted as *technological coefficients*.

We shall use a particular subarray of the $y_{i,s}$ elements in support of our sensitivity analysis. This subarray (a square $m \times m$ matrix) will simply be the inverse of the present basis matrix (i.e., $\mathbf{B}^{-1}$). However, from a physical point of view, it also "summarizes" all the matrix operations that have taken place, up to and including the iteration of interest. Thus, this matrix is sometimes denoted as $\mathbf{T}$, the "transformation matrix."

Referring back to our review of linear algebra of Chapter 3, recall that one way to obtain the inverse of a square matrix, $\mathbf{A}$, is to augment the identity matrix and then perform matrix row operations such that $(\mathbf{A}|\mathbf{I})$ is converted into $(\mathbf{I}|\mathbf{T})$, where $\mathbf{T}$ is simply the inverse of $\mathbf{A}$. When slack variables are added to the linear programming constraints (as with the primal–dual algorithm) to form an initial basis, exactly the same type of thing occurs. That is, with each iteration of the simplex algorithm, we are finding the inverse of the present basis. This inverse appears under the columns of the *original* set of basic variables —the slack variables. Thus, the columns of $\mathbf{B}^{-1}$ may be found by properly arranging the $\mathbf{y}_s$ vectors according to the following rules:

1. Each $\mathbf{y}_s$ column in $\mathbf{B}^{-1}$ must be associated with an original (first tableau) basic variable (i.e., the slack variables).
2. The columns ($\mathbf{y}_s$) must be arranged according to the original (first tableau) order of the original basic variables.
3. The actual values associated with the $\mathbf{y}_s$ vectors in $\mathbf{B}^{-1}$ are those as indicated in the particular tableau under consideration.

Thus,

$$\mathbf{B}^{-1} = (\mathbf{e}_1 \quad \cdots \quad \mathbf{e}_r \quad \cdots \quad \mathbf{e}_m)$$

where $\mathbf{e}_r$ is the $\mathbf{y}_s$ column, from the tableau of interest, under the rth original basic (slack) variable.

As an illustration, consider the tableau of Table 10-2. This tableau is actually the last tableau (Table 9-14) from Example 9-4, as presented earlier. The original set of basic variables were x_3, x_4, and x_5.

Table 10-2

V	4	5	$\mathbf{x}_B$
3	$\frac{9}{8}$	$\frac{5}{8}$	$\frac{23}{4}$
1	$-\frac{5}{8}$	$-\frac{1}{8}$	$\frac{5}{4}$
2	$\frac{7}{8}$	$\frac{3}{8}$	$\frac{21}{4}$
	$\frac{27}{8}$	$\frac{15}{8}$	$\frac{141}{4}$

Examining Table 10-2, we note that the first column of $\mathbf{B}^{-1}$ should be under the x_3 column heading. However, since x_3 is basic and since this is a condensed tableau, this column does not explicitly appear. It is obvious, though, that such a column will have a 1 in the first row (since x_3 is the basic variable for the first row) and zeros elsewhere. The columns for x_4 and x_5 form the final two columns of $\mathbf{B}^{-1}$ and may be read directly from Table 10-2. Thus, we write $\mathbf{B}^{-1}$ as

$$\mathbf{B}^{-1} = \begin{pmatrix} 1 & \frac{9}{8} & \frac{5}{8} \\ 0 & -\frac{5}{8} & -\frac{1}{8} \\ 0 & \frac{7}{8} & \frac{3}{8} \end{pmatrix}$$

Note that, for the original tableau (see Table 9-12), the initial $\mathbf{B}^{-1}$ matrix is simply

$$\mathbf{B}^{-1} = \begin{pmatrix} 1 & 0 & 0 \\ 0 & 1 & 0 \\ 0 & 0 & 1 \end{pmatrix}$$

That is, since we have added slack variables to obtain a set of simultaneous equations, we have also augmented the matrix **A** (the matrix of constraint coefficients) with the identity matrix, **I**.

Discrete Changes

As mentioned earlier, there are two particular types of changes considered in sensitivity analysis. In this section we discuss one type: discrete changes in the original linear programming model structure.

A CHANGE IN c_j

The first discrete change to be considered will be a change in the value of an objective function coefficient (i.e., a c_j). We handle such changes differently according to whether the respective c_j of interest is associated with a basic or nonbasic variable.

Let us first consider a change in a c_j associated with a nonbasic variable (i.e., a change in a $c_{N,s}$). The only impact of such a change is on the single indicator row element, $z_{N,s} - c_{N,s}$. Letting $\hat{c}_{N,s}$ be the *new* value of c_j and $\hat{z}_s - \hat{c}_{N,s}$ be the resultant, new value of $z_s - c_{N\ s}$, we have

$$\hat{z}_s - \hat{c}_{N,s} = \mathbf{c}_B \mathbf{y}_s - \hat{c}_{N,s} \tag{10.1a}$$

or

$$\hat{z}_s - \hat{c}_{N,s} = \sum_{i=1}^{m} y_{i,s} c_{B,i} - \hat{c}_{N,s} \tag{10.1b}$$

Not only can $z_s - c_{N,s}$ change, it could also possibly go negative. This indicates that the present basis is no longer optimal and at least one additional pivot is necessary.

Now consider a change in a c_j associated with a basic variable (i.e., a $c_{B,i}$). Such a change can affect any or all of the $z_s - c_{N,s}$ elements *and* the value of z. Let $\hat{c}_{B,i}$ be the *new* value of c_j, $\hat{z}_s - \hat{c}_{N,s}$ be the *new* value of the indicator row (shadow price) elements, and $\hat{z}$ be the new objective function value. Thus,

$$\hat{z}_s - \hat{c}_{N,s} = \hat{\mathbf{c}}_B \mathbf{y}_s - c_{N,s} \tag{10.2a}$$

or

$$\hat{z}_s - \hat{c}_{N,s} = \sum_{i=1}^{m} y_{i,s} \hat{c}_{B,i} - c_{N,s} \tag{10.2b}$$

and

$$\hat{z} = \hat{\mathbf{c}}_B \mathbf{x}_B \tag{10.3a}$$

or

$$\hat{z} = \sum_{i=1}^{m} x_{B,i} \hat{c}_{B,i} \tag{10.3b}$$

The new $z_s - c_{N,s}$ values should be treated in a manner as discussed above. The new z value simply indicates that the objective function value has changed. Consider the following examples.

Example 10-1: A Change in a Nonbasic c_j

We shall refer to the following example frequently throughout this chapter. The initial problem formulation is given below and the final condensed tableau is shown in Table 10-3.

Table 10-3

	$\mathbf{c}_N \longrightarrow$	0	0	
$\mathbf{c}_B$	V	4	3	$\mathbf{x}_B$
3	2	−1	1	6
5	1	1	0	4
		2	3	38

$$\text{Maximize} \quad z = 5x_1 + 3x_2$$
$$\text{subject to}$$
$$x_1 + x_2 \leq 10$$
$$x_1 \leq 4$$
$$x_1, x_2 \geq 0$$

Notice that x_3 and x_4 are both nonbasic in this final iteration.

Let us assume that there actually is a profit associated with idle resources for the first constraint and that this profit is \$4 per unit of idle resource. Thus, $\hat{c}_{N,2} = \hat{c}_3 = \$4$. From (10-1), this means that

$$\begin{aligned} \hat{z}_2 - \hat{c}_{N,2} &= \mathbf{c}_B \mathbf{y}_2 - \hat{c}_{N,2} \\ &= (3 \quad 5)\begin{pmatrix} 1 \\ 0 \end{pmatrix} - 4 \\ &= -1 \end{aligned}$$

Thus, the indicator row element under x_3 has gone negative and we must continue with the simplex algorithm until optimality is again regained.

Example 10-2: A Change in a Basic c_j

Given the original model of Example 10-1, let us assume that the value of c_2 should really be 2 units rather than 3. Thus, a change in the coefficient associated with a *basic* variable has been made. That is, $\hat{c}_2 = 2 = \hat{c}_{B,1}$. From (10.2) and (10.3) we then obtain

$$\hat{z}_1 - \hat{c}_{N,1} = (2 \quad 5)\begin{pmatrix} -1 \\ 1 \end{pmatrix} - 0 = 3$$

$$\hat{z}_2 - \hat{c}_{N,s} = (2 \quad 5)\begin{pmatrix} 1 \\ 0 \end{pmatrix} - 0 = 2$$

and

$$\hat{z} = (2 \quad 5)\begin{pmatrix} 6 \\ 4 \end{pmatrix} = 32$$

In this example, although the values of the entire bottom row changed, the optimal basis (i.e., x_1 and x_2) did not, and thus no further processing is required.

A Change in b_i

The values on the far right-hand side of the condensed tableau are the $x_{B,i}$ values (i.e., the values of the basic variables for the present iteration). In the *initial* tableau, these values are also the b_i values (i.e., the original right-hand side values of the problem constraints). If a change in a particular b_i is made, there is an impact on both the $\mathbf{x}_B$ vector and the value of z. Recalling that $\mathbf{x}_B$ is given simply by $\mathbf{B}^{-1}\mathbf{b}$ and recalling that $\mathbf{B}^{-1}$ can be found by a proper arrangement of the $\mathbf{y}_s$ column vectors at any iteration, we have

$$\hat{\mathbf{x}}_B = \mathbf{B}^{-1}\hat{\mathbf{b}} \tag{10.4}$$

where $\hat{\mathbf{x}}_B$ = new right-hand side column vector in the tableau of interest

$\mathbf{B}^{-1}$ = inverse of the present basis matrix

$\hat{\mathbf{b}}$ = new set of right-hand side constraint constants

Also,

$$\hat{z} = \mathbf{c}_B \hat{\mathbf{x}}_B \tag{10.5}$$

Formula (10.4) should be studied carefully. The inverse of the basis matrix, $\mathbf{B}^{-1}$, may contain negative elements and thus there is always a possibility that $\hat{\mathbf{x}}_B$ may include some negative elements. This presents no real problem, however, as dual simplex may be implemented so as to regain feasibility. This is illustrated in Example 10-3.

Example 10-3: A Change in a b_i

Consider the problem given in Example 10-1 when b_2 is changed from 4 to 12. That is, $\hat{b}_2 = 12$.

Table 10-3 provides the final tableau for the original problem. We observe that the inverse basis matrix is, from Table 10-3,

$$\mathbf{B}^{-1} = \begin{pmatrix} 1 & -1 \\ 0 & 1 \end{pmatrix}$$

Thus, from (10.4) and (10.5), we have

$$\hat{\mathbf{x}}_B = \begin{pmatrix} 1 & -1 \\ 0 & 1 \end{pmatrix} \begin{pmatrix} 10 \\ 12 \end{pmatrix} = \begin{pmatrix} -2 \\ 12 \end{pmatrix}$$

and

$$\hat{z} = (3 \quad 5) \begin{pmatrix} -2 \\ 12 \end{pmatrix} = 54$$

Our new tableau is then given in Table 10-4. Since the basis is no longer feasible, we use dual simplex (pivoting on x_2 and x_4) and obtain the new final tableau of Table 10-5.

Table 10-4

	$\mathbf{c}_N \longrightarrow$	0	0	
$\mathbf{c}_B$	V	4	3	$\mathbf{x}_B$
3	2	−1	1	−2
5	1	1	0	12
		2	3	54

Table 10-5

	$\mathbf{c}_N \longrightarrow$	3	0	
$\mathbf{c}_B$	V	2	3	$\mathbf{x}_B$
0	4	−1	−1	2
5	1	1	1	10
		2	5	50

A CHANGE IN $a_{i,j}$

Discrete changes in the technological coefficients are relatively easy to handle *if* the $a_{i,j}$ to be changed is associated with a *nonbasic* variable. However, a change in an $a_{i,j}$ associated with a basic variable is considerably more involved and thus, for such a case, we shall resort to simply resolving the problem from the beginning.

Restricting our attention then to $a_{i,j}$ changes for nonbasic variables (i.e., an $a_{i,s}$ in the condensed tableau format), we note that any change in such an $a_{i,s}$ will directly affect the associated $\mathbf{y}_s$ vector (and, indirectly, the value of $z_s - c_{N,s}$). Since, at any iteration, the $\mathbf{y}_s$ column vector is given by $\mathbf{B}^{-1}\mathbf{a}_s$, we have

$$\hat{\mathbf{y}}_s = \mathbf{B}^{-1}\hat{\mathbf{a}}_s \tag{10.6}$$

where $\mathbf{B}^{-1}$ = inverse of the present basic matrix (or $\mathbf{T}$)

$\hat{\mathbf{a}}_s$ = *new* vector of technological coefficients under the sth nonbasic variable

$\hat{\mathbf{y}}_s$ = new vector set of $y_{i,s}$ coefficient in the final condensed simplex tableau under the sth nonbasic variable

The change in $z_s - c_{N,s}$ is given by

$$\hat{z}_s - \hat{c}_{N,s} = \mathbf{c}_B\hat{\mathbf{y}}_s - c_{N,s} \tag{10.7a}$$

or

$$\hat{z}_s - \hat{c}_{N,s} = \sum_{i=1}^{m} \hat{y}_{i,s}c_{B,i} - c_{N,s} \tag{10.7b}$$

and if $\hat{z}_s - \hat{c}_{N,s}$ should go negative, the problem has gone nonoptimal and the simplex algorithm must be applied.

Example 10-4: A Change in $a_{i,j}$

Let us assume, for the problem given in Example 10-1, that $a_{2,2}$ is changed from a zero to a value of, say, 2. Examining the previous final tableau in Table 10-3, we see that the decision variable associated with $a_{2,2}(x_2)$ is basic. Thus, our procedure for examination does not apply to this change, and we should solve the problem over from the beginning.

Consider instead the following problem, for which the final tableau is shown in Table 10-6.

$$\text{Maximize} \quad z = 4x_1 + 3x_2$$
$$\text{subject to}$$
$$5x_1 + 8x_2 \leq 40$$
$$x_1 + x_2 \leq 10$$
$$x_1, x_2 \geq 0$$

Table 10-6

	$\mathbf{c}_N \longrightarrow$	0	3	
$\mathbf{c}_B$	V	3	2	$\mathbf{x}_B$
4	1	$\frac{1}{5}$	$\frac{8}{5}$	8
0	4	$-\frac{1}{5}$	$-\frac{3}{5}$	2
		$\frac{4}{5}$	$\frac{17}{5}$	32

After solving the problem with the results shown in Table 10-6, let us assume that we are informed that an error was made in data collection and that $a_{1,2}$ is really 1 rather than 8. That is, $\hat{a}_{1,2} = 1$ and thus

$$\hat{\mathbf{a}}_s = \hat{\mathbf{a}}_2 = \begin{pmatrix} 1 \\ 1 \end{pmatrix}$$

From (10.6), we have

$$\hat{\mathbf{y}}_2 = \begin{pmatrix} \frac{1}{5} & 0 \\ -\frac{1}{5} & 1 \end{pmatrix} \begin{pmatrix} 1 \\ 1 \end{pmatrix} = \begin{pmatrix} \frac{1}{5} \\ \frac{4}{5} \end{pmatrix}$$

and from (10.7a), we obtain

$$\hat{z}_2 - \hat{c}_{N,2} = (4 \quad 0) \begin{pmatrix} \frac{1}{5} \\ \frac{4}{5} \end{pmatrix} - 3 = -\tfrac{11}{5}$$

The resultant tableau, after the change in $a_{1,2}$, is shown in Table 10-7. The solution is no longer optimal and we must continue the simplex algorithm.

Table 10-7

	$\mathbf{c}_N \longrightarrow$	0	3	
$\mathbf{c}_B$	V	3	2	$\mathbf{x}_B$
4	1	$\frac{1}{5}$	$\frac{1}{5}$	8
0	4	$-\frac{1}{5}$	$\frac{4}{5}$	2
		$\frac{4}{5}$	$-\frac{11}{5}$	32

Addition of a New Constraint

A rather common error in model development is to forget about one or more constraints. It is also not atypical to find that, once the mathematical model has been solved, a new constraint (such as the passage of a new clean air act) comes into being. The consideration of the impact of such new constraints is relatively straightforward.

It is easy to determine whether or not a new constraint has an impact (i.e., without yet evaluating the actual measure of the impact). We simply evaluate the constraint at the present basis ($\mathbf{x}^*$). If the constraint is satisfied, there is *no* impact and we need go no further. However, if the constraint is violated at $\mathbf{x}^*$, the constraint *will* have an impact on the solution and we proceed to the next phase of our analysis.

Having determined that a new constraint will affect the present solution, we proceed to incorporate this new constraint into the previous final tableau. The slack variable for the new constraint will enter the new basis and a new row must be added to the tableau. However, we cannot simply enter the coefficients of the new constraint directly into this new row. Rather, we first must "eliminate" the coefficients of any variables in the new constraint that are basic in the preceding final tableau. We accomplish this through simple matrix row operations that are easier to illustrate than explain.

Example 10-5: Addition of a New Constraint

Returning again to our problem of Example 10-1, let us evaluate the impact of adding the following constraint:

$$x_1 + 3x_2 \leq 15$$

Examining Table 10-3, we see that $\mathbf{x}^* = (4 \quad 6)$. However, when $x_1 = 4$ and $x_2 = 6$, the constraint $x_1 + 3x_2 \leq 15$ is violated and thus has an impact on the solution. Our next step must be to clear the coefficients of any basic variables in this new constraint and then to add it to the tableau. We first rewrite the new constraint as

$$x_1 + 3x_2 + x_5 = 15 \tag{10.8}$$

where x_5 is a zero-cost slack variable. Since x_1 and x_2 are both basic in Table 10-3, the coefficients of both must be set to zero in the new constraint.

Let us first eliminate the coefficient of x_2. The equation for x_2, *from Table 10-3*, is

$$x_2 - x_4 + x_3 = 6$$

Multiplying this equation by 3 and subtracting it from (10.8), we obtain

$$x_1 + 0x_2 + 3x_4 - 3x_3 + x_5 = -3 \tag{10.9}$$

Next, we eliminate the coefficient of x_1. The equation for x_1, *from Table 10-3*, is

$$x_1 + x_4 + 0x_3 = 4$$

Subtracting this last equation from (10.9), we have

$$0x_1 + 0x_2 + 2x_4 - 3x_3 + x_5 = -7 \tag{10.10}$$

At this point we have our new constraint expressed in terms that do not include the coefficients of any basic variables. The new tableau is shown in Table 10-8.

Table 10-8

	$\mathbf{c}_N \longrightarrow$	0	0	
$\mathbf{c}_B$	V	4	3	$\mathbf{x}_B$
3	2	−1	1	6
5	1	1	0	4
0	5	2	−3	−7

Using the fundamental relationships shown below, we may compute the new bottom row for the tableau and this is given in Table 10-9.

$$z_s - c_{N,s} = \mathbf{c}_B\mathbf{y}_s - c_{N,s}$$

$$z = \mathbf{c}_B\mathbf{x}_B$$

Table 10-9

	$\mathbf{c}_N \longrightarrow$	0	0	
$\mathbf{c}_B$	V	4	3	$\mathbf{x}_B$
3	2	−1	1	6
5	1	1	0	4
0	5	2	−3	−7
		2	3	38

The tableau of Table 10-9 is no longer feasible ($x_5 = -7$) and must be returned to feasibility via dual simplex. A pivot on x_5 and x_3 will regain feasibility (the reader is invited to verify this) and we obtain a new final basis of $\mathbf{x}^* = (4, \frac{11}{3})$.

Addition of a New Control Variable

Let us assume that, once we have solved a linear programming problem for a company, they ask us to consider the impact of the introduction of a new product. This new product must be represented by a new control variable (i.e., a new x_j).

The introduction of a new control variable will either affect the optimality of the present solution or else have no affect at all. In the first case, we must bring the new x_j into the basis. In the second, the new x_j stays nonbasic (i.e., we should not introduce the new product into the market).

Actually, we have already learned how to evaluate the impact of the addition of a new variable via the material in the preceding section. Recall that every *variable in the primal* linear programming model coincides with an associated *constraint in the dual.* Thus, to evaluate the impact of a new (primal) variable, we may simply evaluate the resultant impact of the new dual constraint on the associated dual through the technique previously illustrated.

However, if it is desired to work directly with the primal, the approach taken consists of the following two basic phases.

Phase 1. First, determine whether or not the new variable will enter the basis.

(a) Labeling the new variable as x_k, form the kth dual constraint.

(b) Substitute, into the kth dual constraint, the present dual program ($\mathbf{v}^*$).

(c) If the kth dual constraint is feasible for the present dual program, the new (x_k) primal variable will have no affect on the present optimal program and we may cease our analysis. On the other hand, if the kth dual constraint is not satisfied, the new primal variable should enter the basis and we proceed to phase 2.

Phase 2. The addition of a new primal variable, x_k, will first require the addition of a new column to the condensed tableau. For this new column, we must determine the column vector $\mathbf{y}_k$ and the indicator row element $z_k - c_k$. Based on our previous discussion, this is a relatively straightforward procedure.

(a) The new column vector, $\hat{\mathbf{y}}_k$, is given by

$$\hat{\mathbf{y}}_k = \mathbf{B}^{-1}\hat{\mathbf{a}}_k \tag{10.11}$$

where $\mathbf{B}^{-1}$ = transformation matrix as previously defined

$\hat{\mathbf{a}}_k$ = vector of technological (constraints) coefficients for the new variable, x_k

(b) The new indicator row value, $\hat{z}_k - \hat{c}_k$, is given by

$$\hat{z}_k - \hat{c}_k = \mathbf{c}_B\hat{\mathbf{y}}_k - \hat{c}_k \tag{10.12}$$

where $\hat{c}_k$ is the objective function coefficient of the new variable, x_k.

Example 10-6: Addition of a New Variable

The original model of Example 10-1 is as follows:

$$\begin{aligned} \text{maximize} \quad & z = 5x_1 + 3x_2 \\ \text{subject to} \quad & \\ & x_1 + x_2 \leq 10 \\ & x_1 \qquad \leq 4 \\ & x_1, x_2 \geq 0 \end{aligned}$$

The final tableau for this problem was given previously in Table 10-3. The dual program from this table is

$$v_1^* = 3$$
$$v_2^* = 2$$

Let us now assume that a new control variable, x_3 (do not confuse this x_3 with the slack variable x_3 that was used in Example 10-1), is now to be evaluated. We first determine the contribution of x_3 to both the objective function and constraints. We assume that this contribution is as follows:

$$\text{maximize} \quad z = 5x_1 + 3x_2 + 8x_3$$
$$\text{subject to}$$
$$x_1 + x_2 + x_3 \leq 10$$
$$x_1 \quad\quad + 2x_3 \leq 4$$
$$\mathbf{x} \geq \mathbf{0}$$

We now go to phase 1 of the evaluation process and determine whether or not the dual constraint (the third dual constraint) associated with the new primal variable (the third primal control variable) is feasible or not. The third dual constraint for our new model, as given directly above, is simply

$$v_1 + 2v_2 \geq 8$$

Substituting $v_1 = 3$ and $v_2 = 2$, we see that this dual constraint is *not* satisfied. Thus, x_3 should enter the primal basis (recall from the primal–dual space diagrams that, if the dual is infeasible, the primal is not yet optimal).

Since x_3 should enter the basis, we proceed to phase 2. First, we reconstruct Table 10-3, in which we:

1. Add a new column for x_3.
2. Change the subscripts of the two slack variables from 3 and 4 to 4 and 5.

This tableau is shown in Table 10-10.

Table 10-10

	$\mathbf{c}_N \longrightarrow$	0	0	8	
$\mathbf{c}_B$	V	5	4	3	$\mathbf{x}_B$
3	2	-1	1		6
5	1	1	0		4
		2	3		38

Our next step is to determine the $\mathbf{y}$ vector and indicator row value under x_3 in this table. We accomplish this through the use of (10.11) and (10.12), respectively.

$$\hat{\mathbf{y}}_3 = \mathbf{B}^{-1}\hat{\mathbf{a}}_3$$

and

$$\mathbf{B}^{-1} = \begin{pmatrix} 1 & -1 \\ 0 & 1 \end{pmatrix}$$

$$\hat{\mathbf{a}}_3 = \begin{pmatrix} 1 \\ 2 \end{pmatrix}$$

Thus,

$$\hat{\mathbf{y}}_3 = \begin{pmatrix} 1 & -1 \\ 0 & 1 \end{pmatrix}\begin{pmatrix} 1 \\ 2 \end{pmatrix} = \begin{pmatrix} -1 \\ 2 \end{pmatrix}$$

So

$$\hat{z}_3 - \hat{c}_3 = \mathbf{c}_B \hat{\mathbf{y}}_3 - \hat{c}_3$$

where

$$\hat{\mathbf{y}}_3 = \begin{pmatrix} -1 \\ 2 \end{pmatrix}$$

$$\hat{c}_3 = 8$$

Thus,

$$\hat{z}_3 - \hat{c}_3 = (3 \quad 5)\begin{pmatrix} -1 \\ 2 \end{pmatrix} - 8 = -1$$

Our new tableau is thus shown in Table 10-11. Since $\hat{z}_3 - \hat{c}_3$ is negative, the problem is not yet optimal and x_3 must enter the basis. We leave it to the reader to complete this example.

Table 10-11

	$\mathbf{c}_N \longrightarrow$	0	0	8	
$\mathbf{c}_B$	V	5	4	3	$\mathbf{x}_B$
3	2	−1	1	−1	6
5	1	1	0	2	4
		2	3	−1	38

The reader should consider the same problem as given in Example 10-6, wherein the objective function coefficient for the new variable, x_3, is 6 rather than 8. In this case we may terminate the analysis at the end of phase 1.

Parametric Linear Programming

Up to this point we have considered only the impact of *discrete* changes in the structure of a linear programming model. In this section we now consider the impact *over a range* of variation in model structure, but only for two parameters: c_j and b_i. Evaluation of other parameters, over a range, are also possible but tend to be much more involved. Fortunately, for the majority of practical cases, the c_j's and b_i's are the parameters of major interest.

Consider, for example, a mathematical model of a utility company. Now, if such a model were linear (or could be so approximated) and were solved by the simplex method, the results obtained are valid only for conditions reflected by the original mathematical model. Since the utility company's resources include basic energy sources such as oil, coal, and natural gas, and since the availability of such sources varies with time (and other factors), it would be wise to determine the impact of resource availability on the problem solution (i.e., on $\mathbf{x}^*$ and z^*). For such a problem, our interest is centered on the variation of b_i's.

As another example, our interest might be directed toward the selection of an investment portfolio. Obviously, the return from investment opportunities can, in general, only be estimated. Recognizing the inherent risks in such estimates, it is a wise move to evaluate the portfolio choice (and total estimated return) over a variation in individual returns (i.e., a variation in each c_j).

VARIATIONS IN c_j

We consider first an evaluation of the impact, on the original problem solution, of a variation of the objective function coefficients, c_j. This is modeled mathematically by including one or more parameters in the formulation of the objective function and is easiest to explain via examples.

Consider that a company's profit is a linear function of its sales of three products. Letting x_1, x_2, and x_3 be, respectively, the number of units of products 1, 2, and 3 sold, we have

$$z = c_1x_1 + c_2x_2 + c_3x_3$$

where c_1, c_2, and c_3 are the associated per unit profits. Now, let us consider a single parameter that we identify simply as t. Such a parameter might represent calender time, an inflation rate, or so on. If the profit per product can reasonably be expected to vary *linearly* with this parameter, our method of analysis is appropriate. As an illustration, examine the following revised objective function:

$$z(t) = (c_1 + t)x_1 + (c_2 - 2t)x_2 + c_3x_3$$

Reading this function, we see that the profit of x_1 increases linearly with t. The profit of product x_2, on the other hand, decreases—and it does so as a function of two times t. Finally, the profit for x_3 is unaffected by t.

As another example, consider the following objective function:

$$z(t_1, t_2) = (c_1 - t_1)x_1 + (c_2 - t_2)x_2 + (c_3 + t_1)x_3$$

For this case, z is a function of *two* parameters, t_1 and t_2.

Parameters in the objective function appear in the condensed simplex tableau as additional rows *within* the indicator row. Given the model shown below, the associated tableau is given in Table 10-12.

$$\text{Maximize} \quad z(t) = (5 + t)x_1 + (3 + t)x_2$$

subject to

$$x_1 + x_2 \leq 10$$
$$x_1 \qquad \leq 4$$
$$x_1, x_2 \geq 0$$

Table 10-12

$\mathbf{c}_B$	V	1	2	$\mathbf{x}_B$
$\mathbf{c}_N \longrightarrow$		$(5 + t)$	$(3 + t)$	
0	3	1	1	10
0	4	1	0	4
		-5	-3	0
	t	-1	-1	0

Notice carefully that this initial tableau gives us a feasible solution [i.e., $\mathbf{x} = (0, 0, 10, 4)$] and also an optimal solution *if the indicator row is nonnegative*, or if

$$-5 - t \geq 0$$
$$-3 - t \geq 0$$

These two relationships are both satisfied only when $t \leq -5$. That is, at $t \leq -5$, the problem is both feasible and optimal.

There is a rational, physical explanation for our conclusions. Assuming that, in the model above, the c_j's are profits, these profits increase as a function of t. However, at $t = -5$ or less, there is zero or negative total profit. Under such circumstances, it is best to do nothing, as reflected by the fact that both x_1 and x_2 are zero-valued.

We now present a systematic procedure for performing an analysis of the impact of variations in the c_j's.

Step 1. Formulate the problem (including the parameters) and place it into the condensed tableau.

Step 2. Use dual simplex, if necessary, so as to obtain a *feasible* solution (i.e., all $x_{B,i} \geq 0$). The indicator row, which will include the objective function coefficient parameter(s), is ignored at this step.

Step 3. Examine the indicator row. For the problem to be optimal (as well as feasible), *each element* in this row must be *nonnegative*. Determine the parameter ranges over which the tableau is optimal. That is, find

$$L \leq t \leq U$$

where L is the lower (borderline) value and U is the upper (borderline) value for which the tableau is optimal.

Step 4. Set t to its borderline values (if such a value is finite) and develop the new tableaux, giving the alternative optimal solutions.

Step 5. Repeat steps 3 and 4 until all the appropriate ranges of the parameter(s) have been investigated.

The analysis is quite straightforward but, unfortunately, extremely time consuming if done by hand. We illustrate the procedure with the following example.

Example 10-7: Variations in c_j

Let us examine the problem given earlier whose final tableau is shown in Table 10-12. As you may recall, the formulation was:

$$\text{maximize} \quad z(t) = (5 + t)x_1 + (3 + t)x_2$$

$$\text{subject to}$$

$$x_1 + x_2 \leq 10$$

$$x_1 \qquad \leq 4$$

$$x_1, x_2 \geq 0$$

Examining Table 10-12, we see that the present solution, $\mathbf{x}^* = (0, 0, 10, 4)$, is feasible. Thus, we are ready to begin our examination of the indicator row. This row, as mentioned earlier, is optimal as long as

$$-5 - t \geq 0$$

$$-3 - t \geq 0$$

which is true as long as

$$-\infty < t \leq -5$$

The only *finite* bound on t occurs at $t = -5$. Substituting $t = -5$ into the tableau of Table 10-12, the indicator row (which was $-5 - t$, $-3 - t$) becomes 0 and $+2$.

Since there is now a zero under x_1 in the indicator row, we see that there exists an alternative optimal solution for this tableau which may be found, through primal simplex, by bringing in x_1 and letting x_4 depart. The new tableau is then shown in Table 10-13.

Table 10-13

V	4	2	$\mathbf{x}_B$
3	−1	1	6
1	1	0	4
	5	−3	20
t	1	−1	4

Examining Table 10-13, we see that it is optimal if

$$5 + t \geq 0$$
$$-3 - t \geq 0$$

or at $-5 \leq t \leq -3$, $\mathbf{x}^* = (4, 0, 6, 0)$.

There are two finite bounds on t at this point, but we have already examined $t = -5$. Thus, we next look at $t = -3$. When $t = -3$, the tableau of Table 10-13 reveals that an alternative optimal solution exists which may be obtained by bringing in x_2 and removing x_3. That is, when $t = -3$, the indicator row is $+2, 0$. Table 10-14 lists the resulting table.

Table 10-14

V	4	3	$\mathbf{x}_B$
2	−1	1	6
1	1	0	4
	2	3	38
t	0	1	10

Examining Table 10-14, it is optimal only if

$$2 + 0t \geq 0$$
$$3 + \ \ t \geq 0$$

or at $-3 \leq t < \infty$, then $\mathbf{x}^* = (4, 6, 0, 0)$.

At this point, no *finite* bounds on t remain to be examined. Thus, we may summarize the results, for the entire range of t (from $-\infty$ to ∞), as shown in Table 10-15.

Table 10-15

Range of t	*Optimal Program*	*Optimal Objective Value*
$-\infty < t \leq -5$	$\mathbf{x}^* = (0, 0, 10, 4)$	$z^* = 0 + 0t$
$-5 \leq t \leq -3$	$\mathbf{x}^* = (4, 0, 6, 0)$	$z^* = 20 + 4t$
$-3 \leq t < \infty$	$\mathbf{x}^* = (4, 6, 0, 0)$	$z^* = 38 + 10t$

We may, from a physical point of view, interpret these results as follows:

1. If t is less than -5, there is no profit possible and our best course of action is simply to do nothing (i.e., $x_1 = x_2 = 0$).
2. If t lies between -5 and -3, there is some positive profit available through product 1 and thus x_1 enters the basis.
3. When t is greater than -3, both products can provide a positive profit and, in fact, the optimal solution is at $x_1 = 4$, $x_2 = 6$.

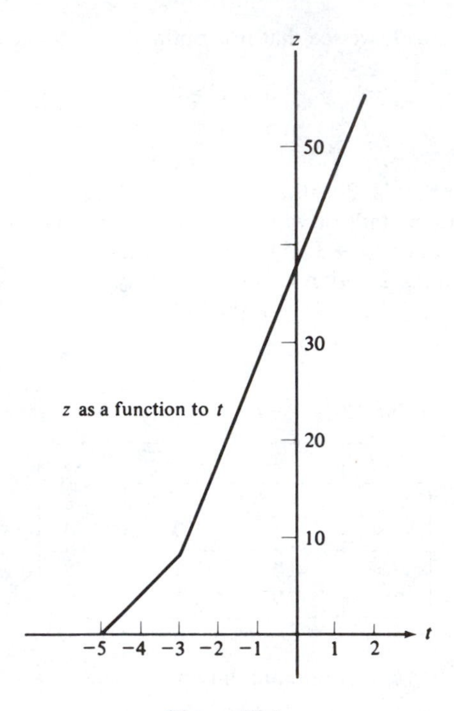

Figure 10-1

Yet another way to represent this parametric analysis is through the graphs of Figure 10-1.

Despite the rather trivial nature of this example problem, it does serve to illustrate the general procedure that is to be carried out for any problem in which variations in c_j are to be analyzed. The use of such an analysis can yield significant information to the organization. Consider, for example, the problem investigated in Example 10-7. Let us assume that, at present, the value of t is -4. At such a value, the optimal program is to produce 4 units of x_1 per time period and none of x_2. Thus, the "production line" for x_2 need not be built. However, if our forecasters predict a likelihood, in the near future, of t exceeding -3, we should plan for the future production of x_2, since it enters the optimal program at that time.

VARIATIONS IN b_i

The only other range variation to be considered is that associated with the right-hand side values (e.g., the units of resource availabilities or lower limits on legal restrictions). Since a change in b_i is equivalent, *in terms of the dual*, to a change in c_j, the process used to analyze a range variation in b_i is the dual of that process used to analyze a range variation in c_j.

Consider the following model.

$$\text{Maximize} \quad z = 5x_1 + 3x_2$$

subject to

$$g_1(t) = x_1 + x_2 \leq 10 - t$$
$$g_2(t) = x_1 \leq 4 - 2t$$
$$x_1, x_2 \geq 0$$

Resource 1 (associated with constraint 1, which is now a function of t as well as $\mathbf{x}$) is being depleted at a rate that is directly proportional to t, while resource 2 is being depleted twice as fast. The questions that we wish to answer then include: What are the optimal program ($\mathbf{x}^*$) and associated profit (z^*) as resources dwindle?

The parameter t on the right-hand side is incorporated into the condensed tableau by simply adding another column inside that column associated with $\mathbf{x}_B$. Table 10-16 illustrates this.

Table 10-16

	$\mathbf{c}_N \longrightarrow$	5	3	$\mathbf{x}_B$	
$\mathbf{c}_B$	V	1	2		t
0	3	1	1	10	−1
0	4	1	0	4	−2
		−5	−3	0	0

The procedure to be used to conduct an analysis of the impact of a variation in the right-hand side vector is as follows:

Step 1. Formulate the problem (including the parameters) and place it into the condensed tableau.

Step 2. Use primal simplex, if necessary, so as to obtain an *optimal* solution (i.e., all $z_s - c_s \geq 0$). The $\mathbf{x}_B$ column, which will include the right-hand side parameter(s), is ignored at this step.

Step 3. Examine the $\mathbf{x}_B$ column. For the problem to be feasible (as well as optimal), each element in $\mathbf{x}_B$ must be nonnegative. Determine the parameter ranges over which the tableau is feasible. That is, find

$$L \leq t \leq U$$

when L is the lower (borderline) value and U is the upper (borderline) value for which the tableau is feasible.

Step 4. Set t to its borderline values (if such a value is finite) and develop alternative new tableaux by pivoting (via dual simplex) on degenerate rows.

Step 5. Repeat steps 3 and 4 until all the appropriate ranges of the parameter(s) have been investigated.

Example 10-8: Variations in b_i

We shall use the model given directly above and shown in Table 10-16 to illustrate the procedure. Since optimality (in terms of all $z_s - c_s \geq 0$) has not yet been achieved, we first use primal simplex so as to achieve this optimality. The departing row will be selected under the assumption that $t = 0$ (i.e., in essence, we ignore t until optimality is reached). Thus, x_1 enters and x_4 leaves the basis so as to achieve the tableau of Table 10-17.

Table 10-17

			$\mathbf{x}_B$	
V	4	2		t
3	−1	1	6	1
1	1	0	4	−2
	5	−3	20	−10

Since Table 10-17 is still not yet optimal, we pivot again via a primal simplex, bringing in x_2 and taking out x_3 so as to obtain Table 10-18. This tableau is optimal and thus we may proceed to the third step in our procedure (i.e., to examine the ranges of t over which the right-hand-side column is feasible). Feasibility will occur when:

$$6 + t \geq 0$$

$$4 - 2t \geq 0$$

or when $-6 \leq t \leq 2$.

Table 10-18

			$\mathbf{x}_B$	
V	4	3		t
2	−1	1	6	1
1	1	0	4	−2
	2	3	38	−7

We are now ready for step 4 of our process. Arbitrarily, we shall first examine the lower limit on ($t = -6$). When $t = -6$, x_2 is zero in the tableau of Table 10-18 and thus, using dual simplex, we pivot on x_2 and x_4 to obtain the tableau shown in Table 10-19.

The basis in the tableau of Table 10-19 is feasible only when:

$$-6 - t \geq 0$$

$$10 - t \geq 0$$

or when $-\infty < t \leq -6$.

At this point, we have examined the parameter t from $-\infty$ to 2. If we return to Table 10-18, we may next attempt to examine t at its borderline value of 2. Thus, in Table 10-18, we now set $t = 2$ and see that x_1 becomes zero. However, we *cannot*

Table 10-19

V	2	3	$\mathbf{x}_B$	t
4	-1	-1	-6	-1
1	1	1	10	-1
	2	5	50	-5

use dual simplex to obtain a new feasible basis as we did earlier because there is no possible dual simplex pivot for the row associated with x_1 in Table 10-18.

This result should not be surprising as we again look at the original problem formulation which, for convenience, is repeated here:

$$\text{maximize} \quad z = 5x_1 + 3x_2$$

$$\text{subject to}$$

$$x_1 + x_2 \leq 10 - t$$

$$x_1 \qquad \leq 4 - 2t$$

$$x_1, x_2 \geq 0$$

Notice in the second constraint that the resource associated with this constraint goes negative at $t > 2$ (i.e., we run out of the resources). This fact is what causes us to be unable to pivot on the x_1 row in Table 10-18 and it is a logical result when interpreted from a physical point of view.

As with variations in c_j, we can now summarize our results in tabular form as shown in Table 10-20.

Table 10-20. SUMMARY OF RESULTS

Range of t	*Optimal Program*	*Optimal Objective Value*
$-\infty < t \leq -6$	$x_1 = 10 - t$ $x_2 = 0$ $x_3 = 0$ $x_4 = -6 - t$	$z = 50 - 5t$
$-6 \leq t \leq 2$	$x_1 = 4 - 2t$ $x_2 = 6 + t$ $x_3 = 0$ $x_4 = 0$	$z = 38 - 7t$
$t > 2$	Problem is infeasible	

RESOURCE VALUES AND RANGES

One particularly significant benefit obtainable through our ability to evaluate a parametric variation on the right-hand side values is that we may use these results to determine the *value* to be placed on (1) either extra units of a resource or (2) the relaxation of a restriction. Before we show how this is done, let us review the interpretation of a dual variable.

Recall that for every primal constraint there is a dual (decision or control) variable. That is, given a primal linear programming problem with, say 12 constraints, there will be 12 dual control variables. Thus, the ith dual control variable is associated with the ith primal constraint. However, not only is there an "association," but there is also an economic interpretation. That is, the value of the ith dual variable represents the improvement (if nonnegative) or degradation (if negative) that may be obtained for every extra unit of resource *i—over a specific range of values for resource i*. Parametric programming, then, may be used to determine this range.

To illustrate the manner in which our analysis proceeds, let us use the following example.

Example 10-9

A chemical company produces two products. Product 1, known as "Crud-Off," is a paint and finish remover for tough refinishing jobs (and is quite thick). Product 2, sold under the brand name "Born Again Wood," is less viscous and intended for easier refinishing (or to be used after one application of "Crud-Off"). Every can of Crud-Off returns a profit of \$0.50, and the return for Born Again Wood is \$0.30 per can.

Since these products can only be shipped out at the end of each week, the total amount produced must fit within the company warehouse, which has a capacity of 400,000 cubic feet. Each container (for either product) consumes 2 cubic feet of warehouse space.

Both products are produced through the same system of pipes, vessels, and processors. Production rates are 2,000 and 2,500 cans per hour for Crud-Off and Born Again Wood, respectively, with a total of 130 hours per week available.

Marketing has determined that, under the present market state and advertising level, the maximum weekly amounts that can be sold are 250,000 cans of Crud-Off and 350,000 cans of Born Again Wood. Finally, because of a previous contract, the company must furnish at least 50,000 cans per week, of Born Again Wood, to a particular customer.

Letting x_1 be the number of cans per week of Crud-Off and x_2 be the number of cans per week of Born Again Wood, we may formulate this problem as the following linear program.

Find x_1 and x_2 so as to

$$\text{maximize} \quad z = 0.5x_1 + 0.3x_2 \tag{10.13}$$

subject to

$$2x_1 + 2x_2 \leq 400{,}000 \tag{10.14}$$

$$\frac{1}{2{,}000}x_1 + \frac{1}{2{,}500}x_2 \leq 130 \tag{10.15}$$

$$(\text{or } 5x_1 + 4x_2 \leq 1{,}300{,}000)$$

$$x_1 \leq 250{,}000 \tag{10.16}$$

$$x_2 \leq 350{,}000 \tag{10.17}$$

$$x_2 \geq 50{,}000 \tag{10.18}$$

$$x_1, x_2 \geq 0$$

where (10.13) is the weekly profit (in dollars) objective, (10.14) is the warehouse capacity limit, (10.15) is the weekly production time limit, (10.16) is the maximum weekly market for x_1, (10.17) is the maximum weekly market for x_2, and (10.18) is the contractual obligation for x_2.

Let us next solve this problem for the optimal, weekly program. The results, shown also in Table 10-21, are

$$\begin{aligned}
x_1^* &= 150{,}000 & v_1^* &= 0.25 \\
x_2^* &= 50{,}000 & v_2^* &= 0 \\
x_3^* &= 0 & v_3^* &= 0 \\
x_4^* &= 350{,}000 & v_4^* &= 0 \\
x_5^* &= 100{,}000 & v_5^* &= 0.20 \\
x_6^* &= 300{,}000 & v_6^* &= 0 \\
x_7^* &= 0 & v_7^* &= 0 \\
z^* &= \$90{,}000 & Z^* &= 90{,}000
\end{aligned}$$

Table 10-21

	$\mathbf{c}_N$	0	0	
$\mathbf{c}_B$	V	3	7	$\mathbf{x}_B$
0.5	1	$\frac{1}{2}$	1	150,000
0	4	$-\frac{5}{2}$	-1	350,000
0	5	$-\frac{1}{2}$	-1	100,000
0	6	0	1	300,000
0.3	2	0	-1	50,000
		0.25	0.2	90,000

where x_3 through x_7 are the slacks for constraints (10.14) through (10.18), respectively, and v_1 through v_5 are the dual variables associated with these same constraints.

Now, if one simply stops at this point and tells the company that their best plan is to produce 150,000 cans of Crud-Off and 50,000 cans of Born Again Wood per week (for a weekly profit of $90,000), he or she has failed to exploit the full potential of the simplex method. Organizations are interested not just in the optimal plan for the present organizational status; they also (and often of more significance) wish to know what they should do in order to grow, increase profits, and so forth. Such information is contained within the simplex tableau.

Notice first that the only positive-valued dual control variables are v_1 and v_5 (note that v_1 through v_5 are dual control variables, while v_6 and v_7 are dual slack variables), associated with the first constraint (10.14) and fifth constraint (10.18), respectively. This means that, under the present optimal program, we may increase profit by $0.25 for every extra cubic foot of warehouse space or by $0.20 for every unit *less* of product x_2 (Born Again Wood) that we ship, per our contract, to our customer.

All other dual control variables are zero, implying that, for the present basis, the worth of any additional units of "resources" 2, 3, or 4 is zero. This is obvious as the slack variables (x_4, x_5, and x_6) associated with these resources are positive-valued in the optimal program (i.e., there are idle units of production capability and we have not yet reached the weekly market limits on x_1 and x_2).

With such information, we have a *portion* of what we need to answer such questions as:

1. What is a reasonable price to pay for additional warehouse space?
2. What is a reasonable penalty to pay for not completely satisfying the contract for 50,000 cans of x_2 (Born Again Wood)?

The partial answer to question 1 is a maximum of \$0.25 per cubic foot of warehouse space rented, per week (i.e., the value of v_1^*). The partial answer to question 2 is a maximum of \$0.20 for every can of x_2 below the 50,000 units for which we have contracted (i.e., the value of v_5^*).

To find the *complete* answer to both questions we must determine the ranges of either b_1 or b_5 over which the values of v_1 and v_5 remain constant. This, of course, calls for a parametric analysis of the right-hand side vector. We now discuss how such an analysis may be performed.

We shall let t_1 be the parameter associated with warehouse capacity (b_1) and t_5 be the parameter associated with contractual requirements on x_2 (b_5). The problem formulation is then:

Find x_1 and x_2 so as to

$$\text{maximize} \quad z = 0.5x_1 + 0.3x_2 \tag{10.19}$$

subject to

$$2x_1 + 2x_2 \leq 400{,}000 + t_1 \tag{10.20}$$

$$5x_1 + 4x_2 \leq 1{,}300{,}000 \tag{10.21}$$

$$x_1 \leq 250{,}000 \tag{10.22}$$

$$x_2 \leq 350{,}000 \tag{10.23}$$

$$x_2 \geq 50{,}000 + t_5 \tag{10.24}$$

$$x_1, x_2 \geq 0$$

At this point, having already solved this problem without the parameters t_1 and t_5, it is not necessary to resolve it from the beginning. Rather, we simply return to Table 10-21, add two columns to $\mathbf{x}_B$ (for t_1 and t_5), and use our technique for determining the impact of a discrete change in **b**. Multiplying $\mathbf{B}^{-1}$ by $\hat{\mathbf{b}}$, we obtain the new right-hand side for Table 10-21 as listed below and as shown in Table 10-22.

Table 10-22

V	3	7	$\mathbf{x}_B$	t_1	t_5
1	$\frac{1}{2}$	1	150,000	$\frac{1}{2}$	−1
4	$-\frac{5}{2}$	−1	350,000	$-\frac{5}{2}$	1
5	$-\frac{1}{2}$	−1	100,000	$-\frac{1}{2}$	1
6	0	1	300,000	0	−1
2	0	−1	50,000	0	1
	0.25	0.2	90,000	$\frac{1}{4}$	−0.2

$$\hat{\mathbf{x}}_B = \mathbf{B}^{-1}\hat{\mathbf{b}}$$

$$= \begin{pmatrix} \frac{1}{2} & 0 & 0 & 0 & 1 \\ -\frac{5}{2} & 1 & 0 & 0 & -1 \\ -\frac{1}{2} & 0 & 1 & 0 & -1 \\ 0 & 0 & 0 & 1 & 1 \\ 0 & 0 & 0 & 0 & -1 \end{pmatrix} \begin{pmatrix} 400{,}000 + t_1 \\ 1{,}300{,}000 \\ 250{,}000 \\ 350{,}000 \\ -50{,}000 - t_5 \end{pmatrix}$$

$$= \begin{pmatrix} 150{,}000 + \frac{1}{2}t_1 - t_5 \\ 350{,}000 - \frac{5}{2}t_1 + t_5 \\ 100{,}000 - \frac{1}{2}t_1 + t_5 \\ 300{,}000 + 0t_1 - t_5 \\ 50{,}000 + 0t_1 + t_5 \end{pmatrix}$$

We next examine the ranges of both t_1 and t_2 over which the program of Table 10-22 remains optimal (i.e., for which v_1 and v_5 do not change). We do this first for t_1 alone by setting t_5 to zero. Thus, the tableau is feasible if

$$150{,}000 + \tfrac{1}{2}t_1 \geq 0$$
$$350{,}000 - \tfrac{5}{2}t_1 \geq 0$$
$$100{,}000 - \tfrac{1}{2}t_1 \geq 0$$
$$300{,}000 + 0t_1 \geq 0$$
$$50{,}000 + 0t_1 \geq 0$$

and this is true if

$$-300{,}000 \leq t_1 \leq 140{,}000$$

Returning to constraint (10.20), we see that this means that as long as b_1 lies between 100,000 and 540,000 cubic feet, the present tableau is optimal. Thus, we may consider renting up to 140,000 extra cubic feet of warehouse space before its associated value (v_1) changes from \$0.25 per cubic foot per week.

For example, if we are given the opportunity of renting, say 100,000 cubic feet of space at a rate of \$0.15 per cubic foot per week, we should take it and expect an increase in weekly profits of

$$(0.25 - 0.15)100{,}000 = \$10{,}000$$

Let us now examine the ranges of t_5 for which our present tableau remains optimal. Setting t_1 to zero, we see that the tableau of Table 10-22 remains feasible if

$$150{,}000 - t_5 \geq 0$$
$$350{,}000 + t_5 \geq 0$$
$$100{,}000 + t_5 \geq 0$$
$$300{,}000 - t_5 \geq 0$$
$$50{,}000 + t_5 \geq 0$$

or the tableau is feasible if

$$-50{,}000 \leq t_5 \leq 150{,}000$$

Examining the fifth constraint, (10.24), we see that the tableau is still optimal if b_5 lies between 0 and 200,000. This means, from a physical point of view, that our maximum, per unit of weekly penalty of $0.20 per can, holds true all the way down to a total elimination (a reduction of 50,000 units of x_2) of this contract. Thus, if the customer is willing to modify this contract, for a penalty of $2,500 per week, so as only to hold the company to 25,000 cans per week of Born Again Wood, we should accept this, as our total weekly profit will increase by

$$0.20(25{,}000) - 2{,}500 = \$2{,}500$$

Summary

This chapter, although relatively lengthy, only begins to illustrate the practical value of the performance of a sensitivity analysis. The problem set at the end of the chapter will provide some additional illustrations. The lesson to be learned, however, is that "solution" of a linear programming problem by simplex is only one phase in problem analysis. The successful analyst will always consider both the sensitivity of the results to possible changes in the model and the potential for extra profit (or reduced cost) that may exist via constraint relaxation. This last analysis is often the most important of all to both the company and the analyst. It shows what can be done to improve the company's position, and it highlights the potential role of the analyst in the actual determination of high-level company policies and plans.

PROBLEMS

Problems 10.1 through 10.6 refer to Example 8-9. Review this example and then answer the questions posed.

10.1. What is the impact of a (discrete) change in the profit per ton of corn from $10 to $8?

10.2. Assume that the predicted output of peanuts (200 tons) is in error and the actual output is only 180 tons. How does this affect the original results?

10.3. What impact would be imposed by a change in output of tobacco from 80 tons to a value of 70 tons?

10.4. Assume that an error was made in the estimated amount of floor space needed for corn. Rather than 40 square feet per ton, we actually need 50 square feet. Does this change the optimal program?

10.5. After solving the problem of Example 8-9, we find that a contract has been signed by our firm obligating us to a delivery of at least 30 tons of tobacco. Does this contract affect our solution?

10.6. We now find that Odum and Odum are considering the soybean market. Each ton of soybeans stored consumes 45 square feet of warehouse floor space and

4 hours per ton for processing. It is estimated that the total output of soybeans, available to Odum and Odum, is 100 tons and the predicted profit is $20 per ton. Should they move into the soybean market? Why or why not?

10.7. Consider Oliver's problem from Example 8-10. What are his optimal strategies if he is willing to consider a *range* in travel hours (previously fixed at 24 hours per month) from 20 up to 40 hours per month?

10.8. Returning to Example 8-10, what is the optimal set of policies if the compensation received from firm *B* is unknown but expected to lie within the range $8 to $20 per hour?

10.9. Given the following problem, evaluate the policy set for a range in profit for variable x_1 from $0 to $5.

$$\text{Maximize} \quad z = 2x_1 + 3x_2$$
$$\text{subject to}$$
$$3x_1 + x_2 \leq 9$$
$$x_1 + 2x_2 \leq 8$$
$$x_1, x_2 \geq 0$$

10.10. For Problem 10.9, examine the impact on the solution for a linear increase (or decrease) in *both* of the constraint right-hand side values. Assume that negative right-hand side values are physically impossible.

10.11. Develop an approach, using visual aids, for the presentation of the results of a sensitivity analysis of a linear programming model to management (where it is assumed that the management has no technical background).

SELECTED APPLICATIONS AND SOME COMPUTATIONAL CONSIDERATIONS

Introduction

Throughout the text we have attempted to demonstrate each algorithm or concept by means of at least one example. A dual purpose of these examples has been to indicate the possible real-world applications of linear programming. In this chapter we present a few additional applications of the linear programming approach. However, these particular examples are less ordinary than those which have been previously cited and, hopefully, they will serve to at least indicate the truly significant range of potential that exists in this field. Also discussed are some additional computational considerations—particularly when using a computer to solve the linear programming model.

An Alternative to Regression

Over the past several years the author has performed an experiment in conjunction with the students enrolled in his linear programming or goal programming classes. These students are either seniors or graduate students from a wide variety of fields and backgrounds and many have attended at least one other university. Despite this, the results of the experiment have never varied.

Each student is assigned to a group consisting of two or three students.

11

Each group is provided with a set of data (often from an actual situation) and a *detailed* problem description. They are all asked to fit a prediction equation to the set of data. Without exception, every group has selected linear regression (i.e., the least-squares method) to solve their problem. In every case they had, already available to them, a much more efficient and appropriate tool (for the particular problems they faced) in linear programming [1].

To further illustrate, consider just one example problem given to these groups. A new firm is engaged in the production of plastic drain pipe. The cost of producing this pipe is considered to be a function of its circumference. Existing data provide cost figures for several pipe diameters (from $\frac{1}{2}$ inch to $3\frac{1}{2}$ inches). The population from which these data have been drawn is unknown with regard to its characteristics (i.e., we do *not* have enough information to determine whether the data has a normal, or any other, specific distribution).

The firm has been approached by the government to produce various sizes of pipe, but sizes never before produced by the firm. They are to respond as soon as possible with a cost estimate. If they receive the contract, they must abide by a penalty clause which says that a fee will be paid to the firm for each size of pipe, *but* a cost penalty will be assigned for any deviation (either *above or below*) from the cost estimates submitted.

Under the assumption that cost is a linear function of pipe circumference, we wish to develop a cost predictive equation [known commonly as a cost estimating relationship or (CER)] of the following form:

$$y_i = ax_i + b$$

where y_i = predicted cost of pipe i

x_i = circumference of pipe i

a and b are *unknown* constants

There are two primary reasons why the least-squares method is not appropriate for the determination of this function. First, the population from which the data were taken is unknown, whereas, with least squares, the associated sensitivity analysis is based on the fundamental assumption of normality of the population. Second, and for this example even more important, is the fact that the least-squares method is based on the minimization of the sum of the *squares* of the residuals (i.e., the difference between the actual data point and the amount predicted by the equation). In this example, the penalty is clearly proportional to the sum of the *absolute deviations* and not the squares of these deviations. That is, we wish to find the values of a and b in the predictive equation so as to minimize

$$\sum_{i=1}^{m} |r_i|$$

where r_i = residual at data point i

m = total number of data points

If we let $C_{A(i)}$ be the actual cost of pipe i and y_i be predicted cost, the residual is simply

$$r_i = C_{A(i)} - (ax_i + b) \tag{11.1}$$

Based on the above, we can establish a mathematical model of the predictive equation problem as follows:

Find a and b so as to

$$\text{minimize} \quad \sum_{i=1}^{m} |r_i| \tag{11.2}$$

subject to

$$C_{A(i)} - (ax_i + b) = 0 \qquad \text{for all } i \tag{11.3}$$

Unfortunately, the model is not in the form of a linear programming problem because of (1) the absolute values in the objective function, and (2) the fact that either a or b (or both) may take on negative values. This may be easily taken care of via the following substitutions. Let

$$a = a_1 - a_2$$
$$b = b_1 - b_2$$

where

$$a_1, a_2, b_1, b_2 \geq 0$$

and modify (11.3) to

$$C_{A(i)} - (a_1 x_i - a_2 x_i + b_1 - b_2) + \eta_i - \rho_i = 0 \qquad \text{for all } i \tag{11.4}$$

where η_i = negative residual (or deviation) at data point i

ρ_i = positive residual (or deviation) at data point i

and both η_i and ρ_i are nonnegative.

Consequently, to satisfy (11.4), we must minimize both η_i and ρ_i, which leads to a new objective:

$$\text{minimize} \quad \sum_{i=1}^{m} (\eta_i + \rho_i) \tag{11.5}$$

As a result, an equivalent linear programming formulation of (11.2) and (11.3) is:

Find $a = a_1 - a_2$ and $b = b_1 - b_2$ so as to

$$\text{minimize} \quad \sum_{i=1}^{m} (\eta_i + \rho_i) \tag{11.6}$$

subject to

$$C_{A(i)} - (a_1 x_i - a_2 x_i + b_1 - b_2) + \eta_i - \rho_i = 0 \qquad \text{for all } i \tag{11.7}$$

$$a_1, a_2, b_1, b_2 \geq 0$$

$$\eta_i, \rho_i \geq 0 \qquad \text{for all } i \tag{11.8}$$

Not only may we solve the foregoing model for the predictive equation, but we may also perform, independent of any assumptions regarding the data population, a sensitivity analysis of the problem using the methods discussed in Chapter 10.

There is yet another reason for employing linear programming rather than least squares. This is because the objective of (11.6) is less affected by outlier data than that of least squares. We demonstrate this aspect through the following example.[a]

---

[a]Based on the paper, which appears as reference [7]. Copyright American Institute of Industrial Engineers, Inc., 25 Technology Park/Atlanta, Norcross, GA 30092. Reprinted with permission.

Example 11-1

In this example we assume that it is our objective to estimate construction costs of electrical line construction efforts. In such line construction tasks, the specific jobs differ mainly in the number of poles, cross-arms, feet of wire, insulators, and so forth, installed by a four-man crew. Table 11-1 summarizes the data taken for eight jobs. Assuming that the present hourly cost of the four-man crew is $70 per hour, the job cost may be calculated as shown in the rightmost column of Table 11-1.

Table 11-1. DATA FOR LINE CONSTRUCTION TASK

	X_1	X_2	X_3	X_4	X_5	T_A	TC_A
Job Number	*Number of Poles*	*Wire (100 Feet)*	*Number of Cross-arms*	*Number of Insulators*	*Number of Guy Wires and Guards*	*Total Time (Hours)*	*Total Cost (Dollars)*
1	1	4	1	2	1	8	560
2	3	10	3	6	1	44	3,080
3	4	24	8	12	2	36	2,520
4	1	5	2	3	0	9	630
5	3	12	3	12	1	25	1,750
6	3	12	3	8	1	27	1,890
7	2	10	4	6	0	16	1,120
8	4	12	8	12	0	28	1,960

A *possible* mathematical form of the CER for this problem is the following linear equation (note that the intercept, b, has been set to zero):

$$\mathrm{TC}_P = a_1X_1 + a_2X_2 + \cdots + a_5X_5 \tag{11.9}$$

where TC_P = predicted cost, dollars, to complete a job

$a_1, \ldots, a_5$ = constants or coefficients of the predicting CER (to be found)

X_j = number of units of component j in the job under construction

Consequently, our problem of CER development for this case simply boils down to fitting a linear expression [equation (11.9)] to the data of Table 11-1.

As discussed, the initial reaction of the typical analyst, when faced with such a problem, is to immediately employ least squares. However, let us suppose that we do not know the characteristics of the population from which the data came. Further, examine the total time or total cost column of Table 11-1 for job 2 and compare this with jobs 5 and 6 (which, as may be seen, are very similar to job 2). The time and cost of job 2 seems "out of line" compared to jobs 5 and 6. This could either be due to a recording error or the fact that a vital factor (such as bad weather) caused the unusual results. Such "outlier" data points occur frequently in real-world problems and, in many cases, it is difficult or impossible to identify precisely either them or their cause. Unfortunately, such outliers can have a severe impact on the resulting CER if least squares is used. This is because the objective of least squares is to find a line or surface with the minimal sum of the squares of the residuals. That is, the objective is to

$$\text{minimize} \quad \sum_{i=1}^{m} (r_i)^2$$

where r_i is the residual of the ith data point. Unfortunately, squaring the residuals (deviations from the data point to the line or surface fitted) will tend to provide outlier data with possible undue influence on the slope and position of the resulting line or surface represented by the CER.

If, instead of least squares, we employ linear programming, our model is: Find $a_1, a_2, \ldots, a_5$ so as to

$$\text{minimize} \quad \sum_{i=1}^{m} (\eta_i + \rho_i) \tag{11.10}$$

subject to

$$\left.\begin{array}{llll} TC_{P(1)} + \eta_1 - \rho_1 & & & = TC_{A(1)} \\ TC_{P(2)} & + \eta_2 - \rho_2 & & = TC_{A(2)} \\ \cdot & & & \cdot \\ \cdot & & & \cdot \\ \cdot & & & \cdot \\ TC_{P(8)} & & + \eta_8 - \rho_8 & = TC_{A(8)} \end{array}\right\} \tag{11.11}$$

where $TC_{P(i)}$ = total predicted cost, from the CER, for job i

$TC_{A(i)}$ = total actual cost for job i

TC_P is in the form shown in (11.9)

Also note that each coefficient ($a_1, \ldots, a_5$) has been replaced by the difference between two strictly nonnegative variables (i.e., $a_j = u_j - v_j$ for $j = 1, \ldots, 5$. See Problem 11.5).

Both least-squares and linear programming were employed to find of the CER of (11.9). The resulting CERs are as follows:

From least squares,

$$TC_P = 1{,}211X_1 + 259X_2 - 588X_3 - 259X_4 - 378X_5$$

From linear programming,

$$TC_P = 533X_1 + 189X_2 + 357X_3 - 84X_4 - 245X_5$$

The two CERs are obviously different, but to see the real implications of these CERs, it is necessary to examine the resulting residuals. These residuals are shown in Table 11-2.

Table 11-2. Residuals

	Residual by:	
Job	*Least Squares*	*Linear Programming*
1	2.9	0
2	8.0	18.5
3	0.5	0
4	1.1	1
5	2.8	0
6	9.0	1.6
7	0.3	0
8	1.3	0

Notice in Table 11-2 that the residual from linear programming for job 2 (the outlier data point) is quite large compared to the other residuals (18.5 as compared to 1.6, 1.0, or 0.0 for all other jobs). This should warn the analyst of a possible problem with this data point (perhaps further investigation may even indicate that this point should be discarded). However, when least squares is used, the residual for the outlier job (job 2) is actually less than that for job 6 (a value of 9 compared to 8) and job 6 is actually a valid data point.

The minimization of the sum of absolute deviations is not the only measure of fit that may be employed in the linear programming approach. We might also consider, for example, the *minimization of the maximum deviation.* Consider, for example, our previous problem involving the determination of a CER for the production of pipes in which we modify the problem slightly. We shall assume that the government contract shall be for a single size of pipe, but although we know the possible range of size (say from 1 to 6 inches in diameter), we have no idea as to the actual size that shall be requested. The company must then submit, as its "bid," the CER itself. However, since there is a penalty associated with either the under- or overestimation of production cost, the firm wishes to develop a CER that minimizes the maximum penalty they might incur.

The general linear programming model for such a measure of fit is as follows: Find the coefficients of the predictive equation so as to

$$\text{minimize} \quad d$$

subject to

$$(C_{A(i)} - C_{P(i)}) - d \leq 0 \qquad \text{for all } i \tag{11.13}$$

$$(C_{A(i)} - C_{P(i)}) + d \geq 0 \qquad \text{for all } i \tag{11.14}$$

$$d \geq 0 \tag{11.15}$$

where d = maximum amount of deviation

$C_{A(i)}$ = actual cost (or value) of data point i

$C_{P(i)}$ = predicted cost (or value) of data point i

and we recall that our predictive equation coefficients are unrestricted in sign.

The reader who goes on to Part Four, which deals with multiple-objective linear programming, will discover that the concepts introduced in the material above (particularly the use of η_i and ρ_i as negative and positive deviation variables) form the basis for one of the first approaches to multiple-objective linear programming, a technique known as weighted goal programming. The methodology of Part Four has, in fact, as just one application, the construction of predictive equations in the environment of several conflicting goals and/or measures of fit.

Input–Output Analysis

Much of the work in input–output (or interindustry) analysis is due to Leontief [9]. The fundamental idea underlying the analysis is that there is a high degree of interdependence among the goods and services produced in an industrial economy. Consequently, an increase or decrease in one sector of production may have an effect on other sectors. There are two basic types of input–output models: the static model and the dynamic model. The static model, which is the one we discuss, deals with but a single time period, whereas the dynamic model investigates changes in the economy over several time periods.

The heart of the input–output model is the transactions table. Each element in this table represents a total sales activity for each industry during the period of interest. Consider, for example, the transactions table of Table 11-3. For simplicity, only three industries are considered to make up the economy: I, II, and III. The first row of the table denotes the activities of industry I. Industry I produced a grand total of 100 (monetary) units of goods of which:

—40 units were sold to the consumer
—10 units were sold to industry II
—30 units were sold to industry III
—20 units were used by I itself for its own production processes

Rows II and III are interpreted in a similar manner. The "households" row is, however, a "catchall" used to represent wages, capital services, and so forth. As such, it is something akin to a fourth "industry." Industries I, II, and III require 50, 70, and 20 units of the households output for their production. The elements of the columns simply represent the total purchase of an industry from the other industries so as to support its operations during the base period.

We shall now refer to the interior of the transactions table (the portion of the table concerned only with industries I, II, and III) and the total output row on the bottom. Dividing each column of this interior matrix by the associated total amount results in a matrix known as the technical coefficients matrix, shown in Table 11-4. To illustrate, the coefficient in the first row and first column

Table 11-3. TRANSACTIONS TABLE

From \ *To*	*I*	*II*	*III*	*Consumers*	*Total*
I	20	10	30	40	100
II	10	20	20	60	110
III	20	10	10	40	80
Households	50	70	20	0	140
Total	100	110	80	140	

Table 11-4. TECHNICAL COEFFICIENTS MATRIX

	I	*II*	*III*
I	0.200	0.091	0.375
II	0.100	0.182	0.250
III	0.200	0.091	0.125

of Table 11-4 is found by dividing 20 by 100 (from the first column in Table 11-3).

The elements of the technical coefficients matrix then represent the per unit contribution of each industry. For example, consider the first column and second row element in this table, which is 0.100. This indicates that, for an output of 1 unit from industry I, an input of 0.100 unit is required from industry II. For convenience we designate the elements of the technical coefficients matrix as $a_{i,j}$, where

$\mathbf{A}$ = matrix of technical coefficients

$a_{i,j}$ = amount of industry i that is necessary to produce 1 unit of commodity j (i.e., the elements of matrix $\mathbf{A}$)

We shall also define the following:

$$x_1 = \text{total output of industry I}$$
$$x_2 = \text{total output of industry II}$$
$$x_3 = \text{total output of industry III}$$

Thus, for the period shown in Table 11-3, the values for x_1, x_2, and x_3 are 100, 110, and 80, respectively.

The variables x_1, x_2, and x_3 make up the *output vector*, which we designate as $\mathbf{X}_k$, where k is the time period under consideration. Thus, for the initial period $k = 0$ and

$$\mathbf{X}_0 = \begin{pmatrix} 100 \\ 110 \\ 80 \end{pmatrix}$$

Next, consider the column under "consumers" in Table 11-3. This is the "demand" for each commodity. We let

y_1 = demand for commodity one (industry I)

y_2 = demand for commodity two (industry II)

y_3 = demand for commodity three (industry III)

Thus, a demand vector, $\mathbf{Y}_k$, may be introduced, where

$$\mathbf{Y}_0 = \begin{pmatrix} 40 \\ 60 \\ 40 \end{pmatrix} = \begin{pmatrix} y_1 \\ y_2 \\ y_3 \end{pmatrix}_{k=0}$$

for our example.

For the base period, we may write the representative equation of each industry (where I = 1, II = 2, and III = 3 in the notation to follow):

$$\begin{aligned} &\text{Industry I:} && x_1 - a_{1,1}x_1 - a_{1,2}x_2 - a_{1,3}x_3 = y_1 \\ &\text{Industry II:} && x_2 - a_{2,1}x_2 - a_{2,2}x_2 - a_{2,3}x_3 = y_2 \\ &\text{Industry III:} && x_3 - a_{3,1}x_1 - a_{3,2}x_2 - a_{3,3}x_3 = y_3 \end{aligned}$$

or for the specific values in our example:

$$\begin{aligned} x_1 - 0.2x_1 - 0.091x_2 - 0.375x_3 &= 40 \\ x_2 - 0.1x_1 - 0.182x_2 - 0.250x_3 &= 60 \\ x_3 - 0.2x_1 - 0.091x_2 - 0.125x_3 &= 40 \end{aligned}$$

This may be rewritten in general matrix form as

$$(\mathbf{I} - \mathbf{A})\mathbf{X} = \mathbf{Y}$$

The matrix $(\mathbf{I} - \mathbf{A})$ is known as the Leontief matrix. Under the assumption that the underlying structure of the economy is linear and that the technical coefficients do not change with time, we can determine a future production vector $\mathbf{X}_k$ given a predicted demand vector $\mathbf{Y}_k$. That is, we solve

$$\begin{aligned} (\mathbf{I} - \mathbf{A})\mathbf{X} &= \mathbf{Y} \\ \mathbf{X} &\geq \mathbf{0} \end{aligned}$$

If the Leontief matrix is nonsingular, then

$$\mathbf{X} = (\mathbf{I} - \mathbf{A})^{-1}\mathbf{Y}$$

Up to this point, the input–output analysis is performed using only very basic linear algebra. However, through linear programming, we may enlarge the scope of the methodology. Consider, for example, that there is a profit, c_i, associated with the output of each unit of industry i. We might then wish to maximize total profit by finding the optimal production output of all industries. That is, find $\mathbf{X}$ so as to

$$\begin{aligned} \text{maximize} \quad & \mathbf{CX} \\ \text{subject to} \quad & \\ (\mathbf{I} - \mathbf{A})\mathbf{X} & \leq \mathbf{Y} \end{aligned}$$

More realistically, the amount produced by each industry has an upper limit, u_i, and thus we wish to find $\mathbf{X}$ to

$$\begin{aligned} \text{maximize} \quad & \mathbf{CX} \\ \text{subject to} \quad & \\ (\mathbf{I} - \mathbf{A})\mathbf{X} & \leq \mathbf{Y} \\ \mathbf{X} & \leq \mathbf{U} \end{aligned}$$

where

$$\mathbf{U} = \begin{pmatrix} u_1 \\ \cdot \\ \cdot \\ \cdot \\ u_m \end{pmatrix}$$

Numerous other variations are possible, and this is why the linear programming approach to input–output analysis is so attractive.

Example 11-2

Given the economy model of Tables 11-3 and 11-4, let us assume that 3 years from now, the following demand and production limit estimates are given:

$$\mathbf{Y}_3 = \begin{pmatrix} 70 \\ 50 \\ 50 \end{pmatrix}$$

$$\mathbf{U}_3 = \begin{pmatrix} 120 \\ 120 \\ 95 \end{pmatrix}$$

Further, it is desired to maximize the amount of units produced by industries I and III.

The associated linear programming model is then:

Find x_1, x_2, and x_3 so as to

$$\text{maximize} \quad x_1 + x_3$$

$$\text{subject to}$$

$$\begin{aligned}
x_1 - 0.2x_1 - 0.091x_2 - 0.375x_3 &\leq 70 \\
x_2 - 0.1x_1 - 0.182x_2 - 0.250x_3 &\leq 50 \\
x_3 - 0.2x_1 - 0.091x_2 - 0.125x_3 &\leq 50 \\
x_1 &\leq 120 \\
x_2 &\leq 120 \\
x_3 &\leq 95 \\
x_1, x_2, x_3 &\geq 0
\end{aligned}$$

Computational Aspects

The solution of linear programming problems by hand can be a time-consuming task even for problems having no more than a few variables and constraints. Although experience indicates that such efforts vastly enhance the student's understanding and appreciation of the solution methodology, problem solving by hand (or even with the aid of pocket calculators) is hardly a rational approach to real-world problems, which may include hundreds or thousands of variables and constraints. For this reason, numerous computerized linear programming algorithms have been developed. Typically, when we speak of a computerized algorithm (i.e., one performed on the digital computer), we denote this as a computer code. The subject of the section is then the various computational aspects one encounters in either developing new linear programming computer codes or in dealing with existing ones.

Computer codes for solving the general linear programming problem, in existence during the writing of this text, can handle problems with several thousands of variables and constraints. The commercial code often used by the author, known as MPSX and developed by IBM, has the capability to solve problems involving up to about 16,000 constraints and as many variables as one may store. In addition, there are methods that sometimes allow us to solve even larger problems than indicated by the rated capacity of the computer. We shall discuss one of these (the method of decomposition) in a subsequent subsection.

In essence, although almost all computer codes for the linear programming model are based on the simplex method, most of the larger and more efficient of these codes are quite different in approach from the algorithms we have presented earlier. The primary reason for this is that these codes are developed to be compatible with and take advantage of the modern digital computer.

Further, problems that are not at all apparent in the solution of linear programming models of small to modest size often become the limiting aspect of the computer code for large problems. A factor that even further complicates the understanding of modern computer codes now arises with the advent of mini- and microcomputers. Rather than solving the problem on a single computer, one often finds a significant advantage in grouping a set of mini- and/or microcomputers into a "distributed computing network" wherein different computers may work on different segments of the algorithm.

We begin our discussion of computational aspects with a discussion of two phenomena, round-off errors and degeneracy, that were discussed in Chapter 7. However, it is probably a good idea to review these aspects (particularly round-off) before proceeding to other details.

Round-off Errors

One of the most common, frustrating, and surprising (at least to the novice) of the computational aspects to be considered is that known as round-off error. Round-off error occurs because the digital computer, despite its astonishing speed and accuracy, still has a finite limitation to the precision by which a number is described. A finite number of digits are available for the representation of any given number. For example, the fraction one-third may be represented by, say, only 10 digits:

$$0.3333333333$$

Although this may seem insignificant at first, the impact of round-off can tend to accumulate and grow with each iteration of the algorithm. The larger the problem, the more likely one is to encounter the round-off problem.

There is, primarily because of round-off, a fundamental rule to be followed in the use of *any* linear programming computer code: "Check your results against your constraints." To the novice's surprise, these may not check out despite the fact that there is nothing "wrong" with the computer code.

There are numerous clever and sophisticated approaches for *reducing* the impact of round-off error. However, one very straightforward approach which will normally improve the performance of the code is to *scale* the functions. Typically, the greater the difference between the smallest and largest coefficient in the linear programming model, the greater the probability of increasing the round-off error. For example, if the right-hand side of a constraint is, say, 3,000,000 gallons, it might be wise to work in terms of thousands (or even millions) of gallons if this can reduce the range between model coefficients.

Degeneracy and Cycling

In theory, a degenerate problem could lead to a computational loop or cycle that is repeated infinitely. In practice this has not occurred on *actual* problems and is highly unlikely to present a problem. However, some codes do include certain rules to avoid this possibility. Another technique used to avoid the problem is perturbation (an aspect that is actually naturally imposed by round-off). The reader interested in the cycling problem is directed to the references [2, 6].

Decomposition

One of the more appealing ways, at least in theory, of dealing with large linear programming models is to decompose the model into smaller submodels whereby the solution to these submodels may somehow be used to generate the solution to the overall problem.

One quite unsophisticated and rather brute-force approach to decomposition is to:

1. Select a portion (k) of the (m) constraints (where $k < m$) and, for the moment, disregard the others. The k constraints for consideration may be selected by guess, intuitively, or on the basis of a somewhat more rational exercise.
2. Form the linear model with only the k constraints and solve. If the solution obtained *also* satisfies the ($m - k$) constraints that were not included in the reduced model, we are finished. Otherwise, go to step 3.
3. Pick some of the most violated of the constraints that were not previously included in the reduced model and place them into the new, reduced model. We might also eliminate some of the constraints from the reduced model that were found to be easily satisfied (e.g., having large slack values).
4. The process continues by repeating steps 2 and 3 until a satisfactory answer is obtained. Although a bit naïve, the approach has been used to solve quite large problems in practice with fairly good results.

The approach listed above is not systematic, nor does it take any advantage of problem structure. One of the best known of the systematic decomposition procedures is due to Dantzig and Wolfe [5] and operates on problems having the following structure:

$$\text{Maximize (or minimize) } z = \sum_{j=1}^{r} \mathbf{c}_j \mathbf{x}_j$$

$$\begin{bmatrix} \mathbf{A}_1 & 0 & 0 & \cdots & 0 \\ 0 & \mathbf{A}_2 & 0 & \cdots & 0 \\ \cdot & & & & \\ \cdot & & & & \\ \cdot & & & & \\ 0 & 0 & 0 & \cdots & \mathbf{A}_r \\ \mathbf{A}_{r+1} & \mathbf{A}_{r+2} & \mathbf{A}_{r+3} & \cdots & \mathbf{A}_{2r} \end{bmatrix} \begin{bmatrix} \mathbf{x}_1 \\ \mathbf{x}_2 \\ \cdot \\ \cdot \\ \cdot \\ \mathbf{x}_r \end{bmatrix} = \begin{bmatrix} \mathbf{b}_1 \\ \mathbf{b}_2 \\ \cdot \\ \cdot \\ \cdot \\ \mathbf{b}_r \end{bmatrix}$$

$$\mathbf{x}_j \geq \mathbf{0} \qquad \text{all } r$$

Note that $\mathbf{c}_j$, $\mathbf{x}_j$, and $\mathbf{b}_i$ are all *vectors* while $\mathbf{A}_i$ are *matrices*. To illustrate, consider the following problem:

Find $\mathbf{x}$ to

$$\text{maximize} \quad z = 3x_1 + 7x_2 + 2x_3 + 5x_4 + 8x_5 + 5x_6 + 10x_7$$

subject to

$$\begin{aligned}
x_1 \qquad\qquad + 2x_3 + 3x_4 + x_5 + x_6 + 4x_7 &\leq 50 \\
2x_1 + x_2 + x_3 + x_4 \qquad\quad + 3x_6 + x_7 &\leq 70 \\
2x_1 + x_2 &\leq 8 \\
x_1 + 4x_2 &\leq 12 \\
x_3 + 2x_4 + 3x_5 &\leq 20 \\
x_3 \qquad + 2x_5 &\leq 7 \\
2x_6 + 3x_7 &\leq 10 \\
4x_6 &\leq 7 \\
x_6 + 6x_7 &\leq 14 \\
x_j \geq 0 \qquad \text{all } j &
\end{aligned}$$

We can observe the special structure of the problem by listing the constraint matrix as shown in Table 11-5. Frequently, we may also be able to interchange various rows and columns and also discover this structure.

Table 11-5. CONSTRAINT COEFFICIENT MATRIX

$$\mathbf{A} = \begin{bmatrix}
1 & 0 & 2 & 3 & 1 & 1 & 4 \\
2 & 1 & 1 & 1 & 0 & 3 & 1 \\
2 & 1 & & & & & \\
1 & 4 & & & & & \\
 & & 1 & 2 & 3 & & \\
 & & 1 & 0 & 2 & & \\
 & & & & & 2 & 3 \\
 & & & & & 4 & 0 \\
 & & & & & 1 & 6
\end{bmatrix}$$

If we find a large problem having such a structure, the use of the Dantzig–Wolfe algorithm may provide a more efficient approach to its solution and, as such, should be considered. Details of this algorithm, together with numerous other approaches to large-scale problems, appear in the text by Lasdon [8].

Upper- and Lower-bound Constraints

Frequently, we find constraints of the following form in a model: $x_j \geq L_j$, where x_j is a problem variable and L_j is the (nonnegative) constant right-hand side value or the *lower bound* on x_j. Letting our surplus variable be s_j, we can convert this lower-bound constraint to

$$x_j - s_j = L_j$$

Alternatively, we may express x_j in terms of L_j and s_j as

$$x_j = L_j + s_j$$

Since both L_j and s_j are nonnegative, we may replace the variable x_j by $L_j + s_j$ wherever it appears in the problem. Thus, we not only get rid of a constraint, but we also reduce the number of variables needed to describe the problem. For example, given the lower-bound constraint $x_3 \geq 7$, we replace x_3 with $7 + s_3$ and eliminate the constraint.

Upper-bound constraints, such as

$$x_3 \leq 12$$
$$x_1 + x_7 + x_8 + x_{10} \leq 40$$

also appear frequently but, unfortunately, are not as straightforward to deal with, although there do exist several approaches. These techniques, known generally as GUB for generalized upper bounding, may be found in the references [3, 4, 8, 10]; again, the text by Lasdon provides a good summary.

Commercial Codes

Most computer manufacturers will furnish a library routine that may be called to solve linear programming problems. Such routines appear, to the user, as "black boxes" to which they provide the input data that describe the problem and from which they receive the answer. In addition, most of these codes have the ability to perform a sensitivity analysis. Unfortunately, there is very little standardization of codes or of their input and output formats. Another criticism that can be leveled at almost all of these codes is that far too little attention seems to have been paid to the user-computer interface. Consequently, the casual user or the user without an extensive background in computers will find that a great deal of time may be required to learn the system if one wishes to take full advantage of its features.

The "typical" large-scale commercial computer code for linear programming will employ the revised simplex method [4] in conjunction with the product form of the inverse (see Chapter 3). Such codes will use a periodic reinversion of the basis matrix so as to minimize round-off errors. Typically, GUB is either used or can be employed as an option.

The first stage of the computerized algorithm will normally involve finding some initial basis (which need not be feasible). Some codes will even permit the user to enter a guess as to the initial basis. Following this, we discover a wide variety of approaches (some which seem rather arbitrary) to conduct the pivoting operations. However, the use of periodic reinversion of the basis (perhaps every 50 or 100 iterations—or as a function of problem size) appears in almost all the codes [12].

One very effective way for the reader to learn more about the commercial code is to read the user's manual for the code resident within the computer that he or she may use. The text by Orchard-Hays, although published in 1968, is also still one of the most complete introductions to advanced computing techniques in linear programming [10].

Codes for the Multiple-objective Problem

All the computational aspects that exist for the single-objective codes will also be present in the computer codes that carry out the multiple-objective linear programming algorithms of Part Four. However, the analyst will find few commercial codes for the latter problem.

Alternative to Simplex

Over the years, numerous investigators have proposed alternative approaches (i.e., rather than the simplex algorithm) for the linear programming problem. The vast majority of these, however, proceed in a relatively conventional manner (e.g., a search of the vertices of the convex solution space). However, more recently considerable publicity has been given to a quite different approach to the linear programming model. This approach has generally been attributed to the Soviets, specifically to Khachian (or Hacijan, Khatchian, Khachiyan, etc.), and a discussion [11] is provided in the appendix to this chapter for the interested reader.

REFERENCES

1. Campbell, H., and Ignizio, J. P. "Using Linear Programming for Predicting Student Performance," *Journal of Education and Psychological Measurement*, Vol. 32, 1972, pp. 397–401.
2. Charnes, A. "Optimality and Degeneracy in Linear Programming," *Econometrica*, Vol. 20, No. 2, 1952, pp. 160–170.
3. Dantzig, G. B. "Notes on Linear Programming, Parts VIII, IX, X: Upper Bounds, Secondary Constraints and Block Triangularity in Linear Programming," Research Memorandum RM-1367, Rand Corporation, Santa Monica, Calif., October 1954.

4. DANTZIG, G. B., AND ORCHARD-HAYS, W. "Notes on Linear Programming, Part V: Alternate Algorithm for the Revised Simplex Method Using Product Form for the Inverse," Research Memorandum RM-1268, Rand Corporation, Santa Monica, Calif., November 1953.
5. DANTZIG, G. B., AND WOLFE, P. "Decomposition Principle for Linear Programs," *Operations Research*, Vol. 8, No. 1, 1960, pp. 101–111.
6. GASS, S. I. "Comments on the Possibility of Cycling with the Simplex Method," *Operations Research*, Vol. 27, No. 4, 1979, pp. 848–852.
7. IGNIZIO, J. P. "The Development of Cost Estimating Relationships via Goal Programming," *The Engineering Economist*, Vol. 24, No. 1, 1978, pp. 37–47.
8. LASDON, L. S. *Optimization Theory for Large Systems*. New York: Macmillan, 1970.
9. LEONTIEF, W. W. *The Structure of the American Economy, 1919–1939*. New York: Oxford University Press, 1951.
10. ORCHARD-HAYS, W. *Advanced Linear-Programming Computing Techniques*. New York: McGraw-Hill, 1968.
11. PEGDEN, C. D., AND IGNIZIO, J. P. "Khachian's Algorithm for Linear Programming," *AIIE News: Operations Research*, Vol. 14, No. 4, Spring 1980, pp. 1–4.
12. WHITE, W. W. "A Status Report on Computing Algorithms for Mathematical Programming," *Computing Surveys*, Vol. 5, No. 3, 1973, pp. 135–166.

PROBLEMS

11.1. A firm wishes to determine its production rate over the next 6 months. Expected demand for their product is 30 units, 80, 60, 50, 80, and 100 units for months 1 through 6, respectively. The product is perishable and thus cannot be stored from month to month. The loss per unit not sold is $8 and the loss per unit short is estimated to be $5. Because of the nature of the production process, production output must be set as some *linear* function over the time period. That is, if we begin with a production rate of, say, 50 units and end with a rate of 100 units, the production rate for any intermediate month must lie on the straight line between 50 and 100. Formulate this problem and determine its solution if we wish to minimize total costs.

11.2. Formulate and solve Problem 11.1 if we desire to minimize the maximum loss in any one month.

11.3. Compare the results obtained in Problem 11.2 with that which would be achieved using the least-squares method to fit the linear production rate line.

11.4. Finish Example 11-2 and discuss your results.

11.5. In Example 11-1, comment on the appropriateness of replacing the a_j coefficients by the difference between two restricted variables.

Appendix to Chapter 11

KHACHIAN'S POLYNOMIAL ALGORITHM FOR LINEAR PROGRAMMING[a] [5]

C. Dennis Pegden and James P. Ignizio
The Pennsylvania State University

Introduction

In the January 1979 issue of *Doklady*, L. G. Khachian published a paper entitled "A Polynomial Algorithm in Linear Programming" [3]. Although the paper went initially unnoticed in the West, it has recently received wide publicity, including front-page articles in the *New York Times* as well as articles in many popular science magazines, such as *Science News*, *Science*, and *Scientific American*. It has been claimed by some that the new algorithm by Khachian is vastly superior to the simplex method and that it will revolutionize the field of mathetical programming. Although we do not question the theoretical importance of Khachian's algorithm, our experiences in programming and running the algorithm on small test problems indicates that there are some practical issues that must be addressed before Khachian's algorithm can be considered a serious competitor for the simplex method. We find that, in its present form, the algorithm cannot be effectively applied to solve even small linear programs (LPs). In this paper we will briefly describe Khachian's algorithm and summarize the key practical issues which we believe must be addressed before the algorithm can be effectively applied to solve real LP problems.

An algorithm is classified as a polynomial time algorithm if the number of computational steps required to solve a problem using it is bounded above by

[a]Permission granted by the American Institute of Industrial Engineers, Inc., 25 Technology Park/Atlanta, Norcross, GA 30092.

a polynomial in terms of the size of the problem. Klee and Minty [4] have constructed pathological examples which clearly demonstrate that the simplex method is not a polynomial algorithm. Hence before Khachian's development it was unclear whether LP problems were in the class of "easy" problems for which a polynomial time algorithm can be constructed. Therefore, the primary theoretical importance of Khachian's work is that it shows that LPs do indeed fall in the class of easy problems.

It should be noted that although the simplex method can require an exponential time on specially constructed problems, the number of pivots required to solve real problems is generally a linear function of the number of constraints. Consequently, the lack of a polynomial bound on the simplex method is more of theoretical interest than of practical importance.

Khachian's Algorithm

Khachian's algorithm solves a system of linear inequalities of the form $Gx \leq h$, where G is a matrix and x and h are column vectors. Note that there are no nonnegativity restrictions on the x and there is no explicit objective function to be minimized or maximized. Thus it is first necessary to convert an LP problem to an equivalent system of inequalities having the same solution set as the LP. This is accomplished by writing the LP in the form:

$$\begin{aligned} &\text{maximize} \quad c^t y \\ &\qquad \text{subject to} \\ &\qquad Ay \leq b \\ &\qquad y \geq 0 \end{aligned} \tag{1}$$

By the weak duality theorem, any solution to the following system of inequalities will solve (1), where u denotes the solution vector to the dual.

$$\left.\begin{aligned} Ay &\leq b \\ -A^t u &\leq -c \end{aligned}\right\} \tag{2}$$
$$\left.\begin{aligned} -c^t y + b^t u &\leq 0 \\ -y \qquad &\leq 0 \\ -u &\leq 0 \end{aligned}\right\}$$

Note that if we have n' variables and m' constraints for the LP given by (1), then there will be $n = n' + m'$ variables and $m = 2n' + 2m' + 1$ inequalities in the system $Gx \leq h$ given by (2). Note also that for LPs having a unique solution, the feasible space for the corresponding inequality system given by (2) consists of a single point.

Khachian's algorithm solves (2) by starting with a Euclidean ball with center $x_0 = (y, u)$ at the origin and sufficiently large radius r to ensure that the (optimal) solution x^* is contained within the ball; i.e., $||x^*|| \leq r$. The algorithm then proceeds by constructing a series of successively smaller ellipsoids whose centers converge to the solution of (2) and hence (1). Each successive ellipsoid is constructed from the previous ellipsoid such that it always contains the solution. The convergence proof consists of showing that the volume of the ellipsoid can be reduced to any desired value within K iterations, where K is a polynomial in terms of n.

The general procedure for constructing the ellipsoid at each iteration k consists of the following steps:

1. Test each inequality in (2) and select a violated constraint. Denote the variable coefficients of the selected constraint row by the column vector g_i. If all inequalities are satisfied, stop since x_k solves (2) and hence (1). Otherwise, continue.
2. Cut the current ellipsoid in half by passing a plane through the current center x_k and parallel to the hyperplane of the violated constraint ($g_i^t x \leq h$), cutting off the region $g_i^t(x - x_k) > 0$. Since x_k violates $g_i^t x \leq h_i$, all x such that $g_i^t(x - x_k) > 0$ also clearly violate the constraint, and therefore may be discarded from consideration.
3. Construct a new ellipsoid of smaller volume with new center given by x_{k+1} which wholly contains the half ellipsoid from step 2 and return to step 1.

The equation of an ellipsoid may be written in the form $\{w \mid w = x + Qz, ||z|| \leq 1\}$ where x is a n-dimensional vector defining the center of the ellipsoid, Q is an n by n transformation matrix, and z is a n-dimensioned vector defining points in a ball of radius 1. The ellipsoid is completely defined by knowledge of the center x and transformation matrix Q. Khachian provides the following equations for updating x and Q at each iteration corresponding to the new smaller ellipsoid.

$$
\begin{aligned}
F_k &= -Q_k^t g_i \\
x_{k+1} &= x_k + \frac{Q_k F_k}{(n+1)||F_k||} \\
Q_{k+1} &= 2^{1/8n^2} Q_k \mathrm{ORT}(F_k) D
\end{aligned}
\tag{3}
$$

where $\mathrm{ORT}(F_k)$ is an orthogonal n by n matrix whose first column is

$$\frac{F_k}{||F_k||}$$

D is the diagonal matrix given by

$$D = \mathrm{diag}\left(\frac{n}{n+1}, \frac{n}{\sqrt{n^2 - 1}}, \ldots, \frac{n}{\sqrt{n^2 - 1}}\right)$$

The foregoing calculations are straightforward, with the exception of finding the orthogonal matrix whose first column is $(F_k)/||F_k||$. An orthogonal matrix is a matrix whose inverse is the transpose of the matrix. The calculation of finding the orthogonal matrix can be done using the Gram–Schmidt orthogonization procedure [1].

Khachian's algorithm requires an initial starting value for x_0 and Q_0. Khachian suggests using $x_0 = 0$ and $Q_0 = rI$, where I is an n by n identity matrix and r is the radius of the starting ball encompassing the solution and is equal to 2^L, where L is given by

$$L = \sum_{i,j} \log_2 (|g_{ij}| + 1) + \sum_i \log_2 (|h_i| + 1) + \log_2 (nm) + 1 \tag{4}$$

When using $x_0 = 0$ and $Q_0 = 2^L I$ as the initial starting ball, it can be shown that the center of the ellipsoid will converge to the solution of (2) within $16n^2L$ iterations, if a solution exists.

Gaćs and Lovaśz [2] have developed a simpler version of the equations definding the ellipsoid at each iteration. However, in attempting to retrace their development we have found several errors and have constructed counter examples for which their version of the ellipsoid equations appear to be invalid.

An Example

Consider the system of inequalities given below:

$$\begin{aligned} -x_1 \quad &\leq -10 \\ x_1 \quad &\leq \ 12 \\ -x_2 &\leq \ 1 \\ x_2 &\leq \ 1 \end{aligned}$$

To illustrate the calculations we will show the computations for the first iteration. For simplicity in graphing we will use a radius of $r = 16$ instead of $r = 2^L$. A radius of 16 for the initial ball is sufficient since all feasible points are within a ball of radius 16.

$$x_{(0)} = \begin{pmatrix} 0 \\ 0 \end{pmatrix} \qquad Q_{(0)} = \begin{bmatrix} 16 & 0 \\ 0 & 16 \end{bmatrix} \qquad g_1 = \begin{pmatrix} -1 \\ 0 \end{pmatrix} \qquad D = \begin{bmatrix} \frac{2}{3} & 0 \\ 0 & 2/\sqrt{3} \end{bmatrix}$$

$$x_{(1)} = \begin{pmatrix} \frac{16}{3} \\ 0 \end{pmatrix} \qquad Q_{(1)} = \begin{bmatrix} 10.91 & 0 \\ 0 & 18.90 \end{bmatrix}$$

This x and Q defines an ellipsoid with center $(\frac{16}{3}, 0)$ and semi-axes $a = 10.91$ and $b = 18.896$. The starting ball, feasible region, and first ellipsoid for this example is depicted below:

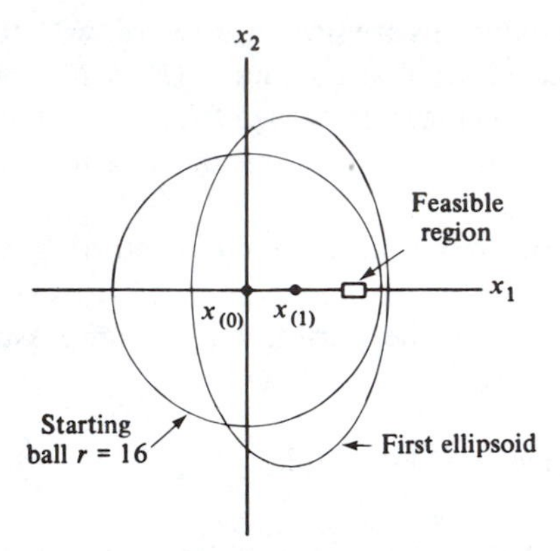

Figure AP11-1

The violated constraint selected for the first iteration is the first constraint. A hyperplane is passed through the current center (0, 0) and parallel to the plane of the first constraint, cutting away the left half of the initial ball. Note that the new ellipsoid is of smaller volume and completely contains the remaining right half ellipsoid and hence the feasible region. Repeated iterations of the algorithm yields smaller and smaller ellipsoids until the centers of the ellipsoids converge to the feasible region.

Practical Issues and Areas for Research

In evaluating the practically of solving LPs using Khachian's algorithm, two primary concerns are the memory requirements and run time of the algorithm. In its present form the algorithm appears to be inferior to the simplex method in both areas. For example, consider a moderate-size LP having 2,000 variables and 300 constraints. A problem of this size is relatively trivial for the simplex method. However, Khachian's algorithm would require a Q matrix dimensioned 2,300 by 2,300 and could require as many as approximately 84.64(L) million iterations, where L is given by equation (4) and is clearly a very large number.

The key element in developing a practical version of the algorithm appears to us to be an improved procedure for selecting values for x_0 and r. A smaller volume for the initial ball generally reduces the number of iterations for the algorithm, particularly when the solution set is a single point as in the case of most LPs. If a good heuristic could be devised for computing x_0 and r such that r is small and $||x^* - x_0|| \leq r$ for the solution x^*, then the computations for the algorithm could be substantially reduced.

Another potentially fruitful research area might be the combining of Khachian's algorithm with other procedures for solving systems of inequalities such as the subgradient method. A hybird approach such as this might prove to be effective.

Although Khachian's algorithm appears to be far less efficient on practical LPs than the simplex method, it does provide a new way of approaching mathematical programming problems. Perhaps its approach can be used to develop new algorithms for other mathematical programming problems for which current algorithms are relatively inefficient.

REFERENCES

1. Birkhoff, G., and MacLane, S. *A Survey of Modern Algebra.* New York: Macmillian, 1953, pp. 192–193.

2. Gaćs, P., and Lovaśz, L. "Khachian's Algorithm for Linear Programming," Computer Science Department Working Paper, Stanford University, 1980.

3. Hacijan, L. G. "A Polynomial Algorithm in Linear Programming," *Doklady Akademii Nauk SSSR*, Vol. 244, No. 5, 1979, pp. 192–194.

4. Klee, V., and Minty, G. "How Good Is the Simplex Algorithm?" in O. Shisha (ed.), *Inequalities III.* New York: Academic Press, 1972, pp. 159–175.

5. Pegden, C. D., and Ignizio, J. P. "Khachian's Algorithm for Linear Programming," *AIIE News: Operations Research*, Vol. 14, No. 4, Spring 1980, pp. 1–4.

part three

SPECIAL MODELS IN SINGLE-OBJECTIVE LINEAR PROGRAMMING

CHARACTERISTICS OF THE "SPECIAL MODEL"

Introduction

The chapters that compose Part Three (Chapters 12 through 15) serve to introduce a *selected* set of rather special single-objective models. Although, by tradition, each of these models is denoted as a linear programming problem, in reality they all involve control variables which must take on only integer values. As such, they are actually all linear *integer* programming problems.

Although this text is not intended to cover the general linear integer model, we have picked these particular subclasses of this model for discussion for the following reasons.

1. Applications involving such models are frequently encountered.
2. Applications in which such models are a part of, or embedded within other, more general models are also fairly frequent.
3. Each special model has its own specially adapted solution technique that provides performance (for this special class) that is, usually, better than that available through more general approaches.

Special Features

There are a number of particular features that tend to identify or isolate the special problem classes that are presented. The recall of such features can be important in practice. For example, if we see a particular real-life problem

12

having such features, or for which such features predominate, it may be possible to find the solution more efficiently than if approached in a more general manner.

The particular features peculiar to the special models (but not necessarily unique) include

—Integer or zero–one variables
—A natural *network* model representation
—And, for *some* of the special models, a condition known as unimodularity

We discuss each of these in turn.

Variables

For all the special classes of problems to be discussed, *all* the variables within each problem must take on only *integer* values. Some problem types will be even more restrictive in that the only permissible values for the variables will be either a zero or a 1.

As mentioned numerous times previously, the simplex method simply searches for extreme points (i.e., basic feasible solutions) and thus any or all of the variables can take on fractional values in the final solution. Consequently, except under rather unique circumstances (to be discussed), the use of simplex on these special problem types is ruled out.

Network Representation

Another feature inherent in the special types of problems to be discussed is that they all find an easy, *natural* representation via *network models*. A network model, for our purposes, consists of a set of two distinct elements: nodes and branches (also known as vertices and edges). These elements form the building blocks from which an unlimited variety of network models may be constructed.

Nodes normally represent points of reception or transmission. Branches serve to connect nodes (every branch connects exactly two nodes) so as to indicate either potential or actual transmission paths. Figure 12-1 indicates an illustrative network model having five nodes and seven branches. The nodes, represented as circles, are designated *A*, *B*, *C*, *D*, and *E*, and the branches (the lines between nodes) are denoted as b_1 through b_7.

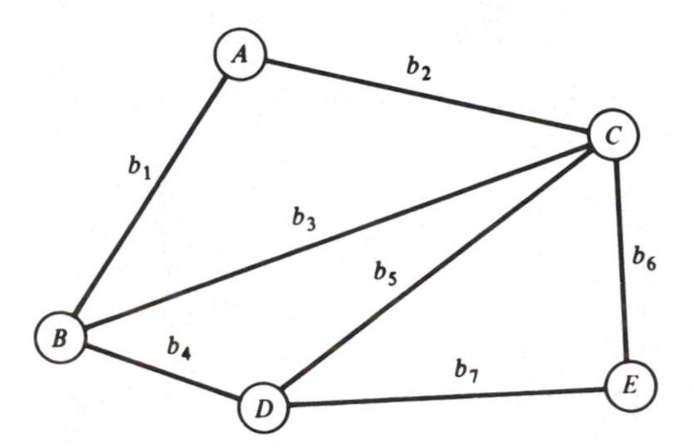

Figure 12-1. Illustrative network model (nonoriented).

Notice that a *direct* path exists between some nodes and not others. For example, nodes *A* and *D* are not directly connected, whereas *A* and *C* or *A* and *B* are directly connected.

The network in Figure 12-1 is an example of an undirected or nonoriented network. That is, its branches are bilateral (i.e., transmission in either direction is possible). An oriented network is then one composed of unilateral branches wherein such branches allow transmission in only a single direction. Graphically, this is indicated by an arrowhead on the branch pointing in the direction of permissible transmission or movement. Figure 12-2 illustrates an oriented network.

There are numerous other topological properties of networks [3], but we shall conclude this brief introduction at this point and cite, where necessary, various other pertinent features in the chapters that immediately follow.

Unimodularity

The property known as unimodularity is not characteristic of *all* the special problem classes to be discussed. However, it does exist in some of the better

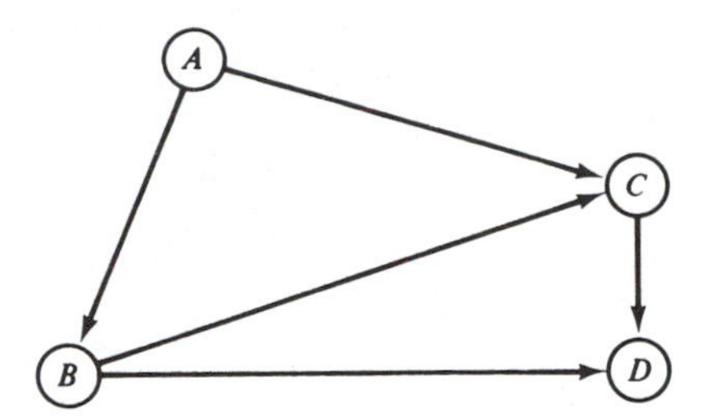

Figure 12-2. Illustrative oriented network.

known of these problem types, particularly among a class known generally as "distribution" or "assignment" problems.

Before we explain exactly what unimodularity is, let us first consider the impact of this property. A problem for which unimodularity exists will always provide a strictly integer solution when solved by simplex. That is, such a problem has optimal extreme points (basic solutions) that are all strictly integer-valued. However, even though the regular simplex method of previous chapters could be used on such problems, we shall observe that even simpler approaches exist.

A problem for which unimodularity exists can be identified, formally, as follows. Observe the coefficient matrix (the **A** matrix) of the problem and determine whether or not all elements are 0's or 1's (plus or minus). If so, proceed to the next step. If not, the problem is not unimodular.

We next observe all square submatrices of **A**. If the determinant of all of these is either 0, +1, or −1, then **A** is unimodular. We shall observe, in the chapters to follow, several classes of problems exhibiting this property.

Computational Complexity

All of the special problem types of Part Three have, as mentioned, strictly integer-valued variables. As such, they are problems in combinatorics. A naive, brute-force approach to their solution would be to attempt to evaluate every possible combination of integer values for the problem variables. The hopelessness of such an approach can be readily demonstrated for problems of even quite small sizes. For example, if we have a problem involving just 10 variables and if, in turn, each variable could take on only the integer values 0 through 9, it can be seen that total number of solutions to be examined is 10^{10}. Obviously, more intelligent approaches are necessary.

Although there are numerous approaches, ranging from brute-force evaluation through highly esoteric methods, the majority of investigation in the solution of these special problem classes (or problems in combinatorics in general has been directed at either "exact" methods or "heuristic" methods. The former, however, has received far more attention and exposure (through the various professional journals) than has the latter.

Exact methods of solution are those which, rather obviously, promise to yield the exact, optimal solution to a problem in combinatorics and employ various techniques so as to reduce the number of solutions to be searched. (Simplex itself is one example of an exact technique.) Mathematicians and academicians generally prefer and are much more comfortable with such approaches than they are with heuristics.

A heuristic method is one that has no formal mathematical basis, is developed more or less through intuition, and which cannot guarantee an exact optimal solution. A "good" heuristic, however, can normally find "good" solutions (often near optimal) in a minimal amount of time. As such, although heuristics find acceptance and wide employment among the practitioner, they are often looked down upon by the more mathematically sophisticated.

Despite the dim view of heuristics held by some, recent investigations (together with massive past experimental results) in the area of "computational complexity" have shown that, for many problems in combinatorics, our best if not our only answer lies in the use of the heuristic approach [1, 2]. The reasons for this are as follows.

Algorithms for problem solution in combinatorics fall into two classes: exponential growth algorithms and polynomial growth algorithms. The solution time for an exponential growth algorithm increases exponentially at, for example, a rate of 2^n, where n is the number of variables involved. However, for a polynomial growth algorithm, this time increases at a dramatically reduced rate, say at a rate of n^2 or n^3. Steen [5] has provided an illustrative example of this difference. Given a problem with 50 variables and an exponential growth, we assume that the rate is 2^{50}. An extremely fast computer might take over 30,000 years to solve such a problem. On the other hand, if the growth rate was polynomial, say at a rate of n^3 (50^3), one could solve such a problem in no more than about 2 minutes.

At present, virtually all informed observers have concluded that many problems in combinatorics cannot be solved (by exact methods) in polynomial time. This means that, for large problems of such types, they simply *cannot be solved* by any exact algorithm.

The implications of this conclusion are rather brutal. Many investigators have simply been wasting their time (together with numerous pages in the professional journals) in futile attempts to find exact solutions to numerous large problems in combinatorics. This time might have been much better spent in developing good heuristics.

For this reason, among others, we have included a few heuristic methods in our presentation of the material in Part Three. However, since this text serves as an introduction to linear programming, the main emphasis is on the more conventional, exact methodologies. Those readers wishing to learn more about the heuristic methods of solution are directed to the references [1, 2, 4, 5].

Coverage of Part Three

Since it is beyond the scope of an introductory text to present all the so-called special case problems, we concentrate on a limited but representative sample. Chapter 13 deals with one very well known class, known as the distribution or transportation problem. The traditional example of such a class is given by the situation involving "warehouses" and "retail centers." We wish to supply each retail center with its periodic order of goods by selecting a warehouse to retail center routing that minimizes the sum total of transport costs. Numerous variations about such a base will also be discussed.

Chapter 14 introduces a class of problem that has been designated as the assignment problem. In actuality, this particular problem class is really just a special subclass of the distribution problem of Chapter 13, but it lends itself to some rather unique approaches. The traditional assignment problem example is that of finding the most effective assignment of machines to jobs (or workers to tasks, etc.). However, numerous more-interesting applications exist and are discussed.

Chapter 15 deals with what are termed network models. Actually, the problem types in Chapters 13 and 14 are, themselves, representable by such network models. However, the problems discussed in Chapter 15 are somewhat more general in scope and include (1) finding the shortest or longest route through a network, (2) determining the saturation level of a network, and (3) using a network to determine the schedule of a project.

REFERENCES

1. Ignizio, J. P. "Solving Large-Scale Problems: A Venture into a New Dimension," *Journal of the Operational Research Society*, Vol. 31, 1980, pp. 217–225.
2. Ignizio, J. P., Wyskida, R. M., and Wilhelm, M. "A Rationale for Heuristic Programming Selection and Validation," *Industrial Engineering*, January 1972, pp. 16–19.
3. Johnson, D. E., and Johnson, J. R. *Graph Theory*. New York: Ronald Press, 1972.
4. Lewis, H. R., and Papadimitrion, C. H. "The Efficiency of Algorithms," *Scientific American*, January 1978, pp. 96–109.
5. Steen, L. A. "Computational Unsolvability," *Science News*, Vol. 109, May 8, 1976 pp. 109–301.

THE TRANSPORTATION PROBLEM

Introduction

The first "special case" problem to be discussed is known as the "transportation" or "distribution" problem. However, this terminology is more or less a result of tradition and/or convenience, since the problem itself has many other applications in addition to that of the transportation of goods. Such convenient terminology typifies most of the special case problems, and thus the reader should realize that the scope of, for example, the traveling salesperson problem is far broader than is implied by its name [1–4].

Figure 13-1 provides a convenient illustration of the typical transportation problem. This network model can be divided into two particular types of nodes. Those nodes on the left represent "sources," and those on the right are "sinks." Source nodes are nodes used to describe a terminal at which a supply of discrete items exists. For example, a source node could represent a warehouse with its inventory. Sink nodes are those nodes that "consume" or demand the goods stored within the source nodes. Thus, a convenient analogy is that the source nodes represent warehouses while the sink nodes represent customers.

To the left of each source node, or warehouse, in Figure 13-1 is a number in square brackets. This number represents the amount of goods available at

13

this particular source node. For example, A has 10 units, B has 15, and C also has 15. The numbers in square brackets to the right of each sink node, or customer, depicts the units of goods needed by that sink. Node I requires 8 units, II needs 12, III needs 10, and IV requests 10 units.

The branches (12 in total) connecting the source nodes to the sink nodes represent paths of "transmission," or transportation routes between each individual source and sink. In parentheses, beside each of these branches, are the *per unit* "costs" of shipping between the associated source and sink node. A fundamental assumption is that this cost is constant (e.g., there is no discount for large shipments). Consequently, if 1 unit is shipped from A to I, the cost is 2 units. Further, if all 10 units are shipped from A to I, the cost is simply 10×2, or 20 units. Transport cost is then a linear function of the units shipped.

The single objective of the traditional transportation problem is to find the minimal cost pattern of shipment. The rigid constraints are associated with the amounts available at each source and demanded at each sink.

The basic transportation problem can be extended in various ways. As one example, we might permit shipments *between* source nodes and *between* sink nodes. Often, this "transshipment" of goods is allowable and will usually result in an improved solution (i.e., lower total transportation costs). Such a problem is then known as the transshipment problem and is also discussed.

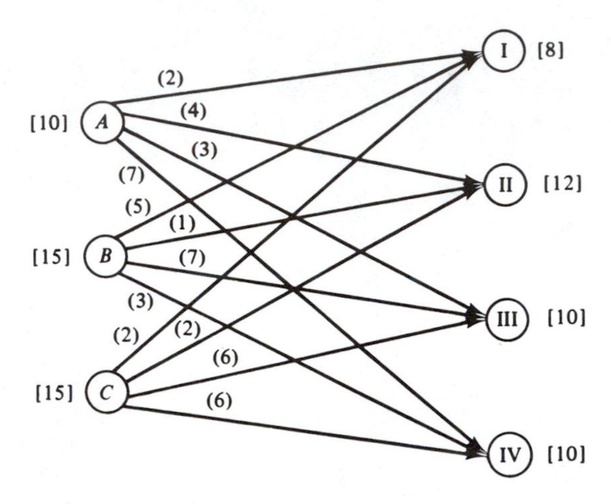

Figure 13-1. Illustrative network model of the transportation problem.

The Transportation Model

In Figure 13-1 we represented one possible model for the transportation problem, the network model. In this chapter we find it easier to deal with yet another representation of this problem. We employ a convenient matrix model [1].

The Matrix Model

Table 13-1 depicts the matrix model for the problem that was illustrated in Figure 13-1. Just as, in earlier chapters, we found the simplex tableau to be a more convenient representation of the general linear programming problem,

Table 13-1. MATRIX MODEL OF THE ILLUSTRATIVE TRANSPORTATION PROBLEM

From \ *To*	*Sink j*				
	I	*II*	*III*	*IV*	*Available*
Source i *A*	[2]	[4]	[3]	[7]	10
B	[5]	[1]	[7]	[3]	15
C	[2]	[2]	[6]	[6]	15
Required	8	12	10	10	40 \ 40

we also find this matrix or table to be a convenient way to both summarize and solve the transportation problem.

Accompanying this matrix representation is a particular set of jargon and concepts that we must know. Notice first that each row in the interior matrix is headed by a source (or warehouse) designator, and each interior column is headed by a sink (or customer) node descriptor. Consequently, at the cell common to row i and column j is a set of information pertaining to the transport of goods from node i to node j. For example, at the node associated with nodes A and I, we find a cell with the per unit transport cost between A and I given in its upper right-hand corner.

Actually, these particular cells can contain additional, pertinent information. This is best illustrated via Figure 13-2, in which a general, interior cell is

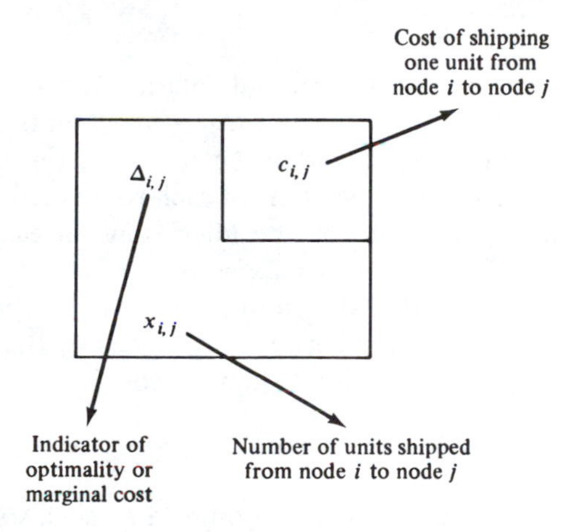

Figure 13-2. The interior cell.

drawn. Notice that this cell contains information not only with regard to node-to-node shipping costs but also with regard to the actual (or proposed) number of units shipped between the particular node pair. Also, in the upper left-hand corner of the cell, we have an "indicator," $\Delta_{i,j}$, element that functions exactly like the indicator row elements of the simplex tableau; that is, it informs us as to whether or not a proposed transportation scheme is optimal and, if not, how to manipulate the present scheme so as to improve the total performance.

The right-most column and bottom-most row are sometimes called the "rims" of the matrix. They provide information as to the number of units of goods available at each source node (the far right-hand side column) and the number of units demanded by each sink node (the bottom row). The intersection of the availability column and demand row is a cell in which the "balance" of the rim conditions may be found (i.e., does supply equal demand?). Noting Table 13-1, we see that the rims are balanced as the total of the supply (40 units) is exactly equal to the total demand (40 units).

Table 13-2. A FEASIBLE SOLUTION

From \ To	I	II	III	IV	Available
A	[2] (8)	[4] (2)	[3]	[7]	10
B	[5]	[1] (10)	[7] (5)	[3]	15
C	[2]	[2]	[6] (5)	[6] (10)	15
Required	8	12	10	10	40 / 40

A *feasible*, but not necessarily optimal solution to the problem of Table 13-1 is given in Table 13-2. Notice that *integer-valued* units of shipment have been allocated (their values circled within the cell) so as to *satisfy* the rim conditions. That is, the sum of the allocations of each row exactly equals the availability of that row, and the sum of the allocations in each column equals column demand.

The total cost of the solution shown in Table 13-2 may be found by multiplying the number of units shipped between node pairs by the per unit shipping cost and then summing. Thus, the total shipping cost is

$$8(2) + 2(4) + 10(1) + 5(7) + 5(6) + 10(6) = 159 \text{ units total}$$

As we shall see, it is relatively easy to determine a feasible solution to a transportation problem. The difficulty lies in determining whether or not the solution is optimal and, if not, how to improve it.

The Mathematical Model

As mentioned in Chapter 12, some of our special models have the property of unimodularity. This is true of the transportation model, as will be revealed via its mathematical formulation [4].

Let us first consider the generalized transportation problem matrix of Table 13-3. In this table we let a_i designate the amount of supply available at source node i and b_j as the amount required at sink node j. As may be seen, this is a matrix of m rows ($i = 1, 2, \ldots, m$) and n columns ($j = 1, 2, \ldots, n$), wherein there are m source nodes and n sink nodes. The total number of interior cells, and thus of control variables, is then m times n.

The goals associated with the transportation problem are then:

1. Determine the number of units shipped from source i to sink j ($x_{i,j}$ values) so as to minimize the total transport costs.

Table 13-3. GENERAL TRANSPORTATION MATRIX

From \ To		Sink j				
		1	2	$\cdots$	n	a_i
Source i	1	$x_{1,1}$ $[c_{1,1}]$	$x_{1,2}$ $[c_{1,2}]$	$\cdots$	$x_{1,n}$ $[c_{1,n}]$	a_1
	2	$x_{2,1}$ $[c_{2,1}]$	$x_{2,2}$ $[c_{2,2}]$	$\cdots$	$x_{2,n}$ $[c_{2,n}]$	a_2
	$\vdots$			$\vdots$		$\vdots$
	m	$x_{m,1}$ $[c_{m,1}]$	$x_{m,2}$ $[c_{m,2}]$	$\cdots$	$x_{m,n}$ $[c_{m,n}]$	a_m
	b_j	b_1	b_2	$\cdots$	b_n	$\sum_i a_i$ / $\sum_j b_j$

2. The supply limits (availability) of each source must be *exactly* satisfied (i.e., we must ship out exactly as many units from a source as are available at that source).
3. The demand requirements of each sink must be *exactly* met.
4. No variables (units shipped) may take on negative values and, further, all shipments must be integer-valued.

Letting $x_{i,j}$ be the number of units shipped from source i to sink j, the baseline model is given as:

$$\text{minimize} \quad z = \sum_{i=1}^{m} \sum_{j=1}^{n} c_{i,j} x_{i,j} \tag{13.1}$$

satisfy

$$\sum_{j=1}^{n} x_{i,j} = a_i \qquad \text{for } i = 1, 2, \ldots, m \tag{13.2}$$

$$\sum_{i=1}^{m} x_{i,j} = b_j \qquad \text{for } j = 1, 2, \ldots, n \tag{13.3}$$

where

$$x_{i,j} = 0, 1, 2, \ldots \tag{13.4}$$

The transformation to the traditional linear programming model is made by assuming that the goals in (13.2) through (13.4) are all absolute and thus that they are rigid constraints. Notice that such a formulation is consistent only if

$$\sum_{i=1}^{m} a_i = \sum_{j=1}^{n} b_j$$

That is, the rim conditions must balance. Although this seems to be highly restrictive, it may be circumvented easily by the employment of what are known as "dummy" sources or "dummy" sinks. This is illustrated later via examples.

The mathematical model for the transportation problem contains two main sets of constraints (ignoring, for the moment, the integer restrictions on $x_{i,j}$): one set associated with the sources and one set associated with the sinks. These may be conveniently summarized in Table 13-4 for the specific example first illustrated in Table 13-1. The interior of this table (ignoring the right-hand side values a_i and b_j) is simply the constraint matrix or matrix **A**. Note that every determinant of order k ($k < m + n$) is either a value of 0, 1, or -1, and thus the problem is unimodular.[a] Also note the particular *pattern* of 0's and 1's within this matrix. Any problem exhibiting such a pattern may be solved via special techniques that we present later.

Properties of the Transportation Problem

The transportation problem exhibits numerous other properties in addition to unimodularity. Some of these are listed below.

Property 1. If the rims of the transportation problem balance (i.e., if $\sum_{i=1}^{m} a_i = \sum_{j=1}^{n} b_j$), then the problem *must* have a feasible (integer) solution.

Property 2. The transportation problem can never be unbounded.

Property 3. If, for any transportation problem, an initial basic feasible solution is found in which all variables are integer, the simplex method will continue to produce all integer basic feasible solutions.

Solving the Transportation Problem

Since the transportation problem is unimodular, we could simply:

1. Add artificial variables to obtain an initial basic feasible solution.
2. Solve by the regular simplex algorithm.

However, the peculiar form of the transportation model may be used to advantage in the solution of the problem via a somewhat simpler approach. This approach consists of two phases in which the matrix model of the transportation problem is employed. Each phase, in fact, is directly analogous to the two steps listed above in the use of simplex.

Phase 1 of the special transportation algorithm consists of a simple heuristic technique by which an initial basic feasible solution is rapidly generated. Notice that, when using the matrix model, the achievement of a basic feasible solution

[a]From our definition in Chapter 12.

Table 13-4. SOURCE AND SINK CONSTRAINTS FOR EXAMPLE

		$x_{1,1}$	$x_{1,2}$	$x_{1,3}$	$x_{1,4}$	$x_{2,1}$	$x_{2,2}$	$x_{2,3}$	$x_{2,4}$	$x_{3,1}$	$x_{3,2}$	$x_{3,3}$	$x_{3,4}$	*Right-Hand Side*
Source Node	1	1	1	1	1	0	0	0	0	0	0	0	0	10
	2	0	0	0	0	1	1	1	1	0	0	0	0	15
	3	0	0	0	0	0	0	0	0	1	1	1	1	15
Sink Node	1	1	0	0	0	1	0	0	0	1	0	0	0	8
	2	0	1	0	0	0	1	0	0	0	1	0	0	12
	3	0	0	1	0	0	0	1	0	0	0	1	0	10
	4	0	0	0	1	0	0	0	1	0	0	0	1	10

corresponds to a solution, with all integer values, that satisfies the rim conditions. The solution shown in Table 13-2 is thus a basic feasible solution for the particular problem under consideration.

There are actually numerous heuristics that may be used to generate the initial basic feasible solution of phase 1 [1]. We shall consider only two of these: the "northwest corner method" and Vogel's approximation method (VAM).

Some of the heuristics available to generate the initial basic feasible solution are extremely fast (they require very few and simple computations) but they also normally give rather poor starting solutions in terms of total transportation cost. Other methods give much better initial solutions but require more lengthy and more elaborate computations. The key to selecting a "best" phase 1 heuristic is to seek a trade-off. That is, find a heuristic that gives at least a moderately good initial solution while requiring a minimal amount of computation. As we shall see, the northwest corner method is very simple but delivers very poor solutions; VAM, on the other hand, represents a very good trade-off between computational requirements and initial accuracy. For that reason, we normally use VAM in phase 1 except in our initial discussion.

Phase 2 of the special method for the transportation problem consists of determining whether or not the solution obtained in phase 1 can be improved. If not, the procedure is terminated. However, if improvement is possible, we employ a procedure, known as MODI [1] to "reshuffle" the present allocation to improve the objective function value. Actually, this reshuffling operation is, in essence, an exchange of a basic variable for a nonbasic variable—the *same procedure used in regular simplex* but made easier (and more obscure) by the nature of the transportation model. We now present these phases, with illustrative examples, in the following discussion. First, however, we consider a preliminary point.

Dealing with the Unbalanced Problem

Prior to discussing the two-phase process, we first note that these procedures are based on the assumption of a balanced (i.e., supply exactly equals demand) problem. This assumption imposes no undue hardship, as any unbalanced problem can be easily converted into an equivalent, balanced model. This is achieved as follows:

1. If the problem is unbalanced wherein *supply exceeds demand* (i.e., $\sum a_i > \sum b_j$), we add a new "dummy" sink column to the matrix. This column has a "demand that is equal to $\sum a_i - \sum b_j$ (i.e., it absorbs the oversupply). Physically, the allocation of goods from a source to a dummy sink simply means that these goods stay at the source node. Consequently, the cost of "shipping" from a source node to a dummy sink should reflect the actual cost associated with holding the goods at the source. Typically, in textbook examples, this cost is given as zero.

2. If the problem is unbalanced wherein *demand exceeds supply* (i.e., $\sum a_i < \sum b_j$), we must add a new "dummy" source row to the matrix. This

row has a "supply" that is equal to $\sum b_j - \sum a_i$ (i.e., it makes up the difference between actual demand and supply). Physically, the allocation of goods from a dummy source to a given sink means that this sink will be *short* these units shown allocated. Consequently, the cost of "shipping" from a dummy source node to a sink node should reflect the actual cost associated with a shortage at the respective sink. Again, in the textbook example, the cost is typically given as zero.

To demonstrate, consider first the problem shown in Table 13-5(a). As may be seen, the problem is unbalanced in that total supply (12 units) exceeds total demand (9 units). The required dummy source column has been added in Table 13-5(b) and, in addition, an arbitrary feasible solution is indicated. Notice that 3 units are allocated from source *B* to the dummy sink. This means that source *B* will retain these 3 units of goods for the period under investigation.

Table 13-5. SUPPLY EXCEEDS DEMAND

From \ *To*	*I*	*II*	a_i
A	[7]	[4]	8
B	[8]	[3]	4
b_j	4	5	12 / 9

(a) Unbalanced Matrix

From \ *To*	*I*	*II*	*Dummy*	a_i
A	[7] (4)	[4] (4)	[0]	8
B	[8]	[3] (1)	[0] (3)	4
b_j	4	5	3	12 / 12

(b) Addition of Dummy Sink

Next, let us examine a problem in which demand exceeds supply, as shown in Table 13-6(a). A dummy source is added, in Table 13-6(b), to balance the problem and, again, an arbitrary allocation scheme is depicted. We see that an allocation of goods, from the dummy source, to both sinks I and II are given. This means, physically, that sink I is short 3 units and sink II is short 4 units.

Table 13-6. DEMAND EXCEEDS SUPPLY

From \ To	I	II	a_i
A	[4]	[8]	5
B	[5]	[6]	6
b_j	14	4	18 \ 11

(a) Unbalanced Matrix

	I	II	a_i
A	[4] (5)	[8]	5
B	[5] (6)	[6]	6
Dummy	[0] (3)	[0] (4)	7
b_j	14	4	18 \ 18

(b) Addition of Dummy Source

Using such approaches, any transportation problem may be balanced so as to proceed with the solution process, which we now consider. Realize that a fundamental assumption of this process is that we begin with a balanced model.

Phase 1: The Initial Basic Feasible Solution

Finding an initial basic feasible solution (i.e., one that satisfies all the rim conditions) is fairly easy by simple inspection for problems of small size, such as that given in Table 13-2. However, when dealing with larger problems it is necessary to employ something a bit more systematic. The northwest corner method is one such approach. Consider the problem shown in Table 13-7. The northwest corner method for a balanced problem proceeds as follows:

Step 1. Given a balanced transportation matrix, examine the cell in the northwest corner of the matrix (i.e., the cell in the uppermost left-hand corner). Call this $N_{i,j}$.

Step 2. Assign, to cell $N_{i,j}$, the minimum of a_i (the row availability) or b_j (the column requirement).

Step 3. Reduce a_i and b_j (for the row and column associated with $N_{i,j}$) by the amount allocated to cell $N_{i,j}$. Thus, either one row or one column (or, in some cases, both) will be satisfied and may be eliminated from the matrix. Cross out this row or column (or, if appropriate, both). Go to step 4.

Step 4. If no rows and columns are left, phase 1 of the procedure is over. Otherwise, return to step 1 and repeat the process with the matrix remaining.

Examining this algorithm reveals its single motivation, to simply find a solution that satisfies the rim conditions. Nothing in the procedure is concerned with finding a *good* solution (i.e., no attention is given to the cost information).

Following the steps of the northwest corner method, we obtain Tables 13-8 and 13-9 for the example provided initially in Table 13-7.

Table 13-7. EXAMPLE FOR NORTHWEST CORNER METHOD

From \ *To*	*D*	*E*	*F*	*G*	a_i
A	[8]	[6]	[4]	[2]	4
B	[10]	[6]	[6]	[2]	3
C	[4]	[2]	[3]	[8]	6
b_j	3	3	3	4	13 \ 13

Table 13-8. APPLICATION OF NORTHWEST CORNER METHOD

(a) Allocation to Northwest Cell (*A–D*)

From \ *To*	*D*	*E*	*F*	*G*	a_i
A	[8] ③	[6]	[4]	[2]	~~4~~ ①
B	[10]	[6]	[6]	[2]	3
C	[4]	[2]	[3]	[8]	6
b_j	~~3~~ ⓪	3	3	4	

(b) Allocation to Northwest Cell (*A–E*) of Reduced Matrix

	E	*F*	*G*	a_i
A	[6] ①	[4]	[2]	10
B	[6]	[6]	[2]	3
C	[2]	[3]	[8]	6
b_j	~~3~~ 2	3	4	

(c) Allocation to Northwest Cell (*B–E*) of Reduced Matrix

	E	*F*	*G*	a_i
B	[6] ②	[6]	[2]	~~3~~ 1
C	[2]	[3]	[8]	6
b_j	~~2~~ 0	3	4	

(d) Allocation to Northwest Cell (*B–F*) of Reduced Matrix

	F	*G*	a_i
B	[6] 1	[2]	~~1~~ 0
C	[3]	[8]	6
b_j	~~3~~ 2	4	

(e) Allocation to Northwest Cell (*C–F*) of Reduced Matrix

	F	*G*	a_i
C	[3] 2	[8]	~~6~~ 4
b_j	~~2~~ 0	4	

(f) Allocation to Northwest Cell (*C–G*) of Reduced Matrix

	G	a_i
C	[8] ④	~~4~~ 0
	4 0	

Table 13-9. RESULTANT SOLUTION

To / From	*D*	*E*	*F*	*G*	a_i
A	[8] ③	[6] ①	[4]	[2]	4
B	[10]	[6] ②	[6] ①	[2]	3
C	[4]	[2]	[3] ②	[8] ④	6
b_j	3	3	3	4	13 / 13

Table 13-9 summarizes the allocation achieved via the implementation of the northwest corner method. The shipping schedule is:

	Number of Units	*From*	*To*
Ship	3	*A*	*D*
Ship	1	*A*	*E*
Ship	2	*B*	*E*
Ship	1	*B*	*F*
Ship	2	*C*	*F*
Ship	4	*C*	*G*

for a total cost of $3(8) + 1(6) + 2(6) + 1(6) + 2(3) + 4(8) = 86$ units of cost. Although this is a feasible allocation, it is not known at this phase whether or not it is the optimal shipping scheme.

An improved algorithm for conducting phase 1 is available through the use of the Vogel heuristic or, as it is typically known, the Vogel approximation method (VAM). VAM is slightly more involved than the northwest corner method, but it usually is worth the extra effort in that, normally, a far better initial solution is obtained. In fact, the only thing that would lead the northwest corner method to a good initial solution is sheer luck.

The reason for the improved performance of VAM is that it pays attention, in its development of a feasible solution, to the cost information contained within the matrix model. It does this through the establishment of "penalty numbers," which indicate the possible cost penalty associated with *not* assigning an allocation to a given cell. A shortened version of the VAM algorithm is given below and is then demonstrated using the example previously solved by the northwest corner method.

Step 1. Given a balanced transportation matrix, determine the VAM penalty numbers for each row (PN_i) and column (PN_j) as follows:

(a) The penalty number for each row (PN_i) is given as the absolute value of the difference between the lowest-cost cell ($c_{i,j}$) and next-lowest-cost cell in each row.

(b) The penalty number for each column (PN_j) is given as the absolute value of the difference between the lowest-cost cell and next-lowest-cost cell in each column.

Step 2. Determine the row or column having the largest penalty number. In the event of ties, break the ties arbitrarily.

Step 3. Assign the maximum number[b] of units possible to the cell with the lowest cost ($c_{i,j}$) in the row or column selected in step 2. Reduce the row and column availabilities and demands by this amount and cross out any row or column (or, if appropriate, both) that has been completely satisfied.

Step 4. Repeat steps 1 through 3 until all units have been allocated.

Various refinements to the algorithm are possible, particularly in additional rules for the breaking of ties (as cited in step 2). For example, in the event of a tie for the maximum penalty number, we could (and will in our examples) break the tie in favor of the row or column having the smallest associated $c_{i,j}$ values.

We now proceed to solve the example given in Table 13-7 by VAM. We begin with Table 13-10(a), wherein the VAM penalty numbers are listed (see the *circled* numbers) to the right of each row and the bottom of each column. Notice that there is a tie for the highest penalty number between the first and second columns and the second row. We break this tie by selecting the second column because it (and row 2) has the lowest $c_{i,j}$ element (a value of 2 for $c_{3,2}$).

Table 13-10(b) is next developed by eliminating the second column and revising the rim conditions. In this new table, the highest VAM penalty number is tied between the second row (i.e., *B*) and the first column (i.e., *D*). The maximum amount possible (3 units) is then allocated to the lowest-cost cell (i.e., $c_{4,3} = 2$). This leads to the third table, Table 13-10(c), wherein 1 unit is allocated to the cell in the third column, first row. Our final table is shown in 13-10(d), where 3 units are assigned to the first column, second row, leaving only one

[b]Always make sure that the rim conditions have not been violated.

possible way to allocate the remaining 3 units. All allocations are summarized in Table 13-11 and the associated cost is 38 units. This is an improvement of 48 units over the allocation developed by the northwest corner method.

Before leaving this example, note that, when dealing with Table 13-10(d), we were able to simultaneously eliminate *both* a row (row *C*) and a column

Table 13-10. TABLES FOR VAM EXAMPLE

(a) First Table

	D	*E*	*F*	*G*	a_i	
A	[8]	[6]	[4]	[2]	4	②
B	[10]	[6]	[6]	[2]	3	④
C	[4]	[2] ③	[3]	[8]	6	①
b_j	3	3	3	4	13 / 13	
	④	④*	①	⓪		

(b) Second Table

	D	*F*	*G*	a_i	
A	[8]	[4]	[2]	4	②
B	[10]	[6]	[2] 3	3	④*
C	[4]	[3]	[8]	3	①
b_j	3	3	4		
	④	①	⓪		

(c) Third Table

	D	*F*	*G*	a_i	
A	[8]	[4]	[2] ①	4	②
C	[4]	[3]	[8]	3	①
b_j	3	3	1		
	④	①	⑥*		

(d) Fourth Table

	D	F	a_i	
A	[8]	[4] 3	3	(4 *)
C	[4] (3)	[3]	3	(1)
b_j	(3)	3		
	(4 *)	(1)		

Table 13-11. RESULTANT SOLUTION

	D	E	F	G	a_i
A	[8]	[6]	[4] (3)	[2] (1)	4
B	[10]	[6]	[6]	[2] (3)	3
C	[4] (3)	[2] (3)	[3]	[8]	6
b_j	3	3	3	4	13 \ 13

(column D). Whenever this occurs (i.e., the simultaneous elimination of a row and column) it is an indication of a *degenerate* solution. Recall from our discussions of linear simultaneous equations (see e.g., Chapter 3) that a degenerate solution is one in which one or more basic variables have a value of zero. This is exactly what has occurred in this example; however, because of the matrix format, it is not as apparent as it was in the simplex tableau.

In general, given a transportation matrix with m sources (rows) and n columns (sinks), a *nondegenerate* solution will have exactly $m + n - 1$ allocations. No solution should contain more and, if any solution has less than $m + n - 1$ allocations, the solution is degenerate. Our previous example had $m = 3$ and $n = 4$, and thus a nondegenerate solution should have $3 + 4 - 1$ or 6 allocations. The solution by the northwest corner method (see Table 13-9) is then nondegenerate, whereas the solution by VAM is degenerate (see Table 13-11; there are only 5 allocations). An optimal solution may well be degenerate and thus there is nothing wrong with having a degenerate solution, except that it poses some added difficulty when we move to the second phase of the solution process.

As stated, a nondegenerate solution to the transportation problem must have exactly $m + n - 1$ allocations. Let us now examine why this is so. Recall from our earlier discussion of the mathematical model of the transportation problem that such problems have exactly $m + n$ constraints. That is, there is a constraint associated with each row and column of the matrix (i.e., with each sink and source). However, of these $m + n$ constraints, only $m + n - 1$ are independent (and can form a basis, as discussed in Chapter 3). This is because once we have satisfied $m + n - 1$ of the sources and sinks, the allocation to the remaining source or sink is fixed.

Dealing with Degeneracy

As mentioned above, there is nothing wrong with a degenerate solution and, in fact, it is possible that the optimal solution to a transportation problem may be degenerate. The only difficulty imposed is from a computational point of view. The second phase of our solution procedure can be employed only with a nondegenerate solution. We can overcome this seemingly severe limitation via a number of fairly straightforward ways. However, for brevity, we discuss only one of these approaches, the addition of ϵ allocations.

Given a transportation problem with a degenerate solution (i.e., the number of allocations is less than $m + n - 1$), we may add artificial allocations to obtain an equivalent solution that the phase 2 algorithm will consider nondegenerate. If, for example, $m = 10$, $n = 7$, and the number of allocations is 13, we must add three ϵ allocations to the matrix to have a "nondegenerate" solution. Our assumption is that ϵ is a very small positive number, nearly zero in value. Once we have finished phase 2, we may set ϵ to zero and obtain the actual solution.

We have to follow certain rules in assigning these ϵ allocations. Basically, an ϵ allocation can only be assigned to an *independent*, *empty* cell. Such a cell may be identified via the employment of a concept known as the θ *path*. The rules for constructing a θ path are as follows:

Step 1. Start with an empty cell. Assign to this cell a $+\theta$ designation.
Step 2. A loop (the θ path) is to be formed that must *start and end* at the initial empty cell. This loop consists of *successive* horizontal and vertical segments (or moves) whose end points, except for the initial cell, must lie at allocated cells.
Step 3. Cells are assigned an alternating pattern of $+\theta$'s and $-\theta$'s (i.e., $+\theta$, $-\theta$, $+\theta, \ldots, -\theta, +\theta$).

Now, for an *independent* cell, we will be *unable to form a closed* θ *path*. Consequently, we normally identify the independent cells and then make our ϵ allocations to those independent cells having the lowest transport cost.

The discussion may be clarified via the example given in Table 13-11. This solution is degenerate by one allocation. We thus need to identify an independent cell and assign an ϵ allocation to that cell.

Checking first for an independent, empty cell, we examine cell *B–F*. This cell is *not* independent because a θ path may be traced as follows:

Cell	θ *Designation*
B–F	$+\theta$
A–F	$-\theta$
A–G	$+\theta$
B–G	$-\theta$
B–F	$+\theta$

This is shown in Table 13-12.

Table 13-12. θ PATH FOR CELL *B–F*

	D	*E*	*F*	*G*
A	[8]	[6]	$-\theta$ [4] 3	$+\theta$ [2] 1
B	[10]	[6]	$+\theta$ [6]	$-\theta$ [2] 3
C	[4] 3	[2] 3	[3]	[8]

Examining another cell, say cell *C–G*, we can find no closed θ path, and thus this cell is independent. An ϵ allocation may be placed in this cell to obtain a "nondegenerate" solution. The reader is advised to check the other empty cells. They are, with the exception of cell *B–F*, all independent for this problem.

PHASE 2: DETERMINATION OF OPTIMALITY

The process followed in phase 1 simply leads to a feasible solution, that is, an integer-valued set of allocations (except for any ϵ allocations that may be added) that exactly satisfy the rim conditions of the balanced matrix model. However, at the conclusion of phase 1, we do not yet know whether or not this initial solution is optimal or, if not, how far from optimality it might be. The purpose of phase 2 is then:

1. To determine if the given solution is optimal and, if not
2. To construct an improved solution.

Phase 2 ends as soon as we can ascertain that no improvement to a given solution can be obtained. As mentioned earlier, phase 2 is simply a way of conducting the simplex algorithm with the transportation matrix rather than by a conventional tableau.

The procedure that we shall employ to implement phase 2 is known as the MODI (modified distribution) algorithm. The general steps of this algorithm are as follows:

Step 1. Phase 1 must have been completed and we must have a *nondegenerate* basic feasible solution. (ϵ allocations are added to achieve this if necessary.)
Step 2. The present solution is checked for optimality. If the solution is not optimal, we pass to step 3. If optimal, the procedure is terminated.
Step 3. The effects of redistributing the allocations (although always satisfying the rim conditions) to cells not presently used are considered.
Step 4. The redistribution that appears to provide the most solution improvement is selected and implemented.
Step 5. Steps 2 through 4 are repeated until optimality is achieved.

We implement steps 2 through 4 by means of MODI. Normally, we suggest that phase 1 be conducted by VAM but, to better illustrate the MODI method, we start our illustrative example using the northwest corner solution to our previous problem. This solution is shown in the MODI matrix of Table 13-13.

Table 13-13. MODI EXAMPLE

	K_j / R_i	D	E	F	G
A		[8] ③	[6] ①	[4]	[2]
B		[10]	[6] ②	[6] ①	[2]
C		[4]	[2]	[3] ②	[8] ④

Table 13-13 has, as its interior, the northwest corner solution shown in Table 13-9. Added to this are a set of row indicators (R_i) and column indicators (K_j) known as the MODI numbers. The steps of the MODI algorithm are then:

Step 1. Given a nondegenerate solution, assign a zero element to any R_i or K_j position.
Step 2. For each *allocated* cell, the following expression must be satisfied:

$$\Delta_{i,j} = R_i + K_j + c_{i,j} = 0 \qquad \text{(for allocated cells)} \qquad (13.5)$$

(One possible R_i, K_j arrangement is shown in Table 13-14.)

Table 13-14

	K_j / R_i	*D*	*E*	*F*	*G*
		−8	−6	−6	−11
A	0	[8] ③	[6] ①	[4]	[2]
B	0	[10]	[6] ②	[6] ①	[2]
C	3	[4]	[2]	[3] ②	[8] ④

Step 3. For each empty cell (i.e., cells without allocations), determine the value of $\Delta_{i,j} = R_i + K_j + c_{i,j}$. Place this value in the upper left-hand corner of the associated cell (as shown in Table 13-15).

Table 13-15

	K_j / R_i	*D*	*E*	*F*	*G*
		−8	−6	−6	−11
A	0	[8] ③	[6] ①	−2 [4]	−9 [2]
B	0	+2 [10]	[6] ②	[6] ①	−9 [2]
C	3	−1 [4]	−1 [2]	[3] ②	[8] ④

Step 4. If the $\Delta_{i,j}$ sums (or indicators) are nonnegative for all the empty cells, the solution is optimal. However, if any empty cells have negative indicators, then *for each unit allocated to that cell the total cost will decrease by the product of the number of units allocated to the cell times* $\Delta_{i,j}$. If any $\Delta_{i,j}$ are negative for an empty cell, go to step 5.

Step 5. Select the empty cell with the most negative $\Delta_{i,j}$ value (cell *A–G* in Table 13-15).[c] We shall next "reshuffle" our present allocation so as to allocate to this empty cell while maintaining a feasible solution. This is accomplished via the construction of a θ path as discussed in step 6.

Step 6. Construct a θ path through the matrix, beginning and ending at the empty cell having the most negative $\Delta_{i,j}$ value (ties are broken by arbitrarily selecting one of the tied cells).

The θ path for our example will start with a $+\theta$ in cell *A–G* and travels

[c]Actually, there is a tie between cells *B–G* and *A–G*, and the selection of *A–G* was done arbitrarily.

to cells *C–G*, *C–F*, *B–F*, *B–E*, *A–E*, and back to cell *A–G*, closing the loop. This is shown in Table 13-16.

Table 13-16. θ PATH

	D	E	F	G
A	[8] ③	$-\theta$ [6] ①	[4]	$+\theta$ [2]
B	[10]	$+\theta$ [6] ②	$-\theta$ [6] ①	[2]
C	[4]	[2]	$+\theta$ [3] ②	$-\theta$ [8] ④

Step 7. The largest number of units that may be allocated to empty cell (*A–G*) is determined by

$$\min_{-\theta} \{x_{i,j}\} \tag{13.6}$$

That is, we find the minimum allocation to any $-\theta$ cell in the θ path. The allocation to the empty cell cannot exceed this amount, or it will cause one or more of the other allocations to go negative. (In our example of Table 13-16, this is 1 unit.) We then add the amount given by (13.6) to all cells with $+\theta$ designations and subtract it from those with $-\theta$'s.

Step 8. We then repeat steps 1 through 7 until optimality, as indicated in step 4, is is achieved.

The procedure used to solve this example is then concluded as shown in Tables 13-17 through 13-22. Notice that, in Table 13-17, the table depicts the reallocation to cell *A–G*. This means that the variable associated with cell *A–G* ($x_{1,4}$) has come into the basis but, since *both* cells *A–E* and *B–F* now have zero allocations, the new solution is degenerate. Consequently, an ϵ allocation must be employed, and we have assigned it to cell *C–E*. The MODI indicators in Table

Table 13-17. SECOND MODI INDICATOR SET

	K_j / R_i	D	E	F	G
		−8	4	3	−2
A	0	[8] ③	+4 [6]	+7 [4]	[2] ①
B	−10	−8 [10]	[6] ③	−1 [6]	−10 [2]
C	−6	−10 [4]	[2] ϵ	[3] ③	[8] ③

Table 13-18. θ PATH

	D	E	F	G
A	$-\theta$ [8] ③	[6]	[4]	$+\theta$ [2] ①
B	[10]	[6] ③	[6]	[2]
C	$+\theta$ [4]	[2] (ε)	[3] ③	$-\theta$ [8] ③

Table 13-19. THIRD MODI INDICATOR SET

	K_j \ R_i	D	E	F	G
		−8	−6	−7	−2
A	0	0 [8]	0 [6]	−3 [4]	[2] ④
B	0	+2 [10]	[6] ③	−1 [6]	[2] (ε)
C	4	[4] ③	[2] (ε)	[3] ③	+10 [8]

Table 13-20. θ PATH

	D	E	F	G
A	[8]	[6]	$+\theta$ [4]	$-\theta$ [2] ④
B	[10]	$-\theta$ [6] ③	[6]	$+\theta$ [2] (ε)
C	[4] ③	$+\theta$ [2] (ε)	$-\theta$ [3] ③	[8]

Table 13-21. FOURTH MODI INDICATOR SET

	K_j \ R_i	D	E	F	G
		−5	−3	−4	−2
A	0	+3 [8]	+3 [6]	[4] ③	[2] ①
B	0	+5 [10]	+3 [6]	+2 [6]	[2] 3 + ε
C	0	[4] ③	[2] 3 + ε	[3] (ε)	+7 [8]

Table 13-22. FINAL SOLUTION

From \ To	D	E	F	G	a_i
A	[8]	[6]	[4] 3	[2] 1	4
B	[10]	[6]	[6]	[2] 3	3
C	[4] 3	[2] 3	[3]	[8]	6
b_j	3	3	3	4	13 \ 13

13-17 show that improvement is possible and we use cell *C–D* (tied with *B–G* for the most negative $\Delta_{i,j}$, -10) as the cell from which to construct our θ path.

Table 13-18 shows the θ path for cell *C–D*, and the new allocation and MODI indicator table are given in Table 13-19. (The smallest allocation in a $-\theta$ cell was 3 in Table 13-18, so 3 has been added to cells *C–D* and *A–G* and subtracted from *A–D* and *C–G*.) Notice that *another* ϵ allocation had to be added to Table 13-19, because in the reallocation process it went degenerate again. This new ϵ has been put in cell *B–G*.

Unfortunately, the allocation of Table 13-19 is still not optimal and we must construct a θ path for cell *A–F*, as shown in Table 13-20. Notice that 3 is the smallest number of units in the $-\theta$ cells, so our new distribution is shown in Table 13-21. We have again gone degenerate, so yet another new ϵ allocation appears in cell *C–F*.

Thankfully, we at last have an optimal solution, since all $\Delta_{i,j}$ indicators are nonnegative. Letting our ϵ's go to zero, the resultant, optimal allocation is shown in Table 13-22, and although degenerate, this solution is optimal, with a cost of 38 units.

This example, although small in size, showed us several things. First, we see that even if we begin phase 2 with a nondegenerate solution, the solution can, at any time during phase 2, go degenerate. Further, during the conduct of phase 2, the ϵ allocations are treated in exactly the same way as the "real" allocations. For example, if the minimum allocation in the $-\theta$ cells of a θ path is ϵ, then ϵ must be subtracted from all $-\theta$ cells and added to all $+\theta$ cells.

Production Scheduling and Inventory Control

As mentioned, the transportation model encompasses far more applications than the name might imply. One of these exists in certain cases of production scheduling and can best be explained by means of an example.

Example 13-1

A firm is engaged in the manufacture of a deodorant known as "Heavy Duty." Demand for Heavy Duty is higher, or course, in warmer weather, and thus seasonal influences play an important factor in production decisions. The estimated demand, per quarter, of the deodorant can be estimated fairly accurately and is shown in Table 13-23(a). Maximum production rates also vary during the year (due to labor turnover, planned equipment overhauls, etc.) and are given in Table 13-23(b).

Table 13-23. DEODORANT DEMAND AND PRODUCTION RATE

Quarter	*Predicted Demand (Cartons)*
1	100,000
2	140,000
3	200,000
4	80,000

(a) Demand

Quarter	*Maximum Production Rate (Cartons)*
1	130,000
2	150,000
3	150,000
4	130,000

(b) Production Rate

Production costs, per carton, are $5 in quarters 1 and 2 and will increase (due to a salary increase for workers) to $6 in quarters 3 and 4. If a carton of deodorant is not used during the quarter it is produced, it incurs an inventory charge of $1 for every quarter it remains in storage.

The company wishes to determine a production schedule for Heavy Duty so as to satisfy demand at minimum cost. Such a schedule will tell them how many cartons to produce in each quarter.

At first, this problem may not seem to bear much similarity to our transportation problem. What is necessary is to probe past the narrative description and problem context so as to examine the *structure* of the problem model. Once this is done, we will see that the problems are of the same class. To do this, first consider the designation of "sinks" and "sources." Although there is no physical movement of goods from sinks to sources, there is an analogous allocation of goods produced in a given quarter and sold within some specific quarter. That is, we may consider our sources to be the four quarters in which the goods may be produced, while our sinks are the four quarters in which the deodorant is sold.

Table 13-24 depicts the "transportation" model matrix for this problem. Note that:

1. The source availabilities are the maximum number of cartons that can be produced each quarter.

2. The sink demands are the estimated number of cartons that can be sold each quarter.
3. The cost ($c_{i,j}$) information in each cell is given by the sum of (a) production cost for the quarter produced and (b) inventory charges incurred, if any. For example, a carton produced in quarter 1 and not sold until quarter 3 has a cost of \$5 (production) plus 2 × \$1 (inventory charges for quarters 1 and 2).
4. Since it is physically impossible to produce a carton in quarter n and sell it in quarter m if $n > m$, the cost of such an allocation is set at infinity (in practice, some very high positive number may be used).

Table 13-24. PRODUCTION SCHEDULING/INVENTORY MODEL

From \ *To*	*Quarter* 1	2	3	4	a_i (*Thousands*)
1	[5]	[6]	[7]	[8]	130
2	[∞]	[5]	[6]	[7]	150
3	[∞]	[∞]	[6]	[7]	150
4	[∞]	[∞]	[∞]	[6]	130
b_j (*thousands*)	100	140	200	80	560 / 520

Since the supply and demands in Table 13-24 are unbalanced, we add a dummy sink with a "demand" of 40 units, which results in Table 13-25. The optimal solution to this problem is also shown in Table 13-25 and is:

Table 13-25. OPTIMAL PLAN

From \ *To*	*Quarter* 1	2	3	4	Dummy	a_i
1	[5] (100)	[6] (30)	[7]	[8]	[0]	130
2	[∞]	[5] (110)	[6] (40)	[7]	[0]	150
3	[∞]	[∞]	[6] (150)	[7]	[0]	150
4	[∞]	[∞]	[∞] (10)	[6] (80)	[0] (40)	130
b_j	100	140	200	80	40	560 / 560

Produce	*For Sale in Quarter:*
100 units in quarter 1	1
30 units in quarter 1	2
110 units in quarter 2	2
40 units in quarter 2	3
150 units in quarter 3	3
10 units in quarter 4	3 *
80 units in quarter 4	4
40 units in quarter 4	dummy *

Although appearing odd, the solution is correct if interpreted properly. Obviously, the 40 units produced in quarter 4 and allocated to the dummy sink reflect unused capacity in this quarter. This is true also for the 10 units produced in quarter 4 and "used" in quarter 3. This allocation, of course, is physically impossible and represents an additional unused capacity.

Another variation, reflective of actual problems, of the production scheduling problem entails the use of overtime in production. That is, we have a certain production capacity each quarter via regular time and some additional (higher priced) capacity with overtime. One approach to handling this situation is to add additional "sources" (i.e., rows) to reflect the overtime production capabilities.

Complications and Extensions

There are numerous extensions and variations of the basic transportation problem, including the transshipment problem, which we discuss in some detail later. Some of the other common complications/extensions include:

—Conversion to a maximizing problem
—Forbidden allocations
—Perturbation methods
—Alternative optimal solutions

The Maximizing Problem

Sometimes the $c_{i,j}$ "cost" elements may actually reflect profits—or some other measure of effectiveness that we wish to *maximize*. The VAM/MODI procedure is, as presented, a minimizing algorithm, but conversion to a maximizing process is straightforward. We present two ways to achieve this.

The approach mentioned in most textbooks is simply to multiply all $c_{i,j}$ elements by -1. Following the VAM/MODI process, we then maximize the objective. However, most students find this approach (involving working with negative $c_{i,j}$ elements) to be subject to considerable human error when solving a problem by hand. A way to avoid this, a method preferred by this author, is to:

1. Find the largest cost element in the matrix.
2. Subtract all cost elements from the largest cost.
3. Proceed as usual with VAM/MODI.

Regardless of the approach used, we must remember to reconvert the answer obtained to reflect the original problem.

FORBIDDEN ALLOCATIONS

If allocation from a given source to a particular sink is not permitted, we usually just assign a very high cost (if minimizing) to the cell under consideration. Normally, this will give us a final allocation which does not include that cell. However, as noted in the previous production scheduling example, an optimal allocation will, on occasion, exist that includes such a cell and one must interpret the *physical* significance of such a result.

PERTURBATION METHODS

Degeneracy is a troublesome outcome in the transportation problem because of its impact on the computational procedure (i.e., the inclusion of the ϵ allocations). There are ways to avoid this difficulty that do not require the utilization of the ϵ allocators. One fairly common approach is to perturb the a_i values (the availability of each source) slightly. That is, we add some very small amount (say δ) to each row. To balance the rims, we must then add $m\delta$ (i.e., the number of rows times δ) to one of the columns. We then solve the perturbed problem and, at the conclusion, reestablish the solution to the original problem by letting δ go to zero.

ALTERNATIVE OPTIMAL SOLUTIONS

Since the $\Delta_{i,j}$ indicators (i.e., the MODI indicators) are the simplex indicators of marginal contribution, we detect alternative optimal solutions in the same manner as in the tabular simplex. That is, if $\Delta_{i,j} = 0$ for any empty cell (i.e., nonbasic variable), then an alternative optimal solution exists which can be formed by reallocating to that cell.

The Transshipment Model

The transportation model has certain inherent restrictions that may or may not exist in real-world situations. If they do not exist, we can often develop a significantly improved solution (i.e., a substantial reduction in total cost). To illustrate, consider the simple problem depicted in Figure 13-3(a). In this

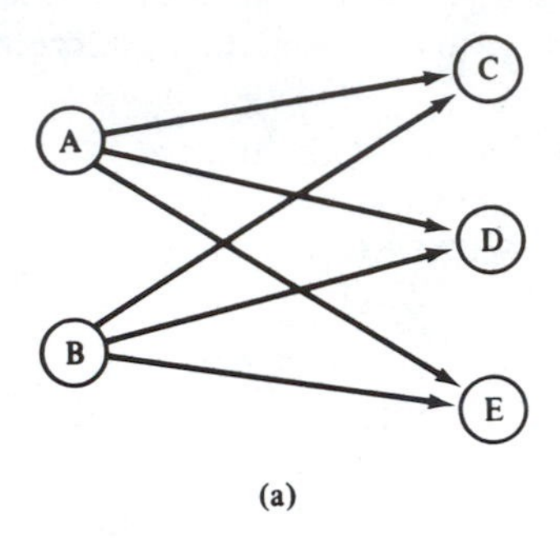

(a)

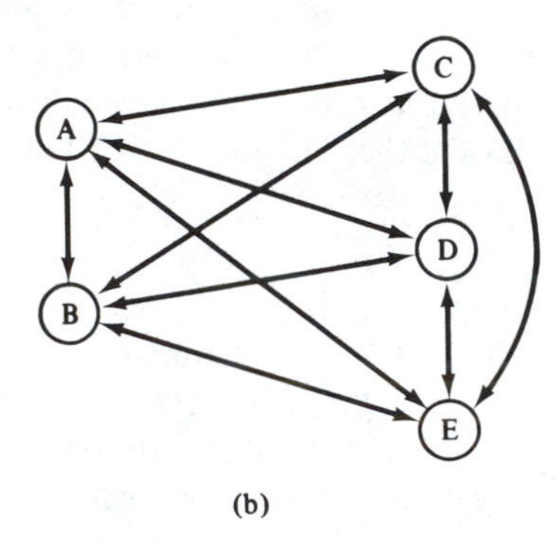

(b)

Figure 13-3. (a) Transportation (bipartite) graph; (b) transshipment graph.

figure, a bipartite graph, all flow proceeds from source nodes to sink nodes. However, it might well be possible to establish a shipping route that includes journeys which include legs that represent either source-to-source, sink-to-sink, or even sink-to-source movements. Consider Figure 13-3(b) in which the graph has now been augmented with additional routing possibilities. It is now possible to, for example, ship goods from source A to sink C via source B (i.e., *to transship*).

As mentioned, the inclusion of additional branches in the network (i.e., transportation legs) can mean that a significant reduction in cost might be achieved. However, the size of the problem is enlarged considerably.

The Transshipment Matrix

The transshipment problem can be placed into a relatively convenient matrix model by extending the notions used in the transportation problem. Table 13-26 depicts the *general* form of such a matrix. As can be seen, the transshipment matrix is made up of four submatrices: (1) source-to-source, (2) source-

Table 13-26. THE GENERAL TRANSSHIPMENT MATRIX

From \ To	Source $1 \ldots i \ldots m$	Sink $1 \ldots j \ldots n$	*Availability*
Source $1 \ldots i \ldots m$	(1) Source-to-source submatrix	(2) Source-to-sink submatrix	$a_1 + U \ldots a_i + U \ldots a_m + U$
Sink $1 \ldots j \ldots n$	(3) Sink-to-source submatrix	(4) Sink-to-sink submatrix	$U \ldots U \ldots U$
Demand	$U \ldots U \ldots U$	$b_1 + U \ldots b_j + U \ldots b_n + U$	

to-sink, (3) sink-to-source, and (4) sink-to-sink. Included within each submatrix are the cells that designate the allocation decision and the transport costs. Notice that submatrix 2 is simply the transportation matrix equivalent (i.e., considering only the source-to-sink flows), which is embedded in the overall problem. The transport cost data for each cell is obtained as usual except that now many more routes will exist. The amount available at each source or sink, and demanded at each source or sink, is not quite so obvious. Notice that all rim availabilities and demands include a constant, U. This constant represents an upper bound on the amount of goods that may flow through any (source or sink) node (i.e., the maximum amount that could be transshipped). To determine U, we first assume a balanced problem. Next, it should be obvious that the maximum number of units that may be transshipped through any given node cannot exceed $\sum_{i=1}^{m} a_i = \sum_{j=1}^{n} b_j$ (i.e., the total amount available or demanded).[d] Thus,

$$U = \sum_{i=1}^{m} a_i = \sum_{j=1}^{n} b_j \tag{13.7}$$

Before proceeding to an example, notice that the cost of shipping from a given node directly to that node is zero. Next, we consider a simple example.

---

[d]If more were transshipped, at least one of these units would have to pass through the same node more than once.

Example 13-2

Consider the transshipment problem depicted in Figure 13-4(a). The amount available at each source is listed, in square brackets, to the left of the source node, and the amount demanded at each sink is given in the brackets to the right of the sink node. The shipping cost between nodes is given in parentheses beside each node-to-node link (for convenience, we have assumed that the costs are the same on a link regardless of the direction of movement, but this is not always so in practice). The embedded transportation matrix, for this problem, is given in Figure 13-4(b).

We shall first, for purpose of comparison, solve for the solution to the embedded transportation problem. The result is given in Table 13-27. The total cost for the optimal solution to the transportation problem is 261 units.

The transshipment matrix for the entire problem is shown in Table 13-28. Since $\sum a_i = \sum b_j = 27$, we have let U be equal to 27 units.

Our next step is to obtain an initial feasible solution to the matrix model (i.e., satisfy the rim conditions). This can be achieved by the use of VAM, just as was done

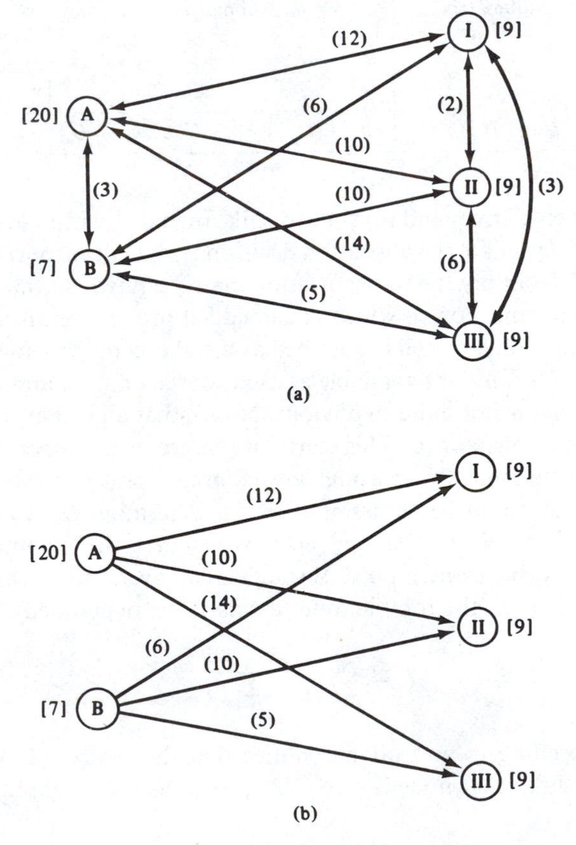

Figure 13-4. (a) Transshipment network; (b) embedded transportation network.

Table 13-27. SOLUTION TO EMBEDDED TRANSPORTATION PROBLEM

From \ To	I	II	III	Available
A	[12] (9)	[10] (9)	[14] (2)	20
B	[6]	[10]	[5] (7)	7
Demand	9	9	9	

Table 13-28. TRANSSHIPMENT MATRIX

	From \ To	Source A	Source B	Sink I	Sink II	Sink III	Available
Source	A	[0]	[3]	[12]	[10]	[14]	20 + 27
Source	B	[3]	[0]	[6]	[10]	[5]	7 + 27
Sink	I	[12]	[6]	[0]	[2]	[3]	27
Sink	II	[10]	[10]	[2]	[0]	[6]	27
Sink	III	[14]	[5]	[3]	[6]	[0]	27
	Demand	27	27	9 + 27	9 + 27	9 + 27	

with the regular transportation matrix. However, since we already have a solution to the embedded transportation problem, we can use this solution (shown in Table 13-27) to obtain an initial feasible, and generally good, solution to the transshipment matrix.

To achieve this initial solution, we first fill in the source-to-sink portion of the submatrix (which is the embedded transportation matrix) with the solution from Table 13-27. Next, we allocate, to each main diagonal cell, the amount U, which, for this problem, is 27 units. The resultant initial allocation is feasible and is shown in Table 13-29.

Our next step is to simply employ phase 2 (i.e., the MODI method). The final, optimal solution is given in Table 13-30. The cost of this solution is just 222 units, compared to 261 units to the embedded transportation model. The savings obtained by considering the transshipment routings is about 15%.

We should now examine this last table to make sure that it is interpreted correctly. First, note that allocations to the main diagonal cells are physically meaningless and may be ignored. Considering the other allocations we begin with node A.

Node A ships out 11 units to node B and 9 to node II. Node B ships 9 units to node I and 9 to node III. Notice that 11 of the units shipped by node B came via node A.

Table 13-29. INITIAL SOLUTION

From \ To	*A*	*B*	*I*	*II*	*III*	*Available*
A	[0] 27	[3]	[12] 9	[10] 9	[14] 2	47
B	[3]	[0] 27	[6]	[10]	[5] 7	34
I	[12]	[6]	[0] 27	[2]	[3]	27
II	[10]	[10]	[2]	[0] 27	[6]	27
III	[14]	[5]	[3]	[6]	[0] 27	27
Demand	27	27	36	36	36	

Table 13-30. FINAL TRANSSHIPMENT SOLUTION

From \ To	*A*	*B*	*I*	*II*	*III*	*Available*
A	[0] 27	[3] 11	[12]	[10] 9	[14]	47
B	[3]	[0] 16	[6] 9	[10]	[5] 9	34
I	[12]	[6]	[0] 27	[2]	[3]	27
II	[10]	[10]	[2]	[0] 27	[6]	27
III	[14]	[5]	[3]	[6]	[0] 27	27
Demand	27	27	36	36	36	

Figure 13-5 denotes the network representation of this solution. The numbers in parentheses at each (utilized) branch are the number of units shipped via that leg of the transshipment network.

This small example had very little transshipment (only the 11 units from source *A* to source *B*) and thus is not typical of the results that will often be obtained in larger problems. Even still, the reduction achieved here, or 15% in total costs, is a substantial savings.

As an aside, the author was engaged a few years ago in a consulting effort

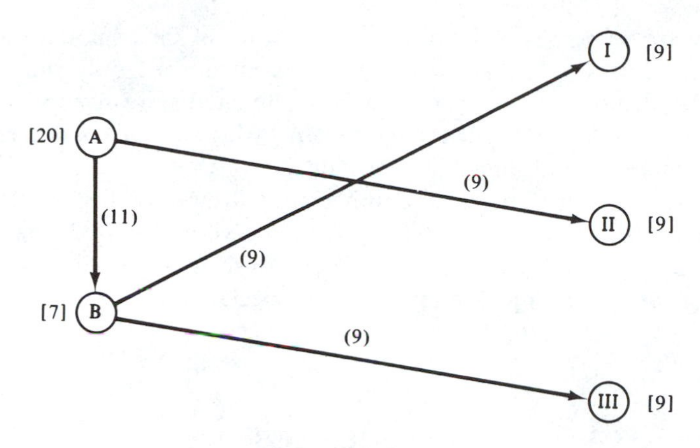

Figure 13-5.

with a moderately large firm in the Southeast. The effort involved the task of determining the location and sizing of several new warehouses. As part of the investigation, we were permitted to examine the firm's computerized dispatching routine. It was a computerized VAM/MODI algorithm for the standard *transportation* model. The firm was genuinely impressed with its performance and we were told that although we could use the routine, we were not to change it. After considerable discussion, the firm finally permitted us to implement a "temporary" modification, which was to remodel the problem as a transshipment matrix. The savings (when compared to several months of past data) varied between 8 and 23%, which translated into several hundreds of thousands of dollars per year.

The main drawback of the transshipment model (assuming that such routings are feasible) is in the size of the final problem to be solved. For example, a problem with 50 warehouses and 300 customers would translate into a transportation problem with 50×300 or 15,000 cells; the transshipment version would have 350×350 or 122,500 cells.

Summary

In this chapter we introduced the special linear integer programming problem known as the transportation problem and demonstrated a two-phase approach to its solution known as the VAM/MODI technique. The technique is directly analogous to conventional simplex, but because of the peculiar structure of the transportation model, it is applied directly to a matrix model of the problem rather than resorting to the simplex tableau. Although the transportation problem's name would indicate a limited applicability, we have seen that other types of problems involve the same mathematical structure and thus are solvable via VAM/MODI.

There are numerous variations and extensions of the common transportation model. However, in this chapter we have introduced only the transshipment model, demonstrating how it may be formulated and solved very much in the same way as the transportation problem, which is, of course, embedded within the more general transshipment model.

In Chapter 14 we introduce another special class of linear integer programming problem, which, in actuality, is only a very restricted version of the transportation problem presented here. We shall see that yet another approach to solution can be developed to apply to this special model.

REFERENCES

1. IGNIZIO, J. P., AND GUPTA, J. N. D. *Operations Research in Decision Making*. New York: Crane, Russak, 1975.
2. LLEWELLYN, R. W. *Linear Programming*. New York: Holt, Rinehart and Winston, 1964.
3. METZGER, R. W. *Elementary Mathematical Programming*. New York: Wiley, 1958.
4. TRUSTUM, K. *Linear Programming*. London: Routledge & Kegan Paul, 1971.

PROBLEMS

Solve Problems 13.1 through 13.3 under the assumption that the interior cell elements represent source-to-sink costs.

13.1.

Source \ *Sink*	*A*	*B*	*C*	*Available*
1	16	8	17	30
2	21	24	37	28
Required	20	10	40	

13.2.

Source \ *Sink*	*A*	*B*	*Available*
1	5	6	5
2	20	2	5
3	8	3	5
Required	10	5	

13.3.

Source \ Sink	A	B	C	Available
1	18	2	5	1
2	9	3	8	1
3	12	6	20	1
Required	1	1	1	

13.4. Repeat Problem 13.1 if the cell values actually represent *profit*.

13.5. Solve the problem shown below wherein (a) the cell entries are costs, and (b) the cell entries are profits.

Source \ Sink	A	B	C	Available
1	14	10	17	42
2	20	17	13	58
Required	20	50	35	

13.6.

Source \ Sink	A	B	C	Available
1	17	25	18	50
2	27	14	30	60
3	21	29	20	90
Required	68	80	52	

Solve if cell entries are costs.

13.7. Given the transportation matrix and (minimum-cost) solution as indicated below, determine via MODI if the distribution scheme is optimal. If not, determine the optimal routing.

Source \ Sink	E	F	G	Available
A	5 (10)	8	3	10
B	7 (1)	4 (3)	5	4
C	2	6 (4)	9	4
D	4	6 (7)	6 (5)	12
Required	11	14	5	

13.8. Solve for the optimal (minimum-cost) distribution scheme, and discuss the physical meaning of the results.

Source \ Sink	I	II	III	Available
A	6	5	7	9
B	5	8	8	3
Required	6	6	5	

13.9. Given that Problem 13.8 is one in which a maximal profit scheme (i.e., the elements represent profits) is desired, determine the solution.

13.10. A firm has three plants that all produce the same product. Their monthly capacities and per unit production costs are given in the table. Four warehouses

Plant	*Regular Capacity*	*Cost per Unit*	*Overtime Capacity*	*Cost per Unit*
I	140	$3.00	60	$5.00
II	110	5.00	40	7.50
III	200	4.25	80	7.00

are supplied from the three plants wherein their monthly demands are: warehouse *A*, 150 units per month; warehouse *B*, 210 units per month; warehouse *C*, 100 units per month; warehouse *D*, 90 units per month. Shipping costs from plant to warehouse are given in the table. What is the optimal production and routing scheme?

From \ *To*	*Shipping Costs per Unit*			
	A	*B*	*C*	*D*
I	$2	$1	$2.5	$4
II	4	3	5	2
III	1	2	4	3

13.11. Red's Rent-Alls specializes in the renting or short-term leasing of heavy equipment. A particularly popular item is the short-term rental of medium-class bulldozers. These bulldozers must be loaded, transported, and unloaded from specially equipped trucks and trailers and, with the cost of fuel and labor, pickup and delivery costs can be substantial. Red O'Hare, the owner of Red's Rent-Alls, has three locations scattered across the central part of Pennsylvania. To be competitive with other renting agencies, Red is reluctant to impose a delivery fee based on distances between his locations and the customers, and thus it is vital that the bulldozer requests be allocated to minimize the overall transportation costs (mileage and labor). Bulldozer availability and requests for the coming month are summarized in the following table:

Rental Location	*Bulldozers on Hand*	*Customer Location*	*Bulldozers Needed*
1	8	*A*	3
2	12	*B*	5
3	20	*C*	12
		D	7

The total estimated delivery costs (dollars) between each rental location and customer is as follows:

	Customer Location			
Rental Location	*A*	*B*	*C*	*D*
1	620	310	200	100
2	230	170	700	630
3	175	300	680	540

Determine the lowest-cost delivery plan.

13.12. "Seldom Seen Slim" operates a small furniture manufacturing enterprise of dubious reputation. Slim's specialty is the production of "antiques," and he is, in fact, one of the reasons why there are now more Colonial Windsor chairs (circa 1725–1825) than there were colonists. Slim has estimated the demand for his "antique" chairs per quarter, as well as the maximum production quantities. He also realizes that he must pay his workers a premium ($2 more per hour) in the spring and summer (quarters 2 and 3) compared to his wage rate in the fall or winter (quarters 1 and 4), because there are additional opportunities for their employment during the warmer weather (in agricultural pursuits). Further, each chair that cannot be sold in the quarter in which it is produced will cost him $7 per quarter for storage. Given that each chair takes 10 hours to produce, at a nonpremium rate of $4 per hour, and the data in the table, deter-

Quarter	*Production Capacity (Chairs)*	*Demand (Chairs)*	*Selling Price per Chair*
1	100	40	$100
2	80	110	120
3	60	120	150
4	120	50	80

mine Slim's optimal production and inventory policy so as to maximize profits.

13.13. The caterer's problem is a well-known example of the transportation problem. The caterer must provide napkins for the next 7 days. Napkins can be either purchased at 30 cents each, or laundered (fast service) at 12 cents each, or laundered (slow service) at 7 cents each. Fast laundry service returns the napkins in 1 day; slow service requires 3 days. The number of napkins required over the next 7 days is 100, 120, 80, 200, 70, 150, and 60. To formulate the problem, consider the source to be each day's supply of dirty napkins and a single supplier of new napkins. The sinks are the daily requirements of napkins. Formulate the problem.

13.14. The networks shown indicate transshipment via all nodes. The numbers in parentheses beside each branch are node-to-node costs; the numbers at each node indicate a demand, if negative, or an availability, if positive. Formulate the problems.

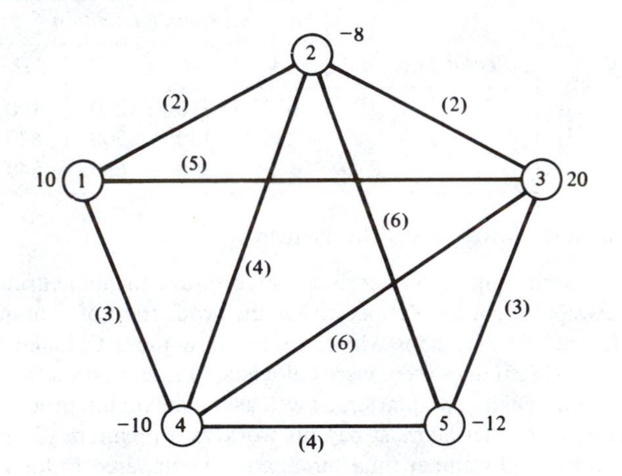

Figure P13-14(a)

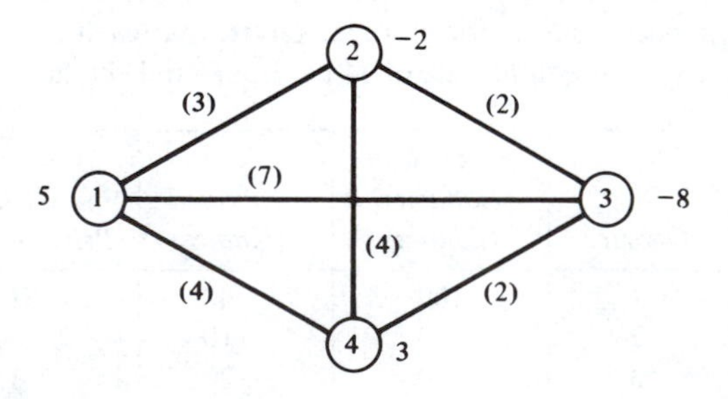

Figure P13-14(b)

13.15. Solve Problem 13.14 by:
(a) VAM only
(b) VAM and MODI
Compare the differences in the results.

THE ASSIGNMENT PROBLEM

Introduction

As mentioned in Chapter 13, the assignment problem is, in structure, simply a special subclassification of the transportation problem. As such, it could be (and sometimes is) treated by the approaches advocated in Chapter 13. However, just as we saw how the transportation problem may be more efficiently handled by a specially tailored methodology, we shall also find that similar advantages may be found in the application of special techniques designed to exploit the particular characteristics of the assignment problem.

Importance and Applications

As we have seen, the transportation problem is a very special subclass of the general, single-objective, linear programming problem. Further, the assignment problem is then a special subclass of the transportation problem. At this point, the reader may wonder if such a very special problem, so far down in the model hierarchy, has but a very limited applicability.

The fact is that there are occasions when the typical practitioner will find a problem that takes on the special structure of the assignment model. However, the number of such encounters will probably be small if we restrict our discussion to the assignment problem under the most rigorous definition. More frequently, we encounter problems in which the assignment problem is embedded (somewhat as the transportation problem is embedded within the transshipment

14

model) within the problem of interest. On other occasions we encounter problems that are not assignment problems but for which we may apply many of the concepts that form the methodology of approach for the assignment model. Consequently, the material to be discussed herein is of use beyond the very specific model under consideration.

The assignment problem gets its name from a particular application in which we wish to assign "individuals" to "tasks (or tasks to machines, etc.). It is assumed that each individual can be assigned to *only one task*, and each task is assigned to *only one individual*. Various personnel assignment problems *may* fit the assumptions and structure of such a model. However, in many cases the strict assumptions listed may not apply. For example, it may well be possible for *more* than one individual to be assigned to a task. Although not satisfying the requirements of the assignment model, we see that extensions to the assignment problem may encompass such variations.

The conventional, single objective of the assignment problem is to minimize the total cost of the assignments made, where this cost may be in terms of dollars required to perform the tasks or, perhaps, the total time[a] required to accomplish all tasks. Consequently, as was the case with the transportation problem, the algorithms developed are normally of the minimizing type.

Consider the problem facing the coach of a swimming team. One of the

---

[a]That is, the *sum* of the times for each task.

events in swimming is the relay race, consisting of four legs: breaststroke, butterfly, backstroke, and freestyle. Each leg is to be assigned to exactly one of the swimmers and the objective, of course, is to win the race by minimizing the total elapsed time. Table 14-1 lists the average time per leg for each of the four relay-team members. A fairly common human tendency is to attempt to assign each individual (swimmer in this case) to the task (event) in which he or she excels. Such an approach will not, in general, lead to either an optimal assignment or even, in many cases, to one that is nearly optimal. The optimal assignment, for this example, is to assign

swimmer *A* to freestyle

swimmer *B* to breaststroke

swimmer *C* to butterfly

swimmer *D* to backstroke

for a total time of 124 seconds.

Table 14-1. SWIMMER/EVENT DATA

	Average Time (Seconds)			
Swimmer	*Butterfly*	*Breaststroke*	*Backstroke*	*Freestyle*
A	31	32	34	29
B	36	30	37	33
C	34	34	37	35
D	30	33	31	28

Problem Complexity

The assignment problem is one of those problems in combinatorics for which intuition generally fails. The initial reaction of many students, when first faced with the assignment problem, is to underestimate severely its inherent difficulty. Consider, for example, the preceding example, with 4 swimmers and 4 events. The total number of ways that n individuals may be assigned to n tasks is n factorial ($n!$). Thus, there are $(4)(3)(2) = 24$ different ways to assign the swimmers. This, then, is a problem that could be solved by brute-force enumeration. However, when $n = 10$, the number of possible assignments is about 3.6 million. In many real-world problems, n has been known to take on values of several hundred or thousand, and thus enumeration is certainly not the way to approach an actual problem.

Coverage

Initially, we discuss the conventional assignment problem, wherein:

1. Each "individual" is assigned to exactly one "task."
2. Each "task" is assigned to exactly one "individual."

Several methods for solving this problem are presented and illustrated by examples.

We then consider two variations on the conventional assignment problem, for which the first is an extension (a relaxation of the conditions listed above wherein a task can be assigned to more than one individual) and the second is a special subclass known as the matching problem.

The Conventional Assignment Problem

We may represent the swim-team assignment problem via, at least, three different models: the network, the matrix, and the mathematical model. Figure 14-1 depicts the network model, which, as may be seen, is a bipartite graph with

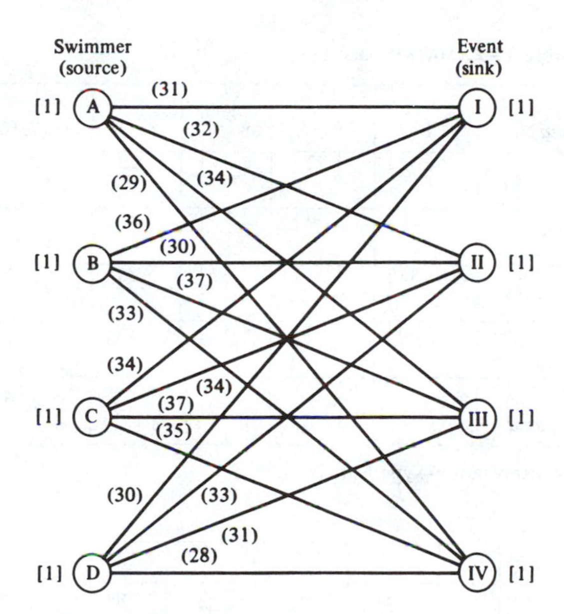

Figure 14-1. Network model.

the "source" nodes on the left to be assigned to the "sink" nodes on the right. Such a model is identical to the transportation network model *except* that:

1. The availability of each source node is exactly 1.
2. The demand at each sink node is exactly 1.
3. The number of source nodes always equals (or is forced to equal) the number of sink nodes.

Note that the expected time to complete each event for a given swimmer is given in parentheses beside each branch (node-to-node link).

The network model may be easily converted to the matrix model of Table 14-2(a) or 14-2(b). Table 14-2(a) is the familiar transportation matrix, which can be used since the assignment problem is but a special class of transportation problem. However, much of the information shown in the transportation type of matrix is unnecessary. Consider, for example, the rim conditions. A 1 will always appear for each availability and demand, and thus the rims may be dropped. Another simplification is concerned with each internal cell. An allocation to a cell will always be either a zero (no assignment) or a 1 (an assignment); consequently, we simply list the cost of each cell as shown in Table 14-2(b). The latter matrix is the simplified or condensed matrix that will normally be used. Note also that, in Table 14-2(a), the optimal assignment has been indicated by simply placing a box about the cost elements of the cells to which an assignment has been made. This, too, will be our normal practice.

Table 14-2. MATRIX MODELS

From \ *To*	1	2	3	4	*Available*
A	31	32	34	29	1
B	36	30	37	33	1
C	34	34	37	35	1
D	30	33	31	28	1
Demand	1	1	1	1	4 \ 4

(a) Transportation-Type Matrix

	1	2	3	4
A	31	32	34	[29]
B	36	[30]	37	33
C	[34]	34	37	35
D	30	33	[31]	28

(b) Condensed (Assignment) Matrix

The final model to be presented is the mathematical model, which is derived in the same way as was the mathematical model of the transportation problem in Chapter 13. This model is, for our specific example:

Find $x_{i,j}$ ($i = 1, 2, 3, 4$ and $j = 1, 2, 3, 4$) to

$$\text{minimize} \quad z = \sum_{i=1}^{m} \sum_{j=1}^{n} c_{i,j} x_{i,j} \tag{14.1}$$

subject to

$$\left.\begin{aligned} \sum_{j=1}^{n} x_{i,j} &= 1 \qquad \text{all } i \\ \sum_{i=1}^{m} x_{i,j} &= 1 \qquad \text{all } j \end{aligned}\right\} \tag{14.2}$$

$$x_{i,j} = 0, 1 \qquad \text{all } i \text{ and } j \tag{14.3}$$

where m = number of source nodes (= 4 in example)

n = number of sink nodes (= 4 in example)

$c_{i,j}$ = "cost" (time in the example) of assigning source node i to sink node j

Notice in particular the restrictions implied in (14.3). In the assignment problem

$$x_{i,j} = \begin{cases} 1 & \text{if source node } i \text{ is assigned to sink node } j \\ 0 & \text{otherwise} \end{cases}$$

Thus, the decision variables within the assignment problem are all zero–one-valued and the problem itself is a linear, zero–one programming problem. This contrasts with the more general linear, integer programming model associated with the transportation problem.

Recall in Chapter 13 that the transportation model possessed the property of unimodularity. This is also true of the assignment model, which may be seen if the structure of the constraint matrix is examined. This means, then, that if we start with any integer feasible solution, the simplex method will converge to the integer-valued optimal solution (wherein the integer values themselves will be restricted to either zero or 1). However, just as we did with the transportation problem, we consider more efficient, special solution methods.

Selected Solution Techniques

There are numerous ways to solve the conventional assignment problem, ranging from the general simplex algorithm to very specific techniques. We restrict our attention to just a few of these approaches: (1) VAM/MODI, (2) the Hungarian algorithm, and (3) branch-and-bound.

VAM/MODI approach. Since the assignment problem is but a special subclass of the transportation problem, we can form the transportation matrix model and solve via the VAM/MODI algorithm presented in Chapter 13.

Unfortunately, the assignment problem will always have a degenerate solution, and thus this approach turns out to not be particularly efficient. Consider, for example, the assignment problem represented in Table 14-3.

Table 14-3. ASSIGNMENT PROBLEM IN TRANSPORTATION MATRIX FORMAT

From \ To	I	II	III	IV	a_i
A	[2]	[10]	[3] 1	[17]	1
B	[5]	[3]	[9]	[10] 1	1
C	[8]	[2] 1	[5]	[14]	1
D	[3] 1	[5]	[10]	[16]	1
b_j	1	1	1	1	4 \ 4

The solution shown in Table 14-3 is found via VAM, based on the assumption that the $c_{i,j}$'s are costs (i.e., that we are minimizing). Note two things:

1. There must always be a *single* allocation to any row or column (this is known as finding an independent set of assignments).
2. The solution is always degenerate.

Given an assignment problem with $m = n$ (i.e., equal number of rows and columns), a nondegenerate solution must have (see Chapter 13) exactly $m + n - 1$ assignments. However, since only a single assignment can appear in any row or column, the assignment problem *always* will have exactly m $(= n)$ assignments. This means that the number of ϵ allocations necessary for phase 2 (MODI) is: $(m + n - 1) - m = n - 1 = m - 1$. For example, an assignment problem with 100 rows (and columns) will require *99 ϵ allocations*. Consequently, we find VAM/MODI to not be an attractive approach to the assignment problem, particularly for problems of modest to large size.

Hungarian algorithm. The so-called Hungarian algorithm [3] provides us with a relatively effective and rather novel approach to the assignment problem. It is particularly appealing when solving by hand problems of small to modest size because it involves the use of visual pattern recognition, an ability in which the human being can often outperform the computer.

Consider an assignment problem in which we are minimizing and *all cell elements are nonnegative*. The lowest possible cost in any cell is thus zero. Further, the lowest *possible* total cost is also zero, which could occur only if it were possible to make a selection of independent assignments to cells for which the individual costs were zero. Now, it is extremely unlikely that we would ever encounter such a problem in actual practice. However, the basic concept implied above is the foundation of the Hungarian algorithm.

Any assignment problem may be transformed into an equivalent problem in which the smallest cost value in any row or column is zero. This is done by means of the following procedure:

Step 1. Select the smallest element in each row (of the assignment matrix) and subtract that element from each element in that row. This will generate at least one zero element in every row of the matrix.

Step 2. Select the smallest element in each column and subtract that element from each element in that column. This step will generate at least one zero element in every column of the matrix.

The solution to the converted matrix must correspond to that of the original. Applying these two steps to the assignment matrix for the problem given in Table 14-3 [or Table 14-4(a)] will result in Table 14-4(b). Notice that the assignment given through VAM for Table 14-3 is also depicted in Table 14-4(b), wherein each assigned cell is indicated by a box about the (converted) cost. Since this assignment is to cells that each have a cost of zero, no better assignment is possible. Consequently, our solution previously arrived at by VAM was actually optimal, with a total cost of $3 + 10 + 2 + 3 = 18$ units.

Normally, it will not be nearly so easy to solve the assignment problem and we must establish a systematic process (the Hungarian algorithm in this particular

Table 14-4. CONVERTED ASSIGNMENT MATRIX

2	10	3	17
5	3	9	10
8	2	5	14
3	5	10	16

(a) Original Matrix

0	8	[0]	8
2	0	5	[0]
6	[0]	2	5
[0]	2	6	6

(b) Converted Matrix

case) to identify the final, optimal solution. Given a matrix that has been converted as above, the steps of the Hungarian algorithm begin by first determining if there is a feasible assignment that may be made by using only cells with zero elements. If so, the problem is solved. If not, we incorporate a process that will, in essence, generate additional zeros in the matrix. The steps of the Hungarian algorithm are given below and are demonstrated on the example problem shown in Table 14-5 (wherein the previous conversion process has already taken place).

Step 1. Place a "box" around a zero in any row that has only one zero in it and cross off all other zeros in the column containing the box. If all rows contain a box, you are finished. Otherwise, go to step 2. (See Table 14-6.)

Step 2. Place a box around a zero in any column that has only one zero in it and cross out all other zeros in the row containing the box. If every column has one box in it, you are finished. Otherwise, go to step 3. (See Table 14-7.)

Step 3. If steps 1 and 2 have not led to a feasible solution, it is necessary to attempt to generate additional zeros in the matrix by drawing the *minimum* number of lines (either horizontal or vertical, or some mixture) that will cover (pass through) *all* the zeros in the matrix. A procedure to aid in drawing such lines follows:

(a) Check the rows having no boxed zeros (row *C* in the example).
(b) Check the columns having a zero in a checked row. (Check column I in the example.)
(c) Check the rows having a boxed zero in a checked column. (Check the second row.)
(d) Repeat steps (b) and (c) until no further rows or columns can be checked.
(e) Draw lines through all unchecked rows and all checked columns. (See Table 14-8.)

Step 4. All elements with lines through them are termed "covered." Elements without lines passing through them are "uncovered." Select the *smallest uncovered element* in the matrix (the number 1 in row *B*, column II) and subtract this element from *all uncovered elements*. Add this element to all *covered cells that occur at the intersection of two covering lines*. (The resultant matrix is shown in Table 14-9.)

Step 5. Repeat steps 1 through 4 until an optimal assignment is found. (See Table 14-10.)

Table 14-5. EXAMPLE FOR HUNGARIAN ALGORITHM

From \ *To*	*I*	*II*	*III*
A	2	0	1
B	0	1	4
C	0	5	5

Table 14-6

	I	*II*	*III*
A	2	0	0
B	[0]	1	3
C	~~0~~	5	4

Table 14-7

	I	*II*	*III*
A	2	[0]	~~0~~
B	[0]	1	3
C	~~0~~	5	4

Table 14-8

	I	*II*	*III*	
A	2	[0]	0	
B	[0]	1	3	✓
C	0	5	4	✓
	✓			

Table 14-9

	I	*II*	*III*
A	3	0	0
B	0	0	2
C	0	4	3

Table 14-10

	I	*II*	*III*
A	3	~~0~~	[0]
B	~~0~~	[0]	2
C	[0]	4	3

The final solution to our example is shown in Table 14-10, wherein we assign:

A to III

B to II

C to I

for a total cost of 2 units (which must be obtained from the *original* table, Table 14-5).

Branch-and-bound algorithm. The branch-and-bound method is a quasi-enumerative approach to problem solving that is fairly efficient for modest-size problems. The general methodology forms an important part of the set of (exact) solution algorithms for the general class of linear integer programming problems, and thus the discussion here, via the assignment problem, serves somewhat as an introduction to generalizations of this approach [1, 5].

Our discussion will be based on the assumption of objective function minimization. Extension to the maximization case is transparent.

We shall designate the optimal (i.e., minimal total cost) solution to the assignment problem as z^*, where this optimal solution lies somewhere between the two values z_U and z_L. That is,

$$z_L \leq z^* \leq z_U$$

where z_U = upper bound on z^*; this is normally the value of the best feasible solution thus far obtained (in the event no feasible solution is known, an obvious upper bound is infinity).

z_L = *lower bound* on z^*; that is, z^* cannot be less than z_L.

Termination of the branch-and-bound process occurs whenever it can be shown that there is no lower bound less than z_U.

The branching portion of the branch-and-bound procedure consists of partitioning the space of all feasible solutions into smaller, mutually exclusive subsets. Consider, for example, the solution space to an assignment problem consisting of three individuals (*A*, *B*, and *C*) and three tasks (1, 2, and 3). The total solution space consists of the following six assignments:

A–1, *B*–2, *C*–3

A–1, *C*–2, *B*–3

B–1, *A*–2, *C*–3

B–1, *C*–2, *A*–3

C–1, *A*–2, *B*–3

C–1, *B*–2, *A*–3

One possible partition of this space, based on the assignments to job 1, consists of the three partitions in Table 14-11. Partition 1 consists of all solutions for which *A* has been assigned to job 1. Partition 2 includes all solutions for which *B* has been assigned to job 1. Partition 3 is the set of solutions associated with the assignment of *C* to job 1.

Table 14-11

Partition 1 (A–1)	*Partition 2 (B–1)*	*Partition 3 (C–1)*
A–1, *B*–2, *C*–3 *A*–1, *C*–2, *B*–3	*B*–1, *A*–2, *C*–3 *B*–1, *C*–2, *A*–3	*C*–1, *A*–2, *B*–3 *C*–1, *B*–2, *A*–3

Each partition is represented by a "node." For each node, the lower bound of the subset associated with that node is computed. If any node has a lower bound that is no better than z_U (the present value of the best feasible solution), no further partitioning of that node is necessary. That is, no better solution can be found via branching from that node. Further, if the lower bound of any node is associated with a *feasible* solution for the subsets associated with that node, that node also is excluded from further partitioning. When these two rules are used to exclude nodes, we say that such nodes have been "terminated" (or, alternatively, "fathomed").

We repeat the partitioning process by branching from any unterminated node (which has not already been partitioned). Normally, we select the node with the smallest lower-bound value. Bounds are again computed and, if possible, nodes are terminated.

The procedure is repeated until all nodes have been terminated or until the value of the best lower bound is no better than that of the best feasible solution.

Consider the problem depicted in Table 14-12. The branch-and-bound process begins by finding a good lower bound (where the closer to the true value of z^*, the better or "tighter" the lower bound is said to be) to the initial, unpartitioned set of solutions. This lower bound can be obtained by simply summing the minimum-cost element in each row (regardless of whether or not the associated assignment is feasible). This gives us a lower bound of $3 + 2 + 10 + 5 = 20$ for our initial node (designated as node 0). This lower bound does not correspond to a feasible solution and thus we branch from node 0.

Table 14-12

	I	*II*	*III*	*IV*
A	15	3	30	20
B	20	2	20	30
C	40	20	10	40
D	5	30	40	40

The branching from node 0 forms the four partitions associated with the four possible decisions associated with task I: *A*–I, *B*–I, *C*–I, and *D*–I. These are indicated in Figure 14-2 as nodes 1, 2, 3 and 4, respectively.

We now determine the lower bounds for the four new nodes (these are

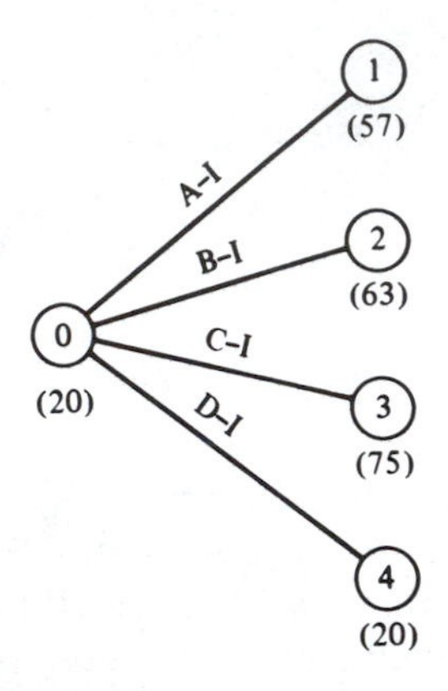

Figure 14-2. Branching on Node 0.

already shown in Figure 14-2 as the number in parentheses beneath each node). These lower bounds are found in the same manner as before *except* that we now have a fixed cost associated with the partial assignment as specified on the branches leading to each node. For example, at node 1 (*A*–I assigned), the lower bound may be found by striking out row *A* and column I and then summing the minimum row cost elements in the remaining submatrix. This gives us $2 + 10 + 30 = 42$. To this, we must add the cost of the specified assignment of *A* to I (15 units) for a total of $15 + 42 = 57$ units. The lower bound on node 1 is thus 57, and it does *not* correspond to a feasible assignment. The lower-bound calculations for all four new nodes is shown in Table 14-13.

Table 14-13. LOWER BOUNDS FOR NODES 1, 2, 3, AND 4

Node	*Lower Bound* (*Specified Assignment Cost*) *plus* (*Sum of Row Minimum Costs*)
1	$(15) + (2 + 10 + 30) = 57$
2	$(20) + (3 + 10 + 30) = 63$
3	$(40) + (3 + 2 + 30) = 75$
4	$(5) + (3 + 2 + 10) = 20$

Our best lower bound is thus 20 units at node 4. We do not yet have a feasible solution. We branch from node 4 according to task II assignments: *A*–II, *B*–II, and *C*–II and the results are shown in Figure 14-3, while the lower-bound computations of the three new nodes (nodes 5, 6, and 7) are given in Table 14-14.

We have now reached two feasible solutions (at nodes 6 and 7). The best of these is at node 6 (37 units), and we may use this best feasible solution to terminate any other (nonpartitioned) nodes having a lower bound no better than

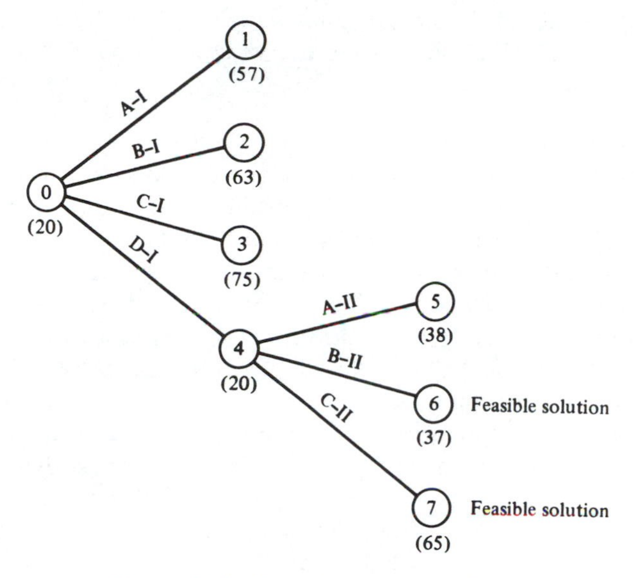

Figure 14-3. Branching from Node 4.

Table 14-14. LOWER BOUNDS FOR NODES 5, 6, AND 7

Node	*Lower Bound* (*Specified Assignment Cost*) *plus* (*Sum of Row Minimum Costs*)
5	$(5 + 3) + (20 + 10) = 38$
6	$(5 + 2) + (20 + 10) = 37$ feasible
7	$(5 + 20) + (20 + 20) = 65$ feasible

37. This results in a termination of *all* other nodes and we are finished. The optimal assignment is thus

D–I

B–II

A–IV

C–III

REFINEMENTS

There are a number of refinements possible with the conventional assignment problem that are of interest. First, consider that we wish to maximize rather than minimize. We may use the minimizing algorithms previously presented so as to accomplish this by modifying the assignment matrix in one of two ways:

1. Multiply all cell elements by minus one (−1); or
2. Subtract all cell elements from the value of the largest cell element in the matrix.

Often, we encounter problems with an unequal number of rows and columns. This is dealt with in a manner similar to that used to "balance the rims" in the transportation problem. That is, we simply add either dummy rows or columns, as required, so as to have an equal number of rows and columns. An assignment to a dummy row or column is interpreted as the inability to either assign a "worker" to a "task," or vice versa. The costs within the dummy row or column are normally zero unless there actually is a cost associated with not staffing a job or assigning a worker. In this case, these associated costs should be used in the dummy row or column.

Occasionally it is forbidden to make a certain assignment. For example, worker *A* may not be permitted to work on task I. To avoid such an assignment, we typically insert a very high cost (compared with those in the other cells) in the cell(s) associated with the forbidden assignment.

Finally, it should be noted that the three methods presented above for solving the assignment problem do not represent all those available. Another method, in particular, has been found quite efficient for the assignment problem. This is the so-called labeling algorithm, which is presented in Chapter 16 for dealing with certain network problems.

A Generalized Assignment Problem

Consider the problem associated with the assignment of individuals to jobs wherein *more than one* individual can be assigned to a single job or, alternatively, where more than one job may be assigned to a single individual. There is a very effective heuristic technique that may be used to solve such problems, even if they are of very large sizes. We discuss this problem and the solution approach herein not only because it represents a practical problem but also because the solution methodology (with minor variations) has been found to apply to several other types of real-world problems [2, 4].

Table 14-15 summarizes the data for an example problem that is used to illustrate the algorithm for solution. In this problem there are nine jobs and four workers. The time required to perform a job is dependent on both the match of the job requirements to worker skills and the individual difficulty of each job. These times are given in the interior cost cells of the table. At the bottom of Table 14-15 is an indication of the maximum amount of time that each worker can devote to his set of jobs (as assigned) during the time period of interest. We wish to assign the nine jobs to the four workers so as to minimize the total time required for all jobs while taking care not to assign any worker to a set of jobs whose time requirements would exceed the limits indicated.

Table 14-15. EXAMPLE MATRIX

Job \ Worker	*A*	*B*	*C*	*D*
1	4	3	12	7
2	8	10	12	6
3	3	5	2	5
4	10	6	2	4
5	10	3	7	9
6	8	10	9	9
7	7	2	10	12
8	5	9	4	14
9	10	8	15	7
Available time (a_j)	15	12	20	14

The mathematical statement of the generalized assignment problem is: Find **x** to

$$\text{minimize } z = \sum_{i=1}^{m} \sum_{j=1}^{n} c_{i,j} x_{i,j}$$

subject to

$$\sum_{j=1}^{n} x_{i,j} = 1 \qquad \text{for all } i$$

$$\sum_{i=1}^{m} c_{i,j} x_{i,j} \leq a_j \qquad \text{for all } j$$

$$x_{i,j} = 0, 1 \qquad \text{for all } i \text{ and } j$$

where $x_{i,j} = \begin{cases} 1 & \text{if job } i \text{ is assigned to worker } j \\ 0 & \text{otherwise} \end{cases}$

$c_{i,j}$ = time required to perform job i by worker j

a_j = total amount of time that worker j can be assigned

The algorithm for solving this problem is very similar to the VAM algorithm presented in Chapter 13. We determine penalty numbers for each row based on the difference between the lowest and second-lowest times in the row. Assignments are made one at a time, to the job having the largest penalty number—*if* such an assignment will not exceed the limit (a_j) on the worker's available time. Specifically, the steps are as follows:

Step 1. Establish the problem in matrix format (see Table 14-15).

Step 2. Set $k = 1$.

Step 3. Compare the available time (a_k) with the smallest time requirement ($c_{i,k}$) in column k. If $\min\{c_{i,k}\} > a_k$, strike out column k.

Step 4. Repeat step 3 for $k = 2, 3, \ldots, n$.

Step 5. Determine the penalty numbers for each row of the remaining matrix as the difference between the lowest and second-lowest times in the row. Place these penalty numbers to the right of each row (see Table 14-16).

Step 6. Make an assignment to the minimum-cost element in the row with the largest penalty wherein such an assignment does not exceed the time available (a_j). (This is job 7 to worker *B* in the example.)

Step 7. Strike out the row associated with the assignment of step 6 and reduce the associated available time. (See Table 14-17.)

Step 8. Repeat steps 2 through 7 until all jobs are either assigned or all columns (workers) have been marked off.

Tables 14-15 through 14-24 illustrate the procedure as applied to the example problem. Note, in particular, the application of step 3 in Tables 14-21 and 14-22. Worker *B* (in Table 14-21) has 4 hours left, but there is no unassigned job that requires 4 or less hours. In Table 14-22, worker *D* has 1 hour left but, again,

Table 14-16

Job \ *Worker*	*A*	*B*	*C*	*D*	*Penalties*
1	4	3	12	7	1
2	8	10	12	6	2
3	3	5	2	5	1
4	10	6	2	4	2
5	10	3	7	9	4
6	8	10	9	9	1
7	7	(2)	10	12	5 (maximum)
8	5	9	4	14	1
9	10	8	15	7	1
Available time (a_j)	15	12	20	14	

Table 14-17

Job \ *Worker*	*A*	*B*	*C*	*D*	*Penalties*
1	4	3	12	7	1
2	8	10	12	6	2
3	3	5	2	5	1
4	10	6	2	4	2
5	10	(3)	7	9	4 (maximum)
6	8	10	9	9	1
7	7	(2)	10	12	
8	5	9	4	14	1
9	10	8	15	7	1
a_j	15	~~12~~ 10	20	14	

Table 14-18

Job \ Worker	A	B	C	D	Penalties
1	4	3	12	7	1
2	8	10	12	(6)	2 (tied for max.)
3	3	5	2	5	1
4	10	6	2	4	2
5	10	(3)	7	9	
6	8	10	9	9	1
7	7	(2)	10	12	
8	5	9	4	14	1
9	10	8	15	7	1
	15	~~12~~ ~~10~~ 7	20	14	

Table 14-19

Job \ Worker	A	B	C	D	Penalties
1	4	3	12	7	1
2	8	10	12	(6)	
3	3	5	2	5	1
4	10	6	(2)	4	2 (max.)
5	10	(3)	7	9	
6	8	10	9	9	1
7	7	(2)	10	12	
8	5	9	4	14	1
9	10	8	15	7	1
a_j	15	~~12~~ ~~10~~ 7	20	~~14~~ 8	

Table 14-20

Job \ Worker	*A*	*B*	*C*	*D*	*Penalties*
1	4	③	12	7	1 (tied for max.)
2	8	10	12	⑥	
3	3	5	2	5	1
4	10	6	②	4	
5	10	③	7	9	
6	8	10	9	9	1
7	7	②	10	12	
8	5	9	4	14	1
9	10	8	15	7	1
a_j	15	~~12~~ ~~10~~ 7	~~20~~ 18	~~14~~ 8	

Table 14-21

Job \ Worker	*A*	*B*	*C*	*D*	*Penalties*
1	4	③	12	7	
2	8	10	12	⑥	
3	3	5	2	5	1
4	10	6	②	4	
5	10	③	7	9	
6	8	10	9	9	1
7	7	②	10	12	
8	5	9	4	14	1
9	10	8	15	⑦	3 (max.)
a_j	15	~~12~~ ~~10~~ ~~7~~ 4	~~20~~ 18	~~14~~ 8	

Table 14-22

Job \ Worker	A	B	C	D	Penalties
1	4	(3)	12	7	
2	8	10	12	(6)	
3	3	5	(2)	5	1 (tied for max.)
4	10	6	(2)	4	
5	10	(3)	7	9	
6	8	10	9	9	1
7	7	(2)	10	12	
8	5	9	4	14	1
9	10	8	15	(7)	
a_j	15	~~12~~ ~~10~~ ~~7~~ 4	~~20~~ 18	~~14~~ ~~8~~ 1	

Table 14-23

Job \ Worker	A	B	C	D	Penalties
1	4	(3)	12	7	
2	8	10	12	(6)	
3	3	5	(2)	5	
4	10	6	(2)	4	
5	10	(3)	7	9	
6	(8)	10	9	9	1 (tied for max.)
7	7	(2)	10	12	
8	5	9	4	14	1
9	10	8	15	(7)	
a_j	15	~~12~~ ~~10~~ ~~7~~ 4	~~20~~ ~~18~~ 16	~~14~~ ~~8~~ 1	

Table 14-24

Worker / *Job*	*A*	*B*	*C*	*D*	*Penalties*
1	4	③	12	7	
2	8	10	12	⑥	
3	3	5	②	5	
4	10	6	②	4	
5	10	③	7	9	
6	⑧	10	9	9	
7	7	②	10	12	
8	5	9	④	14	1
9	10	8	15	⑦	
a_j	~~15~~ 7	~~12~~ ~~10~~ ~~7~~ 4	~~20~~ ~~18~~ 16	~~14~~ ~~8~~ 1	

all unassigned jobs require more than the time available. The final assignment is then

worker *A* to job 6

worker *B* to jobs 1, 5, and 7

worker *C* to jobs 3, 4, and 8

worker *D* to jobs 2 and 9

for a total time of 37 units.

The Matching Problem

The preceding problem represented a generalization of the assignment problem, whereas the matching problem is actually an interesting special case of the conventional assignment problem. Consider the bipartite graph of the assignment problem in Figure 14-4. The equivalent matrix model is given in Table 14-25. Notice that, in the graph, there is a branch between a "worker" node (on the left) and "job" node (on the right) only if the worker is capable of performing that job. In Table 14-25, this is noted in the matrix cells by either 1's or 0's. Thus, since worker *A* can do job 1, a 1 is placed in the corresponding cell. Since the worker is not qualified for job 2, a 0 appears in that cell.

Table 14-25. MATCHING MATRIX

Worker \ Job	*1*	*2*	*3*	*4*
A	1	0	1	0
B	0	1	0	0
C	1	0	1	1

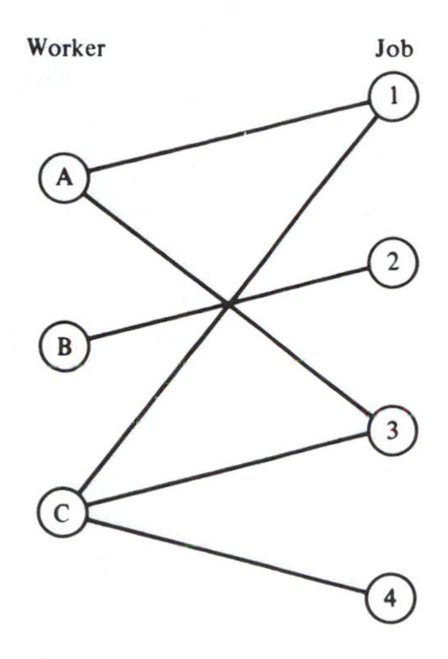

Figure 14-4. Matching graph.

The mathematical model for the general matching problem is then:
Find **x** to

$$\text{maximize} \quad \sum_{i=1}^{m} \sum_{j=1}^{n} c_{i,j} x_{i,j}$$

subject to

$$\sum_{j=1}^{n} x_{i,j} = 1 \qquad \text{for each } i$$

$$\sum_{i=1}^{m} x_{i,j} = 1 \qquad \text{for each } j$$

$$x_{i,j} = 0, 1$$

This is identical to the maximizing version of the assignment problem *except* that $c_{i,j}$ will only take on 0 or 1 values. As such, any technique that may solve the conventional assignment problem can also be used to solve the matching problem. Our interest lies not so much in these methods as in the problems that may be represented by the matching model.

Example 14-1

The top six international trade lawyers have stock in five large firms, as shown in Table 14-26. Can the five firms each hire one of these lawyers so that no conflict of interest arises (i.e., a lawyer cannot be counsel to a firm in which he has stock)? Figure 14-5 has been drawn to represent the resulting matching problem. Note carefully that a branch between a lawyer and firm exists only if that lawyer has *no* stock in that firm.

Table 14-26

Lawyer	*Holds Stock in Firm?*				
	1	*2*	*3*	*4*	*5*
A	No	Yes	Yes	No	Yes
B	No	Yes	No	Yes	No
C	Yes	No	Yes	Yes	Yes
D	Yes	Yes	Yes	No	Yes
E	Yes	Yes	Yes	Yes	Yes
F	Yes	No	Yes	Yes	No

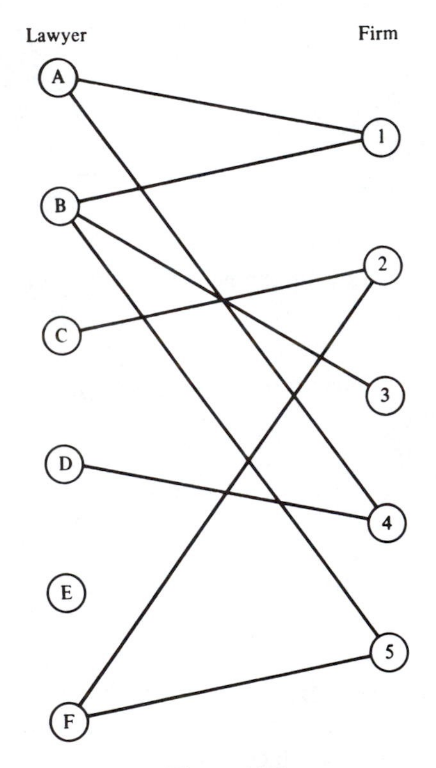

Figure 14-5

Solving via any of the techniques discussed for the assignment problem, we obtain a matching as follows:

Lawyer	*Firm*
A	1
B	3
C	2
D	4
F	5

and lawyer *E* is not assigned to any firm.

Additional examples appear in the problem set at the end of the chapter. Extensions of the matching problem and generalized assignment problem appear in the references [2, 4].

REFERENCES

1. Agin, N. "Optimum Seeking with Branch and Bound," *Management Science*, Vol. 13, No. 4, 1965, pp. 400–412.
2. Ignizio, J. P., and Case, K. E. "A Method for Validating Certain Set-Covering Heuristics," Chap. 13 in P. Brock (ed.), *The Mathematics of Large-Scale Simulation*. La Jolla, Calif.: Simulation Councils, 1972.
3. Kuhn, H. W. "The Hungarian Method for the Assignment Problem," *Naval Research Logistics Quarterly*, Vol. 2, No. 1–2, 1955, pp. 83–97.
4. Kwak, N. K. *Mathematical Programming with Business Applications*. New York: McGraw-Hill, 1973.
5. Lawler, E. L., and Wood, D. E. "Branch and Bound Methods: A Survey," *Operations Research*, Vol. 14, 1966, pp. 699–719.

PROBLEMS

14.1. Solve the swimmer/event problem of Figure 14-1 by:
(a) The VAM/MODI method
(b) The Hungarian algorithm method
(c) The branch-and-bound method

14.2. Solve Problem 13.3 by both the Hungarian algorithm and branch-and-bound methods.

14.3. A quality control check shows that the number of defects in six components produced by the Mho Electronics Company are a function of the workers assigned to produce these components. The average number of defective components, per week, expected for each of six workers is given in the accompanying table. How should we assign workers to components so as to minimize the total expected number of defects?

Worker \ *Component*	*A*	*B*	*C*	*D*	*E*	*F*
1	12	41	17	15	18	13
2	14	20	17	15	53	15
3	22	23	25	40	27	28
4	18	50	20	19	27	25
5	7	8	20	10	10	30
6	23	25	25	30	25	43

14.4. The Midwestern Universal Testing and Nuclear Technology Systems Company, also known as MUTANTS, Inc., intends to subcontract three special valves for its nuclear power plants. Four bids have been received and, according to governmental restrictions, no subcontractor may be permitted to produce more than one valve type. The bids, in terms of thousands of dollars per valve type, and the potential subcontractor information are given in the accompanying table. Solve the problem via both the Hungarian algorithm and branch-and-bound methods.

Value Type \ *Contractor*	*I*	*II*	*III*	*IV*
A	12	13	20	18
B	8	20	10	18
C	20	22	30	25

14.5. Prove that the optimal solution to an assignment problem must include the assignment of at least one source to the sink for which it has minimum cost.

14.6. In Problem 14.3 it is decided that a worker should not be considered for assignment to any component for which his or her average defect rate is above the median rate for that component (e.g., the median rate for component *A* is 16, and thus only workers 1, 2, or 5 should be assigned to that component). Considering *only* this qualification measure (i.e., the screening as based on the median), determine if an assignment is possible. Use the Hungarian algorithm for solution.

14.7. Solve Problem 14.6 by the branch-and-bound method.

14.8. Ten jobs are to be processed through the utilization of four machines. Each machine can process any job, but at different efficiencies. A machine cannot process more than one job at a time, but its available capacity (in terms of minutes available per day) may allow for processing more than one job over

Job \ *Machine*	*I*	*II*	*III*	*IV*
1	5	14	6	10
2	7	3	6	5
3	8	10	6	12
4	12	14	8	10
5	6	17	8	4
6	10	8	8	8
7	9	6	9	10
8	9	10	9	9
9	10	12	9	12
10	3	3	14	6
Available time	22	18	26	30

the planning period. The time requirements per job/machine assignment and total times available for each machine are shown in the accompanying table. Determine the assignment that minimizes the total processing time.

14.9. A distributed processing system consists of an interconnected network of various computational facilities (e.g., computers, minicomputers, microprocessors, intelligent terminals, etc.). One such system contains three different types of computers and there are six "jobs" (algorithms to be processed on the system). No single algorithm can be partially processed on one computer and partially by another (i.e., the algorithm must be completed by the computer it is initially assigned to). The processing times of the algorithms vary according to the computer performing the processing as shown in the accompanying table. How should the algorithms be assigned so as to complete the processing of all algorithms in minimum time?

Algorithm \ *Computer*	*I*	*II*	*III*
1	18	16	12
2	14	21	19
3	23	27	33
4	16	24	23
5	17	24	24
6	25	28	30
Available time (seconds)	47	41	46

14.10. Formulate both the primal and dual for Problem 14.9.

14.11. Find the minimal-time solution to the following generalized assignment problem (without job splitting).

Job \ *Machine*	*A*	*B*	*C*	*D*
1	10	12	8	7
2	8	10	7	7
3	20	10	22	30
4	12	14	20	32
5	17	15	14	15
6	5	10	15	16
7	9	8	12	17
Available time (hr)	20	18	20	22

NETWORK ANALYSIS

Introduction

The transportation and assignment problems (and variations thereof) of Chapters 13 and 14 are not only very special subclasses of linear programming, they are also very special subclasses of (linear) network models. In this chapter we continue our discussion of linear programming problems that may be conveniently modeled as networks. This more general class of network problems is normally solved by a set of techniques that form a methodology termed "network analysis." The specific algorithms of network analysis that we present here are those which, although they may not always be mathematically elegant, are computationally efficient for large-scale problems, the type of problems encountered most often in actual practice.

Included among the types of network problems to be discussed are:

1. Finding the shortest path or paths through a network.
2. Finding the longest path or paths through a network.
3. Network models for project scheduling and management.
4. Network saturation or maximal loading determination.
5. Finding the minimal "spanning tree" of a network.

Sometimes we observe such problems directly. Other times we shall find that they exist embedded within larger, more complex problems such as those that occur in logistics, facility layout, materials handling, and computer architecture.

15

Before proceeding to the models and their methods of solution, we first introduce, briefly, some of the more common notions and concepts associated with the network model [1, 2].

Networks: Descriptions and Definitions

The basic building blocks of any network consist of two items:

1. Nodes (or vertices).
2. Branches (or edges, or links, or arcs).

Nodes provide a convenient way to represent points of storage or transfer, and branches are often used to designate the paths, or perhaps courses of action, that connect the nodes.

Branches can be further classified as either unilateral (directed) or bilateral (undirected). Rather obviously, a unilateral branch implies that the "items" which "flow" through this branch can only do so in one direction, whereas with a bilateral branch, flow in either direction is possible. The transportation model of Chapter 14 was an example of a network with strictly unilateral branches. The flow was of goods from the source nodes to the sink nodes. The transshipment problem, on the other hand, generally included several bilateral branches and, in some cases, possibly all the branches could be bilateral.

Consider Figure 15-1(a), which depicts a bilateral branch between node *A* and node *B*. The number in parentheses above the arrowhead to the right (flowing into node *B*) might represent the per unit cost of items sent from *A* to *B*. The number in parentheses above the arrowhead on the left (flowing into node *A*) could then be the per unit cost of items sent from *B* to *A*. These two costs do not necessarily have to be the same.

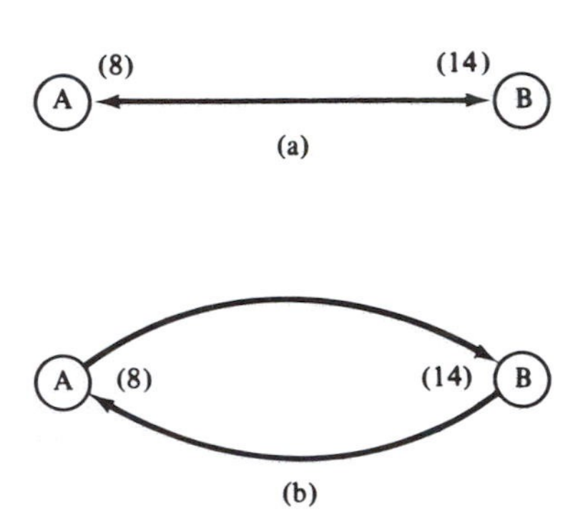

Figure 15-1. (a) Bilateral branch; (b) equivalent network.

Any bilateral branch may be transformed into two equivalent, unilateral branches. This is shown in Figure 15-1(b). Consequently, given any network, we may represent that network with strictly unilateral branches.

Normally, we will not place any arrowheads on bilateral branches. Thus, if a branch is undesignated, it will be assumed that flow in either direction is possible.

Not all nodes within a network necessarily have to be connected to the rest of the network. That is, in some cases, a node is said to be isolated. An example of an isolated node was given in Figure 14-5. Notice that node *E* is isolated in that network.

Another way to think of nodes and branches is that the node might represent a state of the system, whereas the branch depicts the transition between one state and another. Such a concept is particularly useful when one considers stochastic networks (i.e., networks wherein the flow from one node to another is a function of some probability distribution). Consider, for example, the network shown in Figure 15-2. Node *A* might represent the present state of, say, a new product that is about to be marketed. Nodes *B*, *C*, and *D* represent all the possible states of the market for this product 1 month after introduction. Based on various surveys, it is estimated that the transitions to states *B*, *C*, and *D* have probabilities of 30%, 50%, and 20%, respectively. These probabilities are indicated beside each branch.

Some further notions and definitions that help one to describe and understand networks are *chains*, *trees*, *link branches*, and *tree branches*. We list these

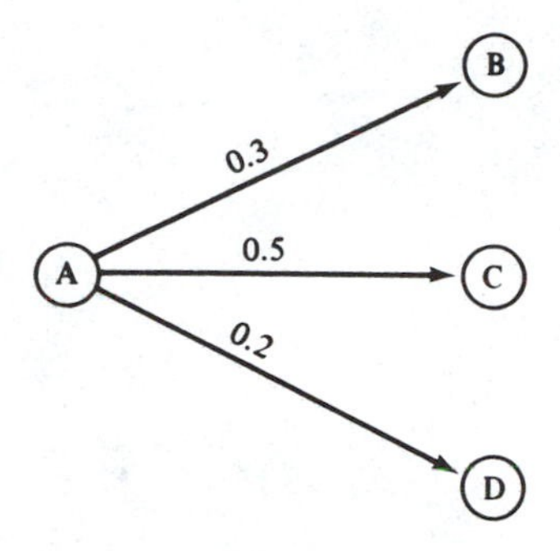

Figure 15-2. Stochastic network.

(and other) definitions below and illustrate via examples. We use the network of Figure 15-3 to demonstrate these ideas.

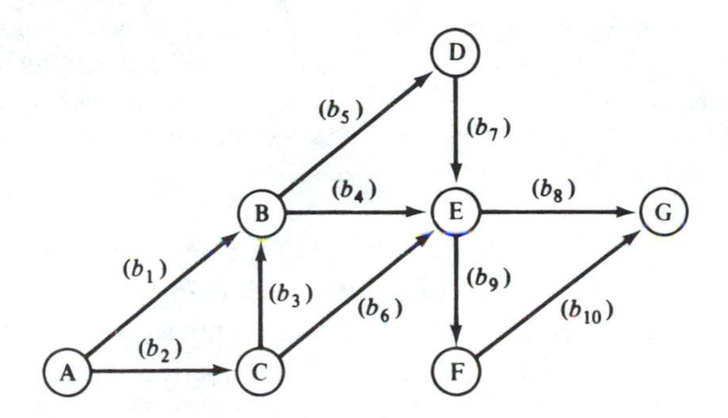

Figure 15-3

CHAIN A chain between nodes i and j is a sequence of branches that connect these two nodes. For example, one of the chains connecting nodes A and G is the sequence of branches b_1, b_4, and b_8, or vice versa.

PATH When the *direction* of travel between two nodes is specified, the associated chain is termed a path. Thus, a path from node A to node F is composed of branches b_2, b_6, and b_9 (in that order).

CYCLE A cycle is a path in which a particular node (or nodes) is (ultimately) connected to itself. There are no cycles possible in Figure 15-3. (Why?) However, in the undirected network of Figure 15-4, one cycle is the chain consisting of b_1, b_4, and b_2.

TREE A network tree is a network in which all nodes are connected by branches (i.e., no isolated nodes) and in which no cycles exist. One possible tree for Figure 15-4 consists of the branches b_1, b_3, b_4, and b_7. A network tree thus provides a connection of all network nodes by a minimal number of branches.

BRANCH NUMBER In a general network, with all nodes connected, the number of branches (b) is related to the number of nodes (n) as

$$b \geq n - 1$$

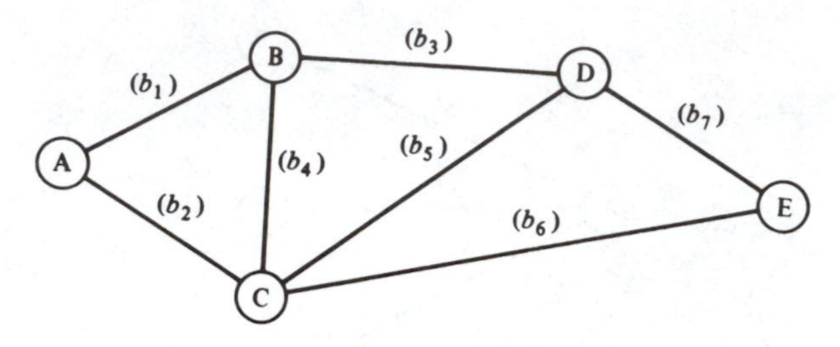

Figure 15-4

However, for a network tree, the number of branches is

$$b = n - 1$$

BRANCH MEASURE The measure of "cost" associated with each network branch is, for convenience in discussion, normally referred to as its "length." However, in general the measure associated with a branch can include (a) its cost, (b) the time required to traverse the branch, (c) its profit or gain, or (d) its capacity.
MINIMAL SPANNING TREE The tree of minimal total length is called the minimal spanning tree.

In our discussion of networks, we restrict our interest solely to those networks in which the flow through the network is measured in *discrete* (rather than continuous) units and for which the branch measure is a *linear function of the flow*. This allows us to formulate all problems as linear integer programming models.

The Shortest-Route Problem

We divide our discussion of the shortest-route problem into two distinct problem subclasses: (1) nondirected networks, and (2) directed networks that contain no cycles. The latter network type is known as an *acyclic-directed network*. If the network is acyclic-directed, the algorithms for solution can be made considerably more efficient than for the nondirected network.

Linear Programming Formulation

To ease the formulation, we assume that our route is to be determined between node 1 and node n, where n also equals the total number of network nodes. Further, it will be assumed that a single branch connects node i to node j (i.e., distances and branches associated with nodes that are not directly connected are ignored). Also, in the event of a directed branch, we do not consider flows in opposition to the direction. Thus, we wish to find the set of $x_{i,j}$ to

$$\text{minimize} \quad \sum_{i=1}^{n} \sum_{j=1}^{n} d_{i,j} x_{i,j} \tag{15.1}$$

subject to

$$\sum_{k=2}^{n} x_{1,k} = 1 \tag{15.2}$$

$$\sum_{k=1}^{n-1} x_{k,n} = 1 \tag{15.3}$$

$$\sum_{i=1}^{n} x_{i,k} - \sum_{j=1}^{n} x_{k,j} = 0 \qquad \text{for } k = 2, \ldots, n-1 \tag{15.4}$$

$$x_{i,j} = 0, 1 \qquad \text{for all } i, j \tag{15.5}$$

where $d_{i,j}$ = "distance" from node i to j (for directly connected nodes)

$$x_{i,j} = \begin{cases} 1 & \text{if the branch from node } i \text{ to node } j \text{ is selected} \\ 0 & \text{otherwise} \end{cases}$$

The first two constraints assure us that exactly one branch is taken from the first node, whereas exactly one branch is taken into the final node. The third constraint simply forces the number of branches *into* an intermediate node to equal those *out of* that node.

Shortest Route through a Nondirected Network

Consider the network shown in Figure 15-5. The "lengths" between nodes are shown in parentheses beside each branch. Since no arrowheads appear on the branches, the network is nondirected. For a network this small, we could easily find the shortest path from, say, node 1 to node 5, by enumeration which is, as shown in Table 15-1, the route from node 1 to node 3 to node 5. However, for problems of any realistic size, a more systematic approach is required.

The approach we present here involves the transformation of the shortest-route problem into an equivalent assignment problem. Any method that can

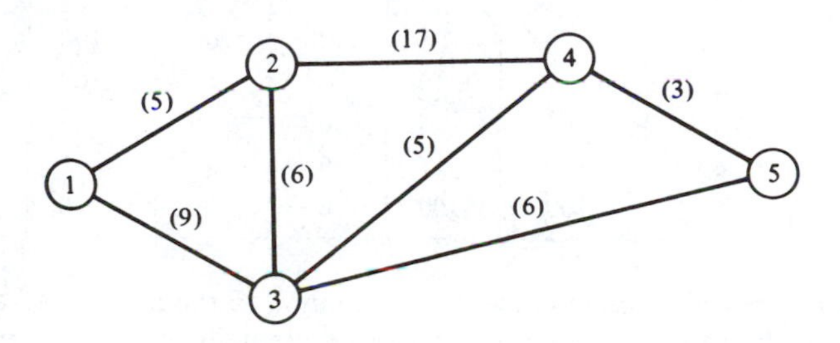

Figure 15-5

Table 15-1. ENUMERATION OF ROUTE

Route	*Distance*
1–2–4–5	25
1–2–3–5	17
1–2–3–4–5	19
1–3–2–4–5	35
1–3–4–5	17
1–3–5	15

solve the assignment problem (see Chapter 14) can then be applied. The algorithm for transformation to the assignment problem is given next.

A matrix of node-to-node distances must first be established. The rules for filling in the cells of this matrix are:

Step 1. Node-to-*same*-node (i.e., *A* to *A*, *B* to *B*, etc.) distances are zero.

Step 2. The path length between nodes not directly connected, with the exception of the distance from the terminal node to source node, is considered infinite. For ease in computation, a very large number with respect to other matrix elements may be used. (The number "100" is used in the example.)

Step 3. To assure a routing from the source node to the terminal node (i.e., the final destination), we assign a very large negative value to this element. The negative of the large number used in step 2 will suffice (i.e., use −100 in the example).

Step 4. All other distances are the same as those indicated on the network.

Table 15-2 depicts the assignment problem matrix that is associated with the network of Figure 15-5. The resultant solution is indicated by the boxed elements. The next phase of solution is to associate the assignment problem solution (as shown in Table 15-2) with the shortest route from node 1 to node 5. We accomplish this via the following steps.

Table 15-2. EQUIVALENT ASSIGNMENT MATRIX

Node \ *Node*	*1*	*2*	*3*	*4*	*5*
1	0	5	[9]	100	100
2	5	[0]	6	17	100
3	9	6	0	5	[6]
4	100	17	5	[0]	3
5	[−100]	100	6	3	0

Step 1. Write down the node number corresponding to the initial (i.e., source) node.

Step 2. Go to the row headed by the previous node designation and write down the node number associated with the boxed element of that row. This is the new node designation.

Step 3. Repeat step 2 until the terminal node has been selected.
Step 4. The routing is then the same as the nodes found in steps 1 through 3 and in the same sequence.

The shortest route from node 1 to node 5 is thus node 1 to node 3 and node 3 to node 5 for a total length of 15 units.

The same approach may be used for determining the shortest route through an acyclic-directed network. We only need note that the distance from one directly connected node to another is infinity (i.e., a large number) if the flow is *against* the direction specified by the arrowhead. However, as mentioned, there is an easier approach.

Shortest Routes in Acyclic-Directed Networks (The Labeling Algorithm)

The acyclic-directed network lends itself to efficient solution techniques for the determination of both:

1. The shortest route between the source node and a single terminal node (or sink node).
2. The shortest routes between the source node and *all other* nodes in the network.

To illustrate, consider the network of Figure 15-6. Note that it is an acyclic-directed network. Also, note the manner in which the nodes have been numbered. A branch must emanate from a node with a lower node number than that to which it is incident (i.e., directed).

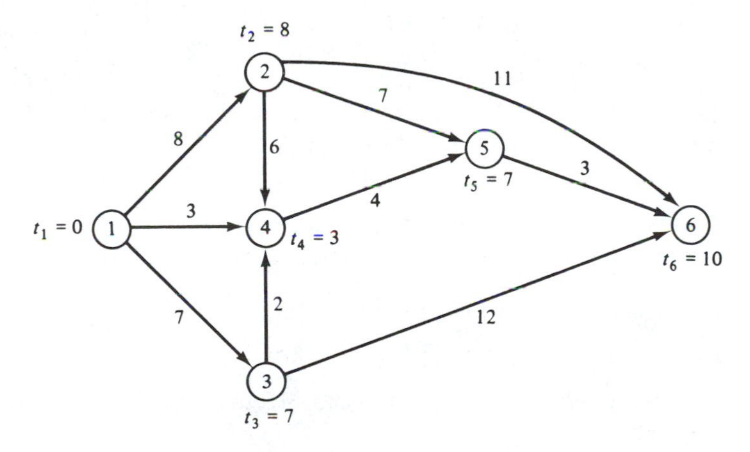

Figure 15-6. An acyclic-directed network.

We shall first consider finding the shortest route from node 1 to node 6. The algorithm for the determination of the shortest route from a source node to a terminal node is given below:

Step 1. Label node 1 with the designation $t_1 = 0$.

Step 2. Label all remaining nodes (in ascending order) according to

$$t_j = \min (t_i + d_{i,j}) \qquad i = 1, 2, \ldots, j - 1 \tag{15.6}$$

where $d_{i,j}$ is the distance from node i to node j. Notice carefully that although node i does not have to be directly connected to node j, the path between i and j must always be according to the specified branch flow directions to be considered.

Step 3. When the last node (node n) has been labeled with its value (t_n), then t_n represents the length of the shortest route through the network. The actual path from source to terminal node is found by tracing backward from node n to all nodes where $t_j = t_i + d_{i,j}$. Those nodes satisfying (15.6) *and* which form a path from source to terminal node represent the shortest route.

The algorithm is best understood through actual demonstration. We shall solve for the shortest route from node 1 to node 6.

Example 15-1: Shortest Route for Figure 15-6

Step 1. $t_1 = 0$

Step 2. $t_2 = \min (t_1 + d_{1,2}) = \min (0 + 8) = 8$

$t_3 = \min (t_1 + d_{1,3}) = \min (0 + 7) = 7$

$t_4 = \min \{(t_1 + d_{1,4}), (t_2 + d_{2,4}), (t_3 + d_{3,4})\}$
$= \min \{(0 + 3), (8 + 6), (7 + 2)\} = 3$

$t_5 = \min \{(t_2 + d_{2,5}), (t_4 + d_{4,5})\}$
$= \min \{(8 + 7), (3 + 4)\} = 7$

$t_6 = \min \{(t_2 + d_{2,6}), (t_3 + d_{3,6}), (t_5 + d_{5,6})\}$
$= \min \{(8 + 11), (7 + 12), (7 + 3)\} = 10$

Step 3. The length of the shortest path is thus $t_n = t_6 = 10$ units. The path itself is found determining those branches that satisfy the equation $t_j = t_i + d_{i,j}$, which include:

branch 5–6[a]	for $t_6 = 10$
branch 4–5	for $t_5 = 7$
branch 1–4	for $t_4 = 3$
branch 1–3	for $t_3 = 7$
branch 1–2	for $t_2 = 8$

Of these branches, only (1–4), (4–5), and (5–6) form a complete path from node 1 to node 6.

We now consider the case in which we desire to determine the shortest routes between the source node and *all other* nodes in the network. With

[a]That is, the branch between nodes 5 and 6.

reference to Example 15-1 and based on node 1 as the initial node, the process is essentially the same as before. For example, the length of the shortest route from node 1 to node 4 is simply $t_4 = 3$, and tracing back we find that it consists of the single branch from node 1 to node 4.

If we wish to find the shortest routes from a node other than node 1, we simply revise the network by dropping all nodes (and the branches that emanate from them) with a node number less than that of the new initial node. For example, if node 3 is to be the initial node, nodes 1 and 2 (and associate branches) are removed and we proceed as before.

Example 15-2: Equipment Replacement

The cost of operating and maintaining a piece of machinery (such as a truck, a computer, or assembly line) will normally increase with time due to wear-out, obsolescence, and so forth. An important problem, then, is to determine the optimal policy for the replacement of this equipment over some given planning horizon. We assume that the costs of purchasing the equipment in year i and replacing it in year j have been computed for all reasonable actions. These are shown on the equivalent network of Figure 15-7, wherein the costs are represented as node-to-node distances. For example, the cost associated with a purchase at the beginning of year 1 and replacement at the beginning of year 3 is given by the "distance" on the branch from node 1 to node 3 (1,200 units in the figure).

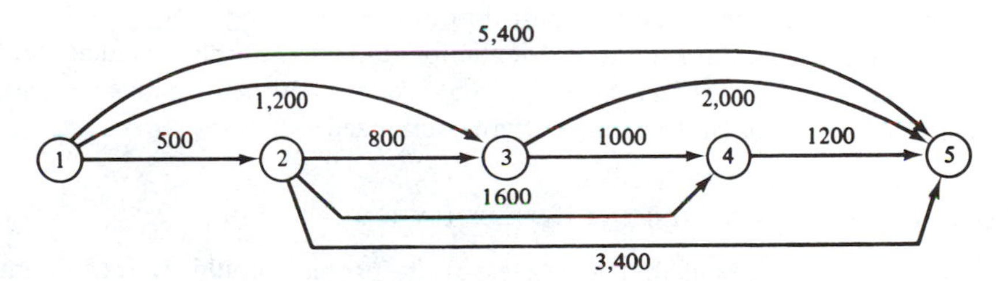

Figure 15-7

The "shortest route" through the network of Figure 15-7 is then actually the set of policies that determine a minimal-cost plan for purchase and relacement over the 4-year planning horizon. (Normally, the planning horizon will be much longer and serve to develop a considerably larger network, but the same approach is appropriate.) Solving by our algorithm for the shortest route, we have:

Step 1. $t_1 = 0$

Step 2. $t_2 = \min (t_1 + d_{1,2}) = \min (0 + 500) = 500$

$$t_3 = \min \{(t_1 + d_{1,3}), (t_2 + d_{2,3})\}$$
$$= \min \{(0 + 1{,}200), (500 + 800)\} = 1{,}200$$
$$t_4 = \min \{(t_2 + d_{2,4}), (t_3 + d_{3,4})\}$$
$$= \min \{(500 + 1{,}600), (1{,}200 + 1{,}000)\} = 2{,}100$$
$$t_5 = \min \{(t_1 + d_{1,5}), (t_2 + d_{2,5}), (t_3 + d_{3,5}), (t_4 + d_{4,5})\}$$
$$= \min \{(0 + 5{,}400), (500 + 3{,}400), (1{,}200 + 2{,}000), (2{,}100 + 1{,}200)\}$$
$$= 3{,}200$$

Thus, the minimal-cost policy results in a total cost of $3,200 and is given by

branch 3–5	for $t_5 = 3{,}200$
branch 2–4	for $t_4 = 2{,}100$
branch 1–3	for $t_3 = 1{,}200$
branch 1–2	for $t_2 = 500$

The equivalent complete path is from node 1 to node 3 to node 5. Thus, we replace the equipment every 2 years.

The Longest-Route Problem

The only network type for which it makes sense to talk about finding the longest route is the acyclic-directed network. The reason for this should be clear if we refer back to Figure 15-4. If we seek the longest route from node *A* to node *E*, in this nondirected network, it is *infinity*. For example, one such route is that from node *A* to node *B* to node *C* to node *A* and continuing forever in this loop. Consequently, all discussion herein is with reference to the acyclic-directed network, in which nodes are numbered so that the node at the end of an arrowhead is of a higher number than that at the beginning.

The mathematical formulation for the longest-route problem is identical to that given in equations (15.1) through (15.5), except that the objective (relationship 15.1) is to be maximized rather than minimized.

Longest Route via Assignment Representation

We previously noted that a shortest-route problem could be transformed into an equivalent assignment problem. The same type of conversion is also possible for the longest-route problem *if* the network is acyclic-directed. The conversion process is identical to that previously presented for the shortest-route problem except that:

1. All path distances between nodes that are directly connected are multiplied by -1.
2. If the matrix cell represents a flow in opposition to the branch direction, the path is considered infinite (in practice we use a large number in relation to the other cell elements).

To demonstrate, consider the network shown in Figure 15-6. The equivalent assignment matrix, with the solution shown in boxes, is given in Table 15-3. The route associated with the solution (of length 21) is nodes 1–2–4–5–6.

Table 15-3

Node \ Node	*1*	*2*	*3*	*4*	*5*	*6*
1	0	[−8]	−7	−3	100	100
2	100	0	100	[−6]	−7	−11
3	100	100	[0]	−2	100	−12
4	100	100	100	0	[−4]	100
5	100	100	100	100	0	[−3]
6	[−100]	100	100	100	100	0

LONGEST ROUTE VIA LABELING ALGORITHM

The labeling algorithm of the previous section (i.e., for the shortest route problem) can also be easily modified so as to be used to find the longest route. We simply revise the formula in step 2 to read

$$t_j = \max (t_i + d_{i,j}) \qquad i = 1, 2, \ldots, j - 1 \tag{15.7}$$

We illustrate this by determining the longest route through the network of Figure 15-6, from node 1 to node 6.

Step 1. $t_1 = 0$
Step 2. $t_2 = \max \{(t_1 + d_{1,2})\} = \max (0 + 8) = 8$
$t_3 = \max \{(t_1 + d_{1,3})\} = \max (0 + 7) = 7$
$t_4 = \max \{(t_1 + d_{1,4}), (t_2 + d_{2,4}), (t_3 + d_{3,4})\}$
$= \max \{(0 + 3), (8 + 6), (7 + 2)\} = 14$
$t_5 = \max \{(t_2 + d_{2,5}), (t_4 + d_{4,5})\}$
$= \max \{(8 + 7), (14 + 4)\} = 18$
$t_6 = \max \{(t_2 + d_{2,6}), (t_3 + d_{3,6}), (t_5 + d_{5,6})\}$
$= \max \{(8 + 11), (7 + 12), (18 + 3)\} = 21$

The longest path has a length then of 21 units and its branches are among those which satisfy $t_j = t_i + d_{i,j}$ or:

branch 5–6 for $t_6 = 21$
branch 4–5 for $t_5 = 18$
branch 2–4 for $t_4 = 14$
branch 1–3 for $t_3 = 7$
branch 1–2 for $t_2 = 8$

The complete path consists then of the branches from node 1–2–4–5–6.

The problem of finding the longest path is inherent in the scheduling of projects. In such a problem, the longest path will correspond to the time required to complete all activities (i.e., to finish the project). Since this problem is so prevalent, we discuss one particular approach to its solution (and with a more thorough analysis than implied in this section) in the section to follow.

PERT/CPM in Network Scheduling and Control

The Program Evaluation and Review Technique (PERT) was developed in the late 1950s by the Navy for controlling the development progress of the Polaris Ballistic Missile Program. It was credited with reducing the completion time of that project by 2 years. PERT has been used to schedule major multimillion and multibillion dollar military and aerospace (particularly NASA) projects, such as the Air Force's Minuteman, Skybolt, and the B-70, the Army's Nike-Zeus, Pershing, and Hawk, and NASA's Saturn/Apollo mission. More recently, PERT has found extensive use in the analysis and scheduling of industrial projects involving many interrelated activities including research and development (R&D) programs, construction projects, computer program development, maintenance planning, and so on [3].

The Critical Path Method (CPM) was developed at about the same time as PERT. However, where PERT found its origin in the military sector, CPM was initially applied to industrial projects. An area in which CPM has found rather extensive use is in construction.

Originally, PERT was intended to introduce the aspects of uncertainty in the estimate of activity times, whereas CPM was more or less a deterministic methodology. Other differences existed within the early versions of both techniques, but with the passage of time and the interchange of ideas, it is now often difficult to discriminate between approaches denoted as PERT and those called CPM. As a consequence, we first introduce a *deterministic* version of PERT in which the main objective is to determine the required time to complete a given project. We then introduce a linear programming model of CPM that examines ways in which one may reduce the project time via additional (financial) resources.

PERT Networks

The primary objectives behind the construction of a PERT network model of a project are to (1) determine the probability of meeting specified deadlines, (2) identify activities that are likely to be bottlenecks, and (3) evaluate the impact of any changes in the project. However, perhaps the most useful aspect of the construction of a PERT network is that it provides a truly significant aid in understanding the overall project and various interrelationships within the project.

The primary features of a deterministic PERT network (other than its typically *much* larger size) are indicated in Figure 15-8. The network consists of its nodes, represented by circles, which are used to indicate the beginning or completion of an activity or activities. The branches each represent a separate activity. The node at the tail of the branch indicates the start of the activity, and the node at the head of the branch designates activity completion. Beside each branch is a letter, which indicates the code assigned to the activity, and a number, which indicates the expected number of days (or other units of time) required to complete the activity. (In the original PERT, three time estimates were used: most optimistic time, most pessimistic time, and the most likely time.) Associated with the PERT network are some additional rules.

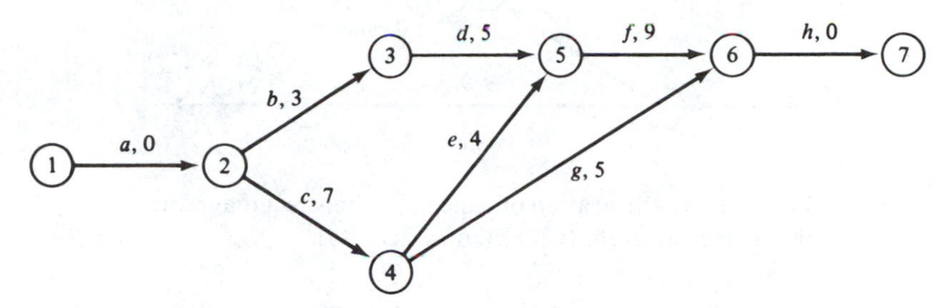

Figure 15-8. An illustrative PERT network.

Three rules that must be followed in network construction are:

1. Each activity (i.e., branch) must be identified with a *unique* combination of nodes.
2. Each node should be numbered using integers where the node number at the end of any activity must be greater than that at the beginning.
3. The network must be acyclic-directed.

We are already familiar with rules 2 and 3, so let us consider rule 1. Consider the *incorrect* PERT network diagram of Figure 15-9(a). The network portion shown indicates that activity *e* cannot be initiated until both activities *c* and *d* are completed. However, the first rule of network construction is violated because activities *c* and *d* both have the same start and end nodes. To avoid this problem, one may use a dummy activity (with zero time requirements) as shown in Figure 15-9(b). The revised network still implies that activity *e* cannot be started until both *c* and *d* are finished, but also it provides a unique node combination for all branches.

Let us next consider a problem in which we wish to represent the following: activity *c* cannot be initiated until activity *a* is completed, while activity *d* cannot be initiated until both activities *a* and *b* are finished. The right and wrong ways of depicting this are shown in Figure 15-10. Note that Figure 15-10(a) really implies that neither *c* nor *d* can start until both *a* and *b* are finished, whereas by the use of a dummy activity, the proper relationships are shown in Figure 15-10(b).

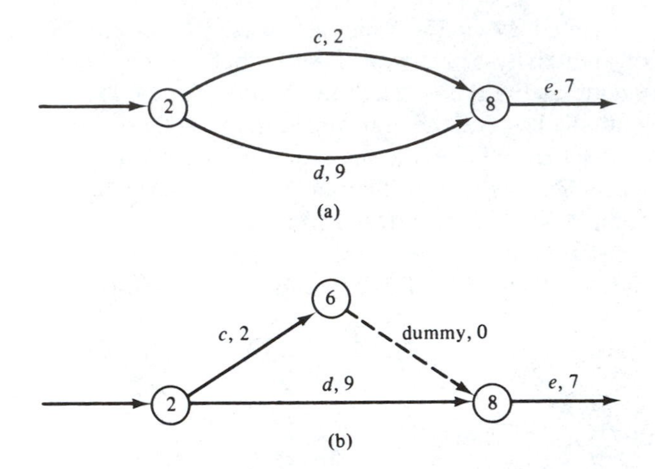

Figure 15-9. Illustration of Rule 1: (a) incorrect diagram; (b) correct diagram (with dummy activity).

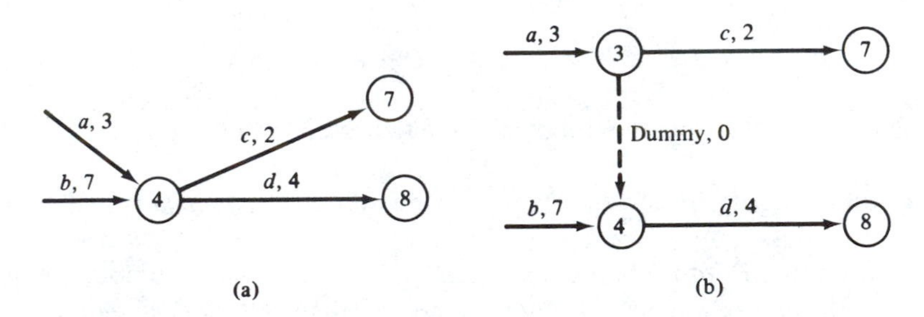

Figure 15-10. (a) Incorrect network; (b) correct network.

We are now ready to demonstrate the construction of a complete network. We shall use the data from Table 15-4, which represents the activities associated with developing a computer program for a firm that specializes in solar heating and cooling systems. The purpose of the program is to determine the required size and orientation of the solar collector for a given building and location. Previously, these calculations were done by hand.

For the moment, the only columns of interest in Table 15-4 are columns *A* through *D*. The associated PERT diagram is given in Figure 15-11.

Our next task is to determine the "critical path" (the longest route) and "slack" times associated with this project. The critical path is the sequence of activities that determines the final completion date of the project. An activity with slack time can be delayed (up to some amount) without any impact on the project completion date, whereas the delay of any activity on the critical path will affect the schedule.

We could solve for the critical path via the methods discussed previously for finding the longest route. However, the method we illustrate has been

Table 15-4. DATA FOR SOLAR HEATING/COOLING PROGRAM

A	*B*	*C*	*D*	*E*	*F*	*G*
					Cost ($ × *100*)	
Activity	*Code Letter*	*Immediate Predecessor(s)*	*Normal Days' Duration*	*Minimum (Crash) Days' Duration*	*Normal*	*Crashing*
Start (assemble work force)	*a*	none	2	2	8	8
Data collection (data book)	*b*	*a*	20	10	40	100
Preliminary design	*c*	*a*	10	6	30	60
Prototype code test	*d*	*c*	30	22	60	100
Installation on system	*e*	*d, b*	8	6	28	50
Final testing	*f*	*e*	10	5	22	60
Delivery	*g*	*f*	1	1	2	2

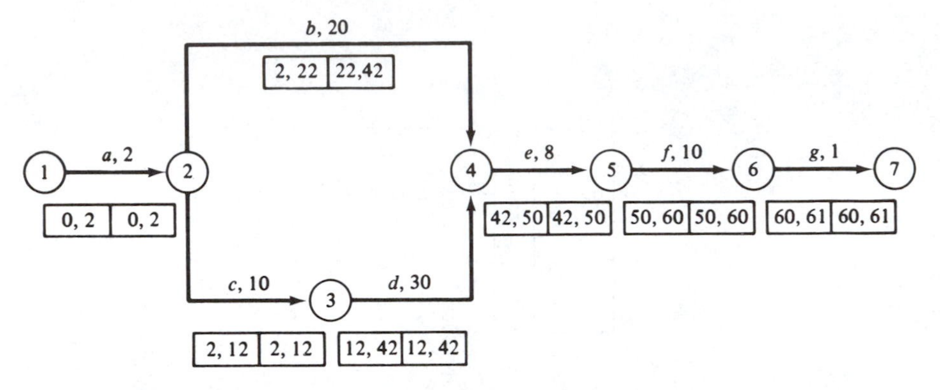

Figure 15-11. PERT diagram for solar heating/cooling program.

designed especially for the PERT application. This technique is a labeling approach in which the following label appears under each branch (including dummy activities) of the PERT network:

ES EF	LS LF

ES stands for early start, or the earliest time that this activity could be started; EF stands for early finish, or the earliest possible time that the activity could be completed. The right-hand side of the label includes LS for latest start and LF for latest finish. The steps to be followed in filling in the values for ES, LS, EF, and LF are:

1. Starting with the first network activity and proceeding from left to right, we first fill in the value for ES. ES for the start activity is always zero. EF is then the sum of the activity time plus ES.
2. The ES values for all other activities will be equal to the largest EF value of all the activities immediately preceding. EF then is activity time plus ES. This procedure is continued until all ES and EF values are filled in.
3. Starting with the final network activity and proceeding from right to left, we next fill in the values for LF and LS. LF is either equal to the value for EF for this last activity or to some predetermined project duration. LS is then equal to LF minus the activity for the branch.
4. The LF values for all other activities will be equal to the smallest LS value for *all* activities that immediately follow the activity. LS is then equal to LF minus activity times for each branch.

The critical path is the path having the least (usually 0) slack time. For each activity, slack time is obtained by taking the difference of either LF–EF or LS–ES. Thus, the path through node 1–2–3–4–5–6–7 is the critical path of 61 days' duration.

CPM AND RESOURCE REASSIGNMENT

If we consider the problem previously discussed (see Figure 15-11), we note that, if we wished to shorten the duration (from 61 days total to something less), the first thing to consider are the critical path activities. It is the change in those activities that will, initially, have an impact on the schedule. For example, if activity *d* could be reduced from 30 days to 10 days, the project could be finished in just 41 days. However, further reduction in activity *d* alone will not affect the 41-day duration because, at that time, the critical path shifts (to branch *b*).

When one also includes the consideration of activity costs as well as duration, the project scheduling problem becomes much more complex. We illustrate a linear programming/CPM approach to deal with such a situation.

The CPM model of the project scheduling and control problem requires the following information:

1. Normal time: the time required to complete the project with either minimal cost or for a certain designated level of funding.
2. Normal cost: the activity cost for a normal time schedule.
3. Crash time: the minimum possible time to complete an activity.
4. Crash cost: the cost associated with completing an activity in minimum possible time.

This information has been included in columns *D* through *G* of Table 15-4 for the preceding example. To illustrate, the data collection task normally would require 20 days at a cost of $4,000. It is possible to reduce this time to 10 days but only at a cost of $10,000 (an increase of $6,000). This information is plotted in Figure 15-12, where it has been assumed that the cost of reducing the activity duration is linear between the normal and crash estimates.

We shall then employ the following notation:

$c_{i,j}$ = normal cost for the activity between nodes i and j

$C_{i,j}$ = crash cost for the activity between nodes i and j

$d_{i,j}$ = normal duration of the activity between nodes i and j

$D_{i,j}$ = crash duration of the activity between nodes i and j

$K_{i,j}$ = intersection of the time–cost curve with the cost axis (this is $16,000 for activity b, as shown in Figure 15-12)

F = desired project completion time

$x_{i,j}$ = duration of the activity between nodes i and j

t_k = earliest time for which activities emanating from node k may be started; note that, for node 1, $t_1 = 0$ and, for the final node (node n), $t_n \leq F$

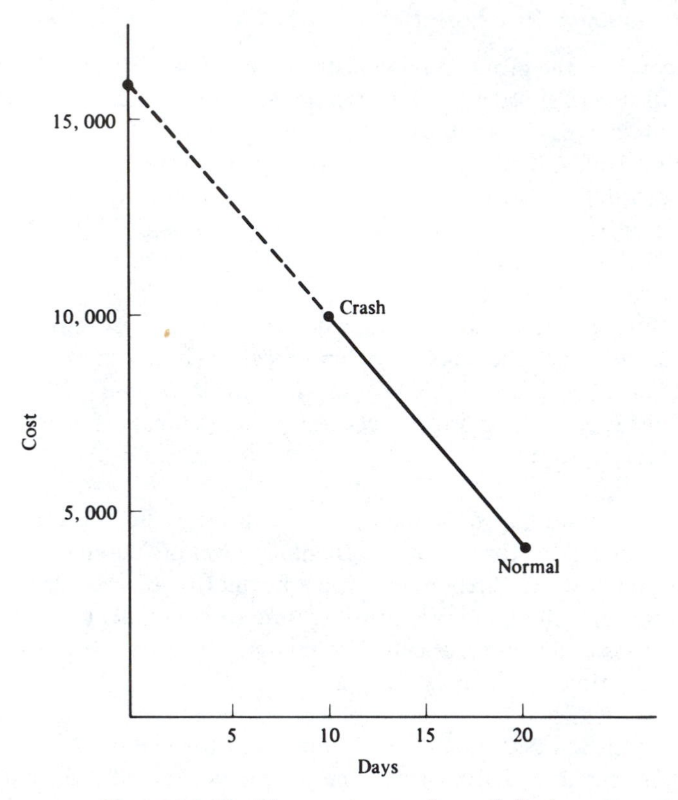

Figure 15-12. Time-cost curve for activity b.

The $x_{i,j}$ and t_k (other than t_1) are then the variables in the model for which we seek to find a value that will minimize the total cost given some desired completion date, F. The cost function we wish to minimize may be developed as follows.

The incremental direct cost of activity (i, j) per unit decrease in $x_{i,j}$ is

$$S_{i,j} = \frac{C_{i,j} - c_{i,j}}{d_{i,j} - D_{i,j}} \tag{15.8}$$

Thus, the direct cost of activity (i, j) is

$$K_{i,j} - S_{i,j}x_{i,j} \tag{15.9}$$

and the *total* direct cost for the entire project is

$$\sum_{(i,j)} (K_{i,j} - S_{i,j}x_{i,j}) \tag{15.10}$$

That is, we sum (15.9) over all activities (i, j).

Notice that $K_{i,j}$ is a constant for all (i, j), and thus the minimization of (15.10) is equivalent to the maximization of

$$\sum_{(i,j)} S_{i,j}x_{i,j} \tag{15.11}$$

The linear programming model for the general CPM project scheduling and control problem is then to find all $x_{i,j}$ and t_k for a given completion time, F so as to maximize (15.11). Or:

$$\text{maximize} \quad z = \sum_{(i,j)} S_{i,j}x_{i,j} \tag{15.12}$$

subject to

$$x_{i,j} \leq d_{i,j} \qquad \text{all } i, j \tag{15.13}$$

$$x_{i,j} \geq D_{i,j} \qquad \text{all } i, j \tag{15.14}$$

$$t_1 = 0 \tag{15.15}$$

$$t_n \leq F \tag{15.16}$$

$$t_i + x_{i,j} - t_j \leq 0 \qquad \text{all } i, j \tag{15.17}$$

Relationship (15.13) indicates that the duration of any activity should not exceed its normal time, while (15.14) specifies that the duration cannot be less than the crash time. Relationship (15.15) simply indicates that the earliest start time for any activities emanating from node 1 is zero [it is not necessary to include (15.15) in the actual formulation], while (15.16) reflects the desire to finish the project by at least time F. Relationship (15.17) indicates that the activity between nodes i and j must be completed before the activities emanating from node j start.

The formulation of the preceding example given a desired completion date of 35 days (F) is:

$$\text{Maximize} \quad z = 7.5x_{2,3} + 6x_{2,4} + 5x_{3,4} + 11x_{4,5} + 7.6x_{5,6}$$

subject to

$$x_{1,2} = 2$$

$$x_{6,7} = 1$$

$$x_{2,3} \leq 10$$

$$x_{2,3} \geq 6$$

$$x_{2,4} \leq 20$$

$$x_{2,4} \geq 10$$

$$x_{3,4} \leq 30$$

$$x_{3,4} \geq 22$$

$$x_{4,5} \leq 8$$
$$x_{4,5} \geq 6$$
$$x_{5,6} \leq 10$$
$$x_{5,6} \geq 5$$
$$t_1 = 0$$
$$t_7 \leq 35$$
$$t_1 + x_{1,2} - t_2 \leq 0$$
$$t_2 + x_{2,3} - t_3 \leq 0$$
$$t_2 + x_{2,4} - t_4 \leq 0$$
$$t_3 + x_{3,4} - t_4 \leq 0$$
$$t_4 + x_{4,5} - t_5 \leq 0$$
$$t_5 + x_{5,6} - t_6 \leq 0$$
$$t_6 + x_{6,7} - t_7 \leq 0$$

The problem before solving by simplex, may be, rather obviously, simplified in form.

Network Saturation

Consider the network shown in Figure 15-13. The figure represents a communications network in which we may send messages from node 1 to node 6 along a finite number of paths. The numbers in parentheses beside each branch represent the maximal capacity of that branch (i.e., the maximum number of messages that may be sent at a given time).

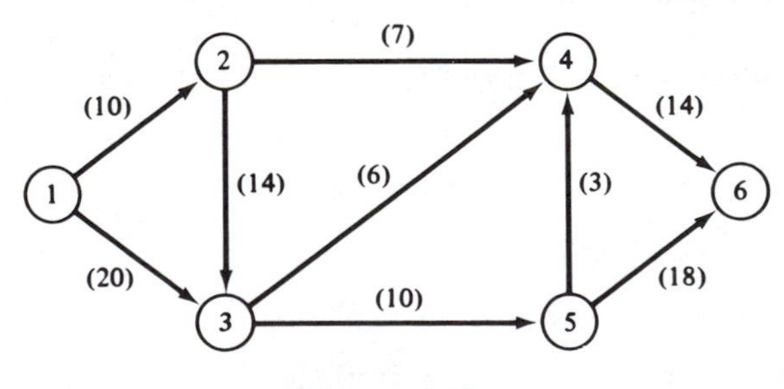

Figure 15-13

There are several questions that may be answered via an analysis of such a network, including:

1. How many messages should be transmitted through each branch so as to maximize the total number of messages received at node 6 at any one time?

2. What is the maximum number of messages that may be sent at any one time? That is, at what level is the network saturated?
3. If it is desired to upgrade the network performance (i.e., increase the amount of messages that may be sent), which branches should first be considered?

We first consider the linear programming formulation of such a problem and then briefly discuss one *heuristic* approach to the solution.

LINEAR PROGRAMMING MODEL

We use the following notation:

$x_{i,j}$ = number of messages routed from node i to node j

$c_{i,j}$ = maximum capacity of the branch from node i to node j

The number of messages received at node 6 is simply $x_{4,6} + x_{5,6}$, and this is the function we wish to maximize. For any intermediate node (nodes 2 through 5), the number of messages in must equal those out. Also, the number of messages out of node 1 must equal those ultimately received at node 6 ($x_{1,2} + x_{1,3} = x_{4,6} + x_{5,6}$). Finally, the branch capacities cannot be exceeded ($x_{i,j} \leq c_{i,j}$). The resulting model is then:

$$\text{maximize} \quad x_{4,6} + x_{5,6}$$

subject to

$$\left.\begin{aligned} & x_{1,2} = x_{2,3} + x_{2,4} \\ & x_{1,3} + x_{2,3} = x_{3,4} + x_{3,5} \\ & x_{2,4} + x_{3,4} + x_{5,4} = x_{4,6} \\ & x_{3,5} = x_{5,4} + x_{5,6} \end{aligned}\right\}\text{balance on intermediate nodes}$$

$$x_{1,2} + x_{2,3} = x_{4,6} + x_{5,6}\}\text{messages out of node 1 equal those into node 6}$$

$$\begin{aligned} x_{1,2} &\leq 10 \\ x_{1,3} &\leq 20 \\ x_{2,3} &\leq 14 \\ x_{2,4} &\leq 7 \\ x_{3,4} &\leq 6 \\ x_{3,5} &\leq 10 \\ x_{4,6} &\leq 14 \\ x_{5,4} &\leq 3 \\ x_{5,6} &\leq 18 \end{aligned}$$

and all $x_{i,j}$ are integers.

Heuristic Approach

There are some relatively efficient exact methods for solving the network saturation problem. However, we discuss one particular heuristic approach that seems to work fairly well in practice (but does not gurantee an optimal solution). This heuristic proceeds by arbitrarily assigning flow units to each path that exists between the source and terminal node. Each time such a flow is assigned, we subtract the flow assigned from the capacity of the branches employed. Obviously, the branch of minimal capacity in the path determines the upper limit of the flow through that path, and this is the amount that we select in each path flow assignment. At some point, we discover that no path exists any longer, and we have then obtained our solution. We demonstrate the approach in Figure 15-13.

1. Let us choose the path from 1 to 2 to 4 to 6. The minimal branch capacity in this route is seven. Thus, our first routing is for 1–2–4–6 for 7 units. We subtract the number seven from each branch in our path.
2. Let us now choose the path 1–3–5–6. The minimal capacity for that path is 10 units. Thus, our route is 1–3–5–6 for 10 units. We then subtract 10 from each branch capacity in this path.
3. Our next path must utilize the branch 3 to 4, since it is now the only branch connecting the left side of the network with the right. Let us choose the path 1–3–4–6. The minimum capacity branch value is six. Our path is then 1–3–4–6 for 6 units.
4. There is no way to proceed from node 1 to 6, and thus our total network capacity is 23 units.

We were lucky in this example, as we actually obtained the optimal solution. Notice that, for the flows assigned, the individual branches that are saturated are: 2 to 4, 3 to 4, and 3 to 5. Any improvement in the system should (at least initially) be directed at these branches.

Minimal Spanning Tree

As mentioned previously, a tree is a network in which all nodes are connected by branches and no cycles exist. The tree of minimal total length is called the minimal spanning tree. For example, consider a distributed computing network in which the nodes represent either minicomputers or the customers who use these computers. The branches between computers and between customer and computer are bilateral communication links. Sometimes a customer will request a computing task that can be accomplished within a single computer; on other occasions, from several up to *all* the computers may be necessary. As a result, a path must exist between each node. Since the cost of the system is determined by the communication link (branch) leasing cost, we wish to find the minimal spanning tree.

The algorithm for determining the optimal spanning tree is about as simple and straightforward as one could imagine. The steps are as follows:

Step 1. Specify U to be the set of unconnected nodes (initially, all nodes are unconnected) and C to be the set of connected nodes.
Step 2. Select *any* node and place it in set C.
Step 3. Determine that node in set U which is nearest to any node in C and connect those nodes (i.e., add this new node to C).
Step 4. Repeat step 3 until all nodes are connected (i.e., the set U is empty).

To illustrate, we use the network shown in Figure 15-14, wherein the branches shown are all *possible* communication links and the number in parentheses beside each branch represents the cost of that branch. The solution procedure follows:

1. $U = \{1, 2, 3, 4, 5, 6\}$,	$C = \varnothing$
2. Select node 5:	$C = \{5\}$
	$U = \{1, 2, 3, 4, 6\}$
3. Connect node 2 to 5:	$C = \{5, 2\}$
	$U = \{1, 3, 4, 6\}$
4. Connect node 1 to 2:	$C = \{5, 2, 1\}$
	$U = \{3, 4, 6\}$
5. Connect node 4 to 1:	$C = \{5, 2, 1, 4\}$
	$U = \{3, 6\}$
6. Connect node 3 to 4:	$C = \{5, 2, 1, 4, 3\}$
	$U \quad \{6\}$
7. Connect node 6 to 3:	$U = \varnothing$, stop

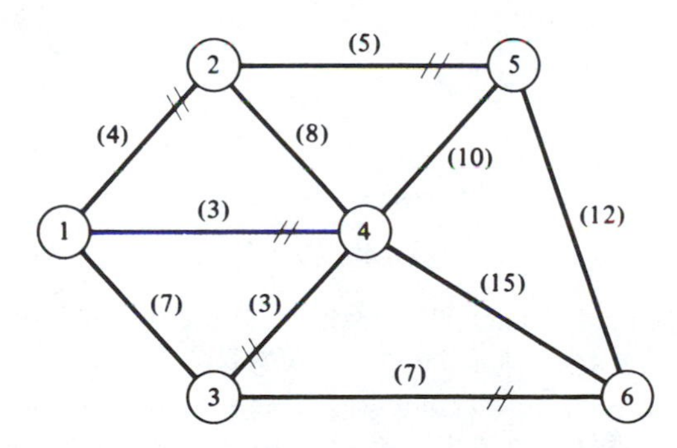

Figure 15-14

The tree selected from the process above is denoted in Figure 15-14 via slashes across the selected branches. The reader may wish to start with another node and see that the same tree results.

Summary

The networks and algorithms presented in this chapter provide only an introduction to the variety of applications and methods that exist in this field. The reader desiring further information is directed to the references.

REFERENCES

1. Hu, T. C. *Integer Programming and Network Flows.* Reading, Mass.: Addison-Wesley, 1970.
2. Weinberg, L. *Network Analysis and Synthesis.* New York: McGraw-Hill, 1962.
3. Weist, J. D., and Levy, F. K. *A Management Guide to PERT/CPM.* Englewood Cliffs, N.J.: Prentice-Hall, 1969.

PROBLEMS

15.1. Draw a network and illustrate the difference between a "chain" and a "path."

15.2. In the directed network shown, list a representative sample of:
(a) Paths
(b) Cycles
(c) Trees

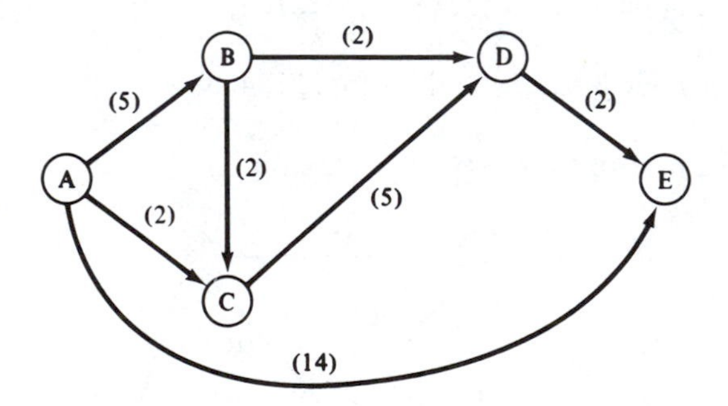

Figure P15-2

15.3. Develop the mathematical formulation for the shortest route from node A to node E for the network given in Problem 15.2.

15.4. Use the various shortest-route algorithms of this chapter to determine the shortest route from node A to node E, for the network given in Problem 15.2, when:
(a) The branches are unilateral (as shown)
(b) The branches are considered to be *bilateral*

15.5. Solve Problem 15.4(a) for the longest path via the algorithms provided in the chapter.

15.6. A small lawn-care firm is faced with an equipment selection and replacement decision with regard to the choice of riding lawnmowers. They have narrowed their choice to two types of mowers: the "Green Machine" and the "Bushwhacker." Operating and replacement costs of both mowers are listed in the table for a 3-year planning horizon. Develop the network model that represents this problem.

	Green Machine		*Bushwacker*	
Year	*Cost at Beginning of Year*	*Operating Cost for the Year**	*Cost at Beginning of Year*	*Operating Cost for the Year**
1990	3,500	200	3,100	300
1991	3,800	240	3,600	360
1992	4,100	260	4,100	400
1993	4,800	300	4,700	450

*As based on a *new* machine. For every year the machine ages, the operating costs increase by $200 for the Green Machine and by $300 for the Bushwhacker.

15.7. Solve Problem 15.6 via the labeling algorithm.

15.8. Construct the (deterministic) PERT network for the project shown in the table. Use the PERT network labeling scheme to determine the critical paths for the projects.

				Cost	
Activity	*Immediate Prede-cessor(s)*	*Normal Days' Duration*	*Minimum (Crash) Duration*	*Normal*	*Crashing*
a	None	4	2	100	300
b	*a*	3	3	200	200
c	*a*	5	2	240	700
d	*b, c*	2	1	200	500
e	*b*	4	2	160	400
f	*d, e*	3	2	240	400

15.9. Develop the mathematical model for the project of Problem 15.8 under the CPM/crashing concept under the assumption that a 20% reduction over normal time is desired.

15.10. If, in Problem 15.2, the branch measures are in terms of branch capacity, formulate the problem as a network saturation model and solve via the heuristic approach of the chapter.

15.11. In Problem 15.10, does the heuristic approach actually give the exact solution? If so, show how one could use the approach and find a less than optimal solution.

15.12. Solve the network in Problem 15.2 for the minimal spanning tree (assume that the branches are *bilateral*).

part four

MULTIPLE-OBJECTIVE LINEAR PROGRAMMING

FORMULATION OF THE MULTIPLE-OBJECTIVE MODEL

Introduction[a]

Part Four, the final portion of this text, concludes the presentation of linear programming by introducing a highly flexible and practical methodology for modeling, solving, and analyzing problems for which we wish to, or must, consider the impact of *multiple*, conflicting objectives. That is, the problems that are so typical of those that are encountered in actual practice are addressed. The existence of such problems was indicated in Chapter 2 with the introduction of the baseline model. However, until now, we have concentrated on an approach that avoids, rather than considers, the multiple-objective environment. Consequently, the thrust of Part Four is to show how one converts the baseline model into a multiple-objective linear programming model, what assumptions are made to do this, and then how to solve the model, analyze the solution, and relate these results to the actual problem. Happily, we shall discover that the methodology of multiple-objective linear programming is really no more difficult from that which was employed in the single-objective problem, and, in fact, there are even some simplifications to be gained. Further, since the method of simplex provides the foundation for both the single- and multiple-objective methods, the amount of new material to be learned is relatively small.

---

[a]The reader may wish to review Chapters 1 and (particularly) 2 at this point.

16

Coverage of the Chapter

In this, our first chapter of Part Four, we focus our attention on two absolutely vital aspects: (1) the terminology and concepts that are necessary to describe and understand our process, and (2) the steps employed to formulate the multiple-objective model. These two factors play a much more important part in the understanding and employment of the multiple-objective process than was the case in single-objective linear programming. Development of the single-objective model is relatively easy, as there are few choices to be considered. That is, we "simply" select one function as our objective and treat the rest as rigid constraints. Formulation of the multiple-objective model requires more understanding of the problem, a fact that we consider to be an advantage even though it will require more thought and time of the analyst.

Unlike single-objective linear programming, there is more than one multiple-objective model (and associated solution technique) from which we may choose. Since we do not have enough time to present all these candidate models and approaches, we use a single multiple-objective model as our basis. This model is known as the *goal programming model* or, more correctly, as *generalized goal programming*. However, the use of the goal programming model does not exclude our consideration of alternative multiple-objective approaches since, as will be shown, the generalized goal programming model may be, through minor changes, extended so as to encompass the facets inherent in the other approaches. Consequently, our choice of the goal programming model for multiple-objective linear programming has been made because:

1. The model development is relatively simple and straightforward.
2. Minor modifications may be employed so as to encompass the alternative approaches (e.g., fuzzy programming, nondominated or efficient solution methods, weighted objectives, etc.) to the multiple-objective linear programming problem.
3. The method of solution is quite simple and is, in fact, just a refinement to the two-phase simplex method presented in Chapter 7.
4. The goal programming model, and variations thereof, have already found extensive implementation in actual problems since the early 1950s.
5. The model and its assumptions seem consistent with typical real-world problems.

Thus, although we are not claiming that the goal programming model and methodology is always the best multiple-objective approach, we note that, as yet, there is no single best approach to the multiple-objective problem and have based our choice of presentation of the goal programming method on its flexibility, efficiency, and ease of use and implementation.

Alternative Approaches

There are a large and growing number of approaches in use and/or proposed for the multiple-objective linear model. As mentioned above, goal programming is just one of these and is described in detail later. However, we now pause to consider just a few of the better-known multiple-objective approaches. (The extension of goal programming to encompass some of these methods is discussed in Chapter 20.)

There are three primary approaches[b] or philosophies that form the basis for nearly all the candidate multiple-objective techniques that have been (and, probably, will be) proposed. These are:

1. Weighting or utility methods
2. Ranking or prioritizing methods
3. Efficient solution (or generating) methods

The weighting method refers to those approaches that attempt to express all the problem objectives in terms of a single measure (such as dollars or "utiles"). The basic thrust of all such methods is to transform a multiple-objective model into a single-objective model. As such, they are attractive from a strictly computational point of view (e.g., conventional simplex may be used if the model

[b]In Chapter 20 we examine an alternative manner by which these philosophies may be categorized.

is linear). However, the obvious drawback to such an approach is that associated with actually developing truly credible weights. For example, if one objective is to reduce inflation and the other is to reduce unemployment, how much more or less important is inflation than unemployment? Even worse, consider the objective of minimizing highway fund outlay and that of minimizing injuries and loss of life through highway accidents. Can we (and should we) really associate a "value," in terms of dollars, to human life? A god-like proclamation that a human life is worth $157,492.57 may provide a number, but does it provide a valid answer?

The ranking or prioritizing methods try to circumvent the heady problems indicated above. Rather than attempting to find a numerical weight for each objective, they simply ask that objectives be ranked according to their perceived importance. Most decision makers can do this and, in fact, ranking is a concept that seems inherent to much of decision making. For example, in determining raises, a procedure used by many organizations is to first rank one's employees in order of their work, productivity, value to the company, and so forth. The problem with the ranking approach is how to associate the results of a given solution to the satisfaction of the ranking.

The final method "avoids" both the problems of finding weights and that of satisfying the ranking. It does this, or at least attempts to, by generating the total set of all the *efficient solutions* (or nondominated solutions, or Pareto optimal solutions). That is, we assume that the decision maker is a rational being. If so, knowing nothing about his or her inherent rankings or weightings of the objectives, we do know that his or her optimal decision must be an efficient solution; that is, one not dominated by any other solution. For example, consider the conflicting objectives mentioned earlier with regard to minimizing inflation and minimizing unemployment. Assume that two alternative economic policies are compared where:

	Policy A	*Policy B*
Reduction in inflation rate (%)	2	2
Reduction in unemployment rate (%)	1	3

Ignoring any other objectives and political factors, policy *B* dominates policy *A* since it is just as good in reducing inflation and better in reducing unemployment. Consequently, we say that policy *A* is dominated, inferior or inefficient and thus of no interest to the decision maker. If we can develop the entire set of efficient (nondominated) solutions we could then present these to the decision maker and let him or her select the one believed most attractive.

The efficient solution approach, like that of weighting, seems attractive on the surface but, like weighting, is subject to severe practical limitations. In most real problems of interest, the number of efficient solutions is so large as to preclude either their development or their presentation to the decision maker.

Actually, the three basic approaches indicated above also represent some of the extremes in multiple-objective approaches. While there are some analysts that adhere strictly to their convenants, it would appear that any robust, practical multiple-objective approach must be developed

1. Either by relaxing the strict interpretation of the methods above.
2. Or by finding a compromise working combination of the above approaches.

The latter is what we shall attempt to obtain via the generalized goal programming approach.

Terminology and Concepts

Terminology, as always, plays an important part in the understanding and appreciation of a methodology. Generalized goal programming has a number of special terms and concepts that we use in the discussions to follow. Included among these are:

OBJECTIVE An objective is a relatively general statement (in narrative or quantitative terms) that reflects the desires of the decision maker. For example, one may wish to "maximize profit" or "minimize labor turnover" or "wipe out poverty."

ASPIRATION LEVEL An aspiration level is a specific value associated with a desired or acceptable level of achievement of an objective. Thus, an aspiration level is used to *measure* the achievement of an objective and generally serves to "anchor" the objective to reality.

GOAL An objective in conjunction with an aspiration level is termed a goal. For example, we may wish to "achieve at least X units of profit" or "reduce the rate of inflation by Y percent."

GOAL DEVIATION Many of us aspire to be independently wealthy, but few will achieve this goal. The difference between what we accomplish and what we aspire to is the deviation from our goal. In all but trivial problems (or in cases where our aspiration levels are unrealistically low), we shall encounter deviations from our goals. Note that a deviation can represent *over*- as well as *under*achievement of a goal.

Goal Formulation

Before proceeding further, let us now examine how one mathematically transforms an objective into a goal within our goal programming framework. We also take careful note that a symmetric procedure is used in the formulation of a "constraint."

Consider the objective function expressed in general terms as $f_i(\mathbf{x})$. Actually, the procedure to be presented is applicable whether $f_i(\mathbf{x})$ is linear or nonlinear, but we shall normally, in the text, assume a linear form of the objective. We then let:

$f_i(\mathbf{x})$ = mathematical representation of objective i as a function of the decision variables $\mathbf{x} = (x_1, x_2, \ldots, x_n)$

b_i = value of the aspiration level associated with objective i

Three possible forms of goals may then result:

1. $f_i(\mathbf{x}) \leq b_i$; that is, we wish to have a value of $f_i(\mathbf{x})$ that is equal to or less than b_i.
2. $f_i(\mathbf{x}) \geq b_i$; that is, we wish to have a value of $f_i(\mathbf{x})$ that is equal to or greater than b_i.
3. $f_i(\mathbf{x}) = b_i$; that is, $f_i(\mathbf{x})$ must exactly equal b_i.

Regardless of the form, we shall transform any of these relations into the goal programming format by adding a negative deviation variable ($\eta_i \geq 0$) and subtracting a positive deviation variable ($\rho_i \geq 0$). Table 16-1 summarizes this statement.

Table 16-1. GOAL FORMULATIONS

Goal Type	*Goal Programming Form*	*Deviation Variables to Be Minimized*
$f_i(\mathbf{x}) \leq b_i$	$f_i(\mathbf{x}) + \eta_i - \rho_i = b_i$	ρ_i
$f_i(\mathbf{x}) \geq b_i$	$f_i(\mathbf{x}) + \eta_i - \rho_i = b_i$	η_i
$f_i(\mathbf{x}) = b_i$	$f_i(\mathbf{x}) + \eta_i - \rho_i = b_i$	$\eta_i + \rho_i$

Next, consider the relationship between the original goal form (i.e., $\leq$, $\geq$, or $=$) and the deviation variables. It should be clear that

1. To satisfy $f_i(\mathbf{x}) \leq b_i$, we must minimize the positive deviation (ρ_i).
2. To satisfy $f_i(\mathbf{x}) \geq b_i$, we must minimize the negative deviation (η_i).
3. To satisfy $f_i(\mathbf{x}) = b_i$, we must minimize *both* η_i and ρ_i.

Again, this is summarized in Table 16-1.

Example 16-1

Let us assume that the profit of a firm may be expressed as a linear function of two variables (products) as:

$$5x_1 + 7x_2 = \text{profit in dollars}$$

If our aspiration is to obtain at least \$1,000 of profit per time period, the goal may be written initially as

$$5x_1 + 7x_2 \geq 1{,}000$$

Adding negative and positive deviation variables, we have

$$5x_1 + 7x_2 + \eta_1 - \rho_1 = 1{,}000$$

Our desire to achieve our goal is then reflected in an attempt to minimize η_i. Now, if $x_1 = 100$ and $x_2 = 100$, we have achieved our goal, wherein

$$\eta_1 = 0$$

$$\rho_1 = 5(100) + 7(100) - 1{,}000 = 200$$

and profit = \$1,200. However, if $x_1 = 100$ and $x_2 = 50$, we have not achieved this goal, because

$$\eta_1 = 1{,}000 - 5(100) - 7(50) = 150$$

$$\rho_1 = 0$$

and the profit = \$850.

Example 16-2

Now consider an objective reflecting the satisfaction of air pollution regulations. Assume that the amount of pollution is a linear function of two variables, such as

$$100x_1 + 200x_2 = \text{air pollution level}$$

Our aspiration is to keep the air pollution level below 5,000 units, and thus we write our goal as

$$100x_1 + 200x_2 \leq 5{,}000$$

or, in goal programming form, as

$$100x_1 + 200x_2 + \eta_1 - \rho_1 = 5{,}000$$

To achieve our goal we must attempt to minimize the positive deviation, ρ_1.

From the examples, it should now be obvious that a goal in equality form $[f_i(\mathbf{x}) = b_i]$ is achieved only when *both* η_i and ρ_i are zero. Thus, we would attempt to minimize their sum.

Let us now consider what we termed a "constraint" in single-objective linear programming. Examining the three forms of goals in the left-hand side of Table 16-1, we may notice that there is no way to tell, by simply looking, that these are goals, formed from objectives, rather than constraints. The inability to distinguish between a goal and a constraint extends beyond simply the mathematical formulation. For example, if one states that he or she wishes to hold the rate of inflation to less than 6%, is that a goal or a constraint? This problem is basically one of semantics and is circumvented by a symmetrical treatment of goals and constraints. Thus, whether the relation shown below is a goal or a constraint, it is transformed and treated in the same way.

$$8x_i + 6x_2 - 4x_3 \leq 100$$

That is, we include the negative and positive deviation variables and minimize the appropriate one (or combination) so as to attempt to achieve the relationship shown.

Such a treatment not only simplifies problem formulation, but it also is far more representative of the real world. Consider, for example, a problem in which a "constraint" is to expend no more than a prespecified level of budget. Treated conventionally, a rigid constraint would not permit a solution that exceeded the budget by even a single cent. Such a solution is *mathematically* infeasible but may well be permissible in the less rigid atmosphere of the real situation. There are a number of other advantages to this symmetric treatment that will become apparent later.

Once every objective and constraint has been transformed, as indicated in Table 16-1, we need to develop a relationship that indicates and *measures the level of achievement* of any solution proposed. Consequently, a natural name for this relationship or function is the "achievement function." Further, since satisfaction of goals (or constraints) is obtained, regardless of original form, by the minimization of various deviation variables, we should indicate this desire in our achievement function.

Evaluating a Solution: The Achievement Function

Given that we have some solution (i.e., *any* solution), $\mathbf{x}$, to a multiple-objective model as represented by the goal formulations described above, the next question is: How do we determine how good the solution is? *The answer to this question depends entirely on one's philosophy with regard to how to measure multiple-objective achievement*, as we see below.

Some of the measures used to evaluate the "goodness" of a solution include:

1. How well does it *minimize the sum of the weighted goal deviations*?
2. How well does it *minimize some polynomial* (or other nonlinear) *form of the goal deviations*?
3. How well does it *minimize the maximum* (i.e., "worst") *goal deviation*?
4. How well does it *lexicographically minimize an ordered* (i.e., ranked or prioritized) *set of goal deviations*?
5. Various combinations of the above.

The achievement function that we shall use (at least initially) is one that combines features from measures 1 and 4. That is, we shall measure achievement in terms of the lexicographic minimization of an ordered set of goal deviations, wherein *within* each set of goals at a particular rank, weights may be used. This achievement function, or vector, then looks as follows:

$$\mathbf{a} = (a_1, a_2, \ldots, a_k, \ldots, a_K) \tag{16.1}$$

where $\mathbf{a}$ = achievement vector for which we seek the lexicographic minimum
k = ranking or priority, where

$$a_k = g_k(\bar{\eta}, \bar{\rho}) \qquad k = 1, 2, \ldots, K \tag{16.2}$$

$g_k(\bar{\eta}, \bar{\rho})$ = linear function of the goal or constraint deviation variables that are to be minimized at rank or priority k

Further, we shall always reserve a_1 (the first term in **a**) for the deviation function associated with any rigid constraints.

Now recall (or be introduced to) what is meant by the lexicographic minimum of an ordered array.

LEXICOGRAPHIC MINIMUM Given an ordered array **a** of nonnegative elements a_k's, the solution given by $\mathbf{a}^{(1)}$ is preferred to $\mathbf{a}^{(2)}$ if

$$a_k^{(1)} < a_k^{(2)}$$

and all higher-order elements (i.e., $a_1, \ldots, a_{k-1}$) are equal. If no other solution is preferred to **a**, then **a** is the lexicographic minimum.

Thus, if we have two solutions, $\mathbf{a}^{(r)}$ and $\mathbf{a}^{(s)}$, where

$$\mathbf{a}^{(r)} = (0,\ 17,\ 500,\ 77)$$
$$\mathbf{a}^{(s)} = (0,\ 18,\ 2,\ 9)$$

$\mathbf{a}^{(r)}$ is preferred to $\mathbf{a}^{(s)}$.

Another term that is used to describe the lexicographic minimum notion is the concept of preemptive priorities. A solution that provides a lexicographic minimum to **a** also satisfies the concept of preemptive priorities. Any goal (or goals) at preemptive priority k (designated by P_k) will always be preferred to (i.e., preempt) any at a lower priority $k + 1, \ldots, K$ regardless of any scalar multiplier associated with these lower priorities.

Although the preemptive-priority concept (or lexicographic minimum) may appear new to some readers, it was already used, implicitly, in single-objective linear programming. Our first (preemptive) priority there was to find a solution that satisfied all constraints. Our next priority is to find a solution that maximizes or minimizes the single objective *without* violating the constraints.

Not only is the preemptive-priority concept implicit in the simplex solution technique, it is also evident in real-world decision making and has been documented in numerous studies. Consider, for example, the decision procedure that is often evident in the purchase of a home. The buyer's first priority may be to consider only a home that is within a 10-mile radius from his or her place of work. All other homes are excluded from consideration. Next, the buyer may desire to limit the purchase price to under \$200,000. Thus, even though homes outside the 10-mile radius may be below \$200,000, they are not considered because they would conflict with the preemptive priority associated with this goal. After narrowing down the homes to those within the 10-mile radius and under \$200,000, the next priority may be that the house have four bedrooms and a double-car garage, and so on. Thus, the preemptive-priority concept is used in the decision analysis as an *iterative screening process*.

Some of the criticisms leveled at goal programming center about the use of the lexicographic minimum (or preemptive priorities). This is unfortunate because, even if the lexicographic minimum does not supply the most desirable measure of achievement for a particular problem, it is generally extremely efficient in finding a good starting solution that may be improved upon by a straightforward relaxation of the strict interpretation of the lexicographic minimum. Further, it has never been intended that it be the rigid, inflexible concept that its critics describe.

Consequently, we employ the lexicographic minimum or preemptive-priority measure primarily because of its flexibility. That is, the notion may be relaxed or extended so as to encompass other measures of achievement. Chapter 20 indicates how this is accomplished. However, for now, we are ready to introduce a set of steps by which one may transform any baseline model into our specific form of the goal programming model.

Steps in Model Construction

The initial phase in the construction of the goal programming model (or any other mathematical programming model) is, or should be, the development of the baseline model, as described in Chapter 2. Once the baseline model has been constructed, we enter the next phase: the conversion of the baseline model into our specific form of the linear goal programming model (i.e., a multiple-objective linear programming model). The assumptions necessary in this conversion are:

1. Aspiration levels may be associated with all objectives so as to transform them into goals.
2. Any rigid constraints (i.e., absolute goals) are ranked at priority 1. All remaining goals may be ranked according to importance.
3. With the exception of priority 1 (i.e., the set of rigid constraints), all goals within a given priority must either be commensurable (i.e., measured in the same units) or, by means of weights, be made commensurable.

These are not particularly restrictive assumptions and, in fact, are generally less severe then those employed to develop the single-objective linear programming model (i.e., the necessity to select a single objective and assume that all remaining functions are rigid constraints).

The steps in the formulation process may then be summarized as follows:

Step 1. Develop the baseline model.
Step 2. Specify aspiration levels for each and every objective.
Step 3. Include negative and positive deviation variables for each and every goal and constraint.

Step 4. Rank the goals in terms of importance. Priority 1 is always reserved for the rigid constraints.

Step 5. Establish the achievement function.

Once these steps have been accomplished, we have a linear goal programming model that takes on the following general form:

Find $\mathbf{x} = (x_1, x_2, \ldots, x_n)$ so as to

$$\text{lexicographically minimize} \quad \mathbf{a} = \{g_1(\eta, \rho), \ldots, g_K(\eta, \rho)\} \tag{16.3}$$

subject to

$$f_i(\mathbf{x}) + \eta_i - \rho_i = b_i \qquad \text{for } i = 1, 2, \ldots, m \tag{16.4}$$

$$\mathbf{x}, \boldsymbol{\eta}, \boldsymbol{\rho} \geq \mathbf{0} \tag{16.5}$$

Since our attention, in this text, is focused on strictly linear models, the form of $f_i(\mathbf{x})$ is given as

$$f_i(\mathbf{x}) = \sum_{j=1}^{n} c_{i,j} x_j \tag{16.6}$$

where $c_{i,j}$ is the coefficient associated with variable j in goal or constraint i.

As mentioned earlier, there is more that must be considered in formulating the multiple-objective model, but the reward is that the resultant model should better reflect the actual problem. Since any solution obtained is only as good as the model it was derived from, the solution to the multiple-objective model should, in general, reflect an improvement over single-objective models.

The best way to obtain proficiency in the development of the linear goal programming model is through practice. Consequently, we present next a few examples of model development and then invite the reader to attempt the exercises given in the problem set at the end of the chapter.

Some Examples

The examples presented below, like those throughout the text, are provided only to indicate the steps involved in the development of a linear goal programming model. In no way can they reflect the size, complexity, amount of work required to collect relevant data, and the frustrations involved in actual practice.

Example 16-3

Time Machine, Inc., is a small cosmetics firm that, until recently, made only nail polish remover. However, when an employee accidently dropped a jar of peanut butter into the remover, it was found that the resultant mixture could temporarily remove facial wrinkles. Thus, the firm now produces two products: Wipe Out, the nail polish remover; and Ageless, the wrinkle remover. The new, improved formulas for each product require differing amounts of two base chemicals (whose names are a company secret), as shown in Table 16-2. The daily availability of these two chemicals cannot be exceeded, as there is only one supplier, who is producing at maximal capacity. Note that, although the daily supply of peanut butter is unlimited, the owner does not want to purchase more than 6 pounds per day, so as to keep secret the fact that this ingredient is going into production.

Table 16-2. DATA FOR EXAMPLE 16-3

Product	*Profit per Gallon*	*Pounds of Chemical A Required per Gallon*	*Pounds of Chemical B Required per Gallon*	*Pounds of Peanut Butter per Gallon*
Ageless wrinkle remover	80	4	4	1
Wipe out nail polish remover	100	5	2	0
Amounts available per day (pounds)		80	48	6

When the president of Time Machine is questioned, the following facts are determined.

1. The daily availability of chemicals A and B cannot, in any way, be exceeded. That is, these are rigid constraints.

The remaining goals and objectives of the firm, in order of preference, are:

2. The firm would like to maintain daily profits at a level above \$800.
3. The daily amount of peanut butter ordered should be kept under 6 pounds (that way it may be disguised as shipments to the company cafeteria).
4. The total number of gallons produced per day of both products should be minimized so as to simplify shipping and handling.

We begin by constructing the baseline model. Letting

x_1 = gallons per day of Ageless wrinkle remover produced

x_2 = gallons per day of Wipe Out nail polish remover produced

then the baseline model is:

$$4x_1 + 5x_2 \leq 80 \tag{16.7}$$

$$4x_1 + 2x_2 \leq 48 \tag{16.8}$$

$$80x_1 + 100x_2 \geq 800 \tag{16.9}$$

$$x_1 \leq 6 \tag{16.10}$$

$$\text{minimize} \quad x_1 + x_2 \tag{16.11}$$

$$x_1, x_2 \geq 0 \tag{16.12}$$

The relationships in (16.7) and (16.8) are associated with the limits on chemicals A and B, respectively. Relationship (16.9) reflects the profit goal, while (16.10) indicates the peanut butter goal. Our desire to minimize the total daily production amount is established in (16.11), while the nonnegativity restrictions are given in (16.12).

The only relationship for which we need an aspiration level is (16.11). We shall assume that the aspired level is 7 gallons or less per day. Coupling this with the rankings already established, we may form the following linear goal programming model:

Find x_1 and x_2 so as to

$$\text{lexicographically minimize} \quad \mathbf{a} = \{(\rho_1 + \rho_2), (\eta_3), (\rho_4), (\rho_5)\} \tag{16.13}$$

subject to

$$4x_1 + 5x_2 + \eta_1 - \rho_1 = 80 \tag{16.14}$$

$$4x_1 + 2x_2 + \eta_2 - \rho_2 = 48 \tag{16.15}$$

$$80x_1 + 100x_2 + \eta_3 - \rho_3 = 800 \tag{16.16}$$

$$x_1 + \eta_4 - \rho_4 = 6 \tag{16.17}$$

$$x_1 + x_2 + \eta_5 - \rho_5 = 7 \tag{16.18}$$

$$\mathbf{x}, \boldsymbol{\eta}, \boldsymbol{\rho}, \geq \mathbf{0} \tag{16.19}$$

The solution to this problem occurs, for the achievement vector specified, at:

$x_1 = 0$ (gallons per day) (i.e., we discontinue the wrinkle remover)

$x_2 = 8$ (gallons per day)

$\mathbf{a} = \{0, 0, 0, 1\}$

Since $a_1 = 0$, all rigid constraints are satisfied. The fact that a_2 is zero indicates achievement of the profit goal, and since a_3 is zero, we have also satisfied the peanut butter limit. The only deviation, 1 unit, is for a_4, which indicates that we could not, for the specifications given, keep the total production below 7 gallons per day (it must be set at 8 gallons per day) without degrading a higher-priority goal.

For this example, exactly the same solution would be obtained from single-objective linear programming if the objective in (16.11) is selected as the single objective to be minimized. We now move to an example for which this will not be the case.

Example 16-4

A firm produces two products. Product *A* has a net return of \$10 per unit and product *B* returns \$8 per unit. Product *A* requires 3 hours per unit in assembly and *B* takes 2 hours per unit produced. Total assembly time is 120 hours per week, but some overtime is possible. However, if overtime is utilized, the net return on both products are reduced by \$1 per unit produced on overtime. Under a present contract, the firm must supply the customer with a minimum of 30 units per week of both products.

Based on conversations with the owners, we establish the following:

1. The contract to the customer must be satisfied and only 120 hours of *regular* time is available.
2. Overtime is to be minimized.
3. Profit is to be maximized.

Letting

x_1 = number of units of product A produced, per week, on regular time

x_2 = number of units of product A produced, per week, on overtime

x_3 = number of units of product B produced, per week, on regular time

x_4 = number of units of product B produced, per week, on overtime

the baseline model may be written as

$$x_1 + x_2 \geq 30 \tag{16.20}$$

$$x_3 + x_4 \geq 30 \tag{16.21}$$

$$3x_1 + 2x_3 \leq 120 \tag{16.22}$$

$$\text{minimize} \quad 3x_2 + 2x_4 \tag{16.23}$$

$$\text{maximize} \quad 10x_1 + 9x_2 + 8x_3 + 7x_4 \tag{16.24}$$

$$\mathbf{x} \geq \mathbf{0} \tag{16.25}$$

Relationships (16.20) and (16.21) reflect the contract obligations for products A and B, respectively. The relationship in (16.22) reflects the fact that regular time is limited to 120 hours, while in (16.23) we indicate our desire to minimize overtime. Relationship (16.24) is the profit objective, while (16.25) indicates the nonnegativity restrictions. To convert the baseline model we must establish aspiration levels for the amount of overtime to be minimized and for the level of aspired profit. We shall assume that management wishes to hold overtime to less than 20 hours per week and that the profit is to be \$800 per week. The final linear goal programming model may then be established as soon as the different goals are ranked. Given that the contracts (for output of A and B) *must* be satisfied and that only 120 hours of regular time is available, the goals of (16.21) through (16.23) all reflect rigid constraints and are set at priority level 1. Then, with the minimization of overtime ranked higher than the maximization of profit, and under the assumption that the two goals are noncommensurable, our resultant goal programming model is:

Find $\mathbf{x} = (x_1, x_2, x_3, x_4)$ so as to

$$\text{lexicographically minimize} \quad \mathbf{a} = \{(\eta_1 + \eta_2 + \rho_3), (\rho_4), (\eta_5)\} \tag{16.26}$$

subject to

$$x_1 + x_2 + \eta_1 - \rho_1 = 30 \tag{16.27}$$

$$x_3 + x_4 + \eta_2 - \rho_2 = 30 \tag{16.28}$$

$$3x_1 + 2x_3 + \eta_3 - \rho_3 = 120 \tag{16.29}$$

$$3x_2 + 2x_4 + \eta_4 - \rho_4 = 20 \tag{16.30}$$

$$10x_1 + 9x_2 + 8x_3 + 7x_4 + \eta_5 - \rho_5 = 800 \tag{16.31}$$

$$\mathbf{x}, \boldsymbol{\eta}, \boldsymbol{\rho} \geq \mathbf{0} \tag{16.32}$$

The goals in (16.27) and (16.28) are rigid constraints (i.e., absolute goals) due to a contract and are achieved by minimizing η_1 and η_2. The goal of (16.29) is also absolute and reflects the maximum amount of regular time available. It is achieved by minimizing ρ_3. The goal in (16.30) is the overtime goal and achieved by minimizing ρ_4 (at priority 2), while that in (16.31) is the profit goal achieved by minimizing η_5 at priority 3. Finally, (16.32) reflects the typical nonnegativity restrictions.

Let us now assume that the overtime goal and profit goal can be placed into the same priority. This can be done if some weighting factor will permit a common measure of effectiveness. For the sake of illustration, we assume that overtime minimization is considered to be three times more important than profit maximization. The only change is to the achievement function, (16.26), which is then expressed as

$$\mathbf{a} = \{(\eta_1 + \eta_2 + \rho_3), (3\rho_4 + \eta_5)\} \tag{16.33}$$

The solution to the model given in (16.26) through (16.32) is as follows:

$$x_1 = 20$$
$$x_2 = 10$$
$$x_3 = 30$$
$$x_4 = 0$$
$$\mathbf{a} = \{0, 10, 270\}$$

That is, 30 units of product *B* are produced on regular time and none on overtime, while 20 units of *A* are produced on regular time and 10 on overtime. The achievement vector indicates that

$a_1 = 0$	all rigid constraints satisfied
$a_2 = 10$	10 hours of overtime above the 20-hour aspiration level were required
$a_3 = 270$	our deviation from the profit goal of \$800 was \$270 (under)

Now, if, for example, the overtime goal had been included as a rigid constraint and the profit were to be maximized, there would be no mathematically feasible solution to the resultant single-objective linear programming model.

Example 16-5

For our final example, consider the blending problem of the Double A Distillery. The Distillery utilizes three grades of whiskey (grades I, II, and III) to produce three different blends (DT, Drunk Tank, and Quiver Liver). The availabilities of the three grades are strictly limited and these availabilities and associated costs are as follows:

Grade I:	1,500 fifths/day at \$6.00/fifth
Grade II:	2,100 fifths/day at \$4.50/fifth
Grade III:	950 fifths/day at \$3.00/fifth

Double A has a quality reputation and, to keep it, they follow their recipes strictly, as shown in Table 16-3.

Table 16-3. DATA FOR EXAMPLE 16-5

Blend	*Recipe*	*Selling Price per Fifth*
DT	Less than 10% of grade II; more than 50% of grade I	$6.00
Drunk Tank	Less than 60% of grade III; more than 20% of grade I	5.50
Quiver Liver	Less than 50% of grade III; more than 10% of grade II	5.00

With the availabilities and recipes as rigid constraints, our remaining objectives and goals, in order of priority, are:

1. Maximize profit.
2. Produce at least 2,000 fifths of the prestige blend, DT, per day.

Our baseline model is formulated by first letting

$x_{1,1}$ = amount of grade I (in fifths) used in blend 1 (DT)

$x_{1,2}$ = amount of grade I used in blend 2 (Drunk Tank)

$x_{1,3}$ = amount of grade I used in blend 3 (Quiver Liver)

$x_{2,1}$ = amount of grade II used in blend 1

$x_{2,2}$ = amount of grade II used in blend 2

$x_{2,3}$ = amount of grade II used in blend 3

$x_{3,1}$ = amount of grade III used in blend 1

$x_{3,2}$ = amount of grade III used in blend 2

$x_{3,3}$ = amount of grade III used in blend 3

The baseline model is then

$$x_{1,1} + x_{1,2} + x_{1,3} \leq 1{,}500 \tag{16.34}$$

$$x_{2,1} + x_{2,2} + x_{2,3} \leq 2{,}100 \tag{16.35}$$

$$x_{3,1} + x_{3,2} + x_{3,3} \leq 950 \tag{16.36}$$

$$\left.\begin{array}{l} \dfrac{x_{2,1}}{x_{1,1} + x_{2,1} + x_{3,1}} \leq 0.1 \\ \dfrac{x_{1,1}}{x_{1,1} + x_{2,1} + x_{3,1}} \geq 0.5 \end{array}\right\} \tag{16.37}$$

$$\left.\begin{array}{l} \dfrac{x_{3,2}}{x_{1,2} + x_{2,2} + x_{3,2}} \leq 0.6 \\ \dfrac{x_{1,2}}{x_{1,2} + x_{2,2} + x_{3,2}} \geq 0.2 \end{array}\right\} \tag{16.38}$$

$$\left.\begin{array}{l} \dfrac{x_{3,3}}{x_{1,3} + x_{2,3} + x_{3,3}} \leq 0.5 \\ \dfrac{x_{2,3}}{x_{1,3} + x_{2,3} + x_{3,3}} \geq 0.1 \end{array}\right\} \tag{16.39}$$

$$\left.\begin{aligned}\text{maximize}\quad & 6(x_{1,1}+x_{2,1}+x_{3,1}) + 5.5(x_{1,2}+x_{2,2}+x_{3,2})\\ & + 5(x_{1,3}+x_{2,3}+x_{3,3}) - 6(x_{1,1}+x_{1,2}+x_{1,3})\\ & - 4.5(x_{2,1}+x_{2,2}+x_{2,3}) - 3(x_{3,1}+x_{3,2}+x_{3,3})\end{aligned}\right\} \tag{16.40}$$

$$x_{1,1}+x_{2,1}+x_{3,1} \geq 2{,}000 \tag{16.41}$$

$$\text{all } x_{i,j} \geq 0 \tag{16.42}$$

Relationships (16.34) through (16.36) reflect the rigid constraints on the availability of the three grades, while (16.37) through (16.39) indicate the rigid constraints on recipes for the three blends. Profit (i.e., sales minus cost) is indicated in (16.40), while the goal to produce at least 2,000 fifths of DT is shown in (16.41).

To convert to the goal programming form we shall assume a profit aspiration of $5,000 per day. The model is then:

Find $x_{i,j}$ (for $i = 1, 2, 3$ and $j = 1, 2, 3$) to

$$\text{lexicographically minimize}\quad \mathbf{a} = \{(\rho_1+\rho_2+\rho_3+\rho_4+\eta_5+\rho_6+\eta_7+\rho_8+\eta_9), (\eta_{10}), (\eta_{11})\}$$

$$\begin{aligned}
x_{1,1} + x_{1,2} + x_{1,3} + \eta_1 - \rho_1 &= 1{,}500\\
x_{2,1} + x_{2,2} + x_{2,3} + \eta_2 - \rho_2 &= 2{,}100\\
x_{3,1} + x_{3,2} + x_{3,3} + \eta_3 - \rho_3 &= 950\\
0.9x_{2,1} - 0.1x_{1,1} - 0.1x_{3,1} + \eta_4 - \rho_4 &= 0\\
0.5x_{1,1} - 0.5x_{2,1} - 0.5x_{3,1} + \eta_5 - \rho_5 &= 0\\
0.4x_{3,2} - 0.6x_{1,2} - 0.6x_{2,2} + \eta_6 - \rho_6 &= 0\\
0.8x_{1,2} - 0.2x_{2,2} - 0.2x_{3,2} + \eta_7 - \rho_7 &= 0\\
0.5x_{3,3} - 0.5x_{1,3} - 0.5x_{2,3} + \eta_8 - \rho_8 &= 0\\
0.9x_{2,3} - 0.1x_{1,3} - 0.1x_{3,3} + \eta_9 - \rho_9 &= 0
\end{aligned}$$

$$6(x_{1,1}+x_{2,2}+x_{3,1}) + 5.5(x_{1,2}+x_{2,2}+x_{3,2}) + 5(x_{1,3}+x_{2,3}+x_{3,3}) - 6(x_{1,1}+x_{1,2}+x_{1,3}) - 4.5(x_{2,1}+x_{2,2}+x_{2,3}) - 3(x_{3,1}+x_{3,2}+x_{3,3}) + \eta_{10} - \rho_{10} = 5{,}000$$

$$x_{1,1}+x_{2,1}+x_{3,1}+\eta_{11}-\rho_{11} = 2{,}000$$

$$\mathbf{x}, \boldsymbol{\eta}, \boldsymbol{\rho} \geq \mathbf{0}$$

Summary

This, our introductory chapter in multiple-objective linear programming, has dealt primarily with the background material necessary in the conversion of the baseline model into a linear goal programming model wherein a lexicographic minimum is sought for an ordered function (the achievement vector) of goal deviation variables. However, it was stressed that this specific multiple-objective model was selected for its flexibility. That is, with only relatively minor changes, the model can be made to encompass the (strictly) weighted goal approach, the efficient solution approach, combinations of these approaches, as well as other concepts in multiple-objective analysis.

Further, although beyond the scope of this text, it should be noted that the goal programming approach is, by no means, limited to strictly linear models. The author has developed a goal programming methodology for the nonlinear model, for the linear model with discrete-valued variables, for the nonlinear model with discrete-valued variables, and for models wherein the achievement function terms themselves are nonlinear [1, 2].

Finally, it cannot be emphasized enough that the solution to the linear goal programming problem (as did the solution to the single-objective linear programming problem) is intended to serve as *a decision aid.* The decision to accept such a solution or to further investigate is one that must be made by an evaluation, rather than an unquestioning acceptance, of the practical appropriateness of the result. In other words, the results obtained through the solution of the mathematical model are not the answer but, rather, provide systematically derived information upon which a policy may be based.

REFERENCES

1. IGNIZIO, J. P. *Goal Programming and Extensions.* Lexington, Mass.: Heath (Lexington Books), 1976.

2. IGNIZIO, J. P. "An Introduction to Goal Programming: With Applications to Urban Systems," *Journal of Computers, Environment and Urban Systems*, Volume 5, 1980, pp. 15–33.

PROBLEMS

16.1. Refer to the discussion of the baseline model in Chapter 2 and its conversion to the conventional, single-objective, linear programming model presented in Chapter 4. Contrast the approaches and assumptions used to convert the baseline model to the single-objective model versus those used in conversion to the goal programming model. Discuss, for both cases, the potential impacts of the assumptions employed on the resultant model and its solution.

16.2. Perform a (limited) literature search (Chapter 7 of reference [1] may be used to initiate such an effort) of the methods used to either weight or rank objectives and goals. Compare and discuss a selected subset of such methods.

16.3. Discuss the differences between the concept of an "objective" and that of a "goal." Why is it so important to draw a distinction between these two notions?

16.4. Given a goal of the general form

$$f_i(\mathbf{x}) + \eta_i - \rho_i = b_i$$

show why $\eta_i \rho_i = 0$.

16.5. How do goal deviation variables (i.e., η_i and ρ_i) differ from the slack and surplus variables used in single-objective linear programming?

16.6. Why is the concept of the lexicographic minimum (i.e., preemptive priorities) used in our goal programming achievement function?

16.7. Discuss how the concept of the lexicographic minimum is actually employed in solving single-objective linear programming problems.

16.8. A critic of goal programming states that it takes more effort and more knowledge of the problem to develop a goal programming model (i.e., versus that needed to construct a conventional linear programming model). Prepare an answer to this criticism so as to defend the use of goal programming.

16.9. A student of conventional linear programming states that goal programming is but a clever extension of linear programming. Prepare an argument that illustrates how linear programming might be more logically and naturally considered as a special case of linear goal programming.

16.10. Formulate the lexicographic linear goal programming model for the following problem. Energistics, Inc., produces a single type of moped (i.e., a small, gasoline-powered motorbike) called the "Spirit of America," or "Spirit" for short. It also imports an Italian moped, called the "Lirasaver," which it simply assembles and checks out. The Spirit sells for $650, and the Lirasaver sells for $725. Demand is such that essentially all the mopeds that the firm could make or import could easily be sold. The cost, to Energistics, of the imported, unassembled Lirasaver is $185 each. Data with regard to production time, assembly time, test time, and labor costs are given in the accompanying table.

	Hours per Unit		
Moped	*Manufacture*	*Assembly*	*Test*
Spirit	20	5	3
Lirasaver	0	7	6
Labor cost, per hour	$12 regular	$8 regular	$10 regular

After discussing the problem in more detail with Luigi Smith, the company president, we conclude that:

1. A profit of at least $3,000 per week is desired.
2. There are 120, 80, and 40 hours of regular time available per week for manufacture, assembly, and testing, respectively.
3. The firm believes that it would be politically wise to sell as many Spirits as possible.
4. The president wishes to minimize worker idle time without resorting, in any case, to overtime.

First formulate the baseline model and then, using your own assumptions as to the ranking of the various objectives and goals, convert to both a lexicographic linear goal programming model and a single-objective linear programming model.

16.11. Develop the goal programming model for Problem 2.10. List and justify any assumptions used in the ranking of the goal set.

16.12. Develop the goal programming models for Problems 2.13 and 2.14. List and justify any assumptions used in the ranking of the goal set.

16.13. Develop the goal programming model for Problems 2.16 and 2.17(a). Assume that considerations due to adverse publicity on mileage are at a higher priority than profit.

16.14. Formulate the goal programming model for Problem 2.20 wherein:

(a) With the exception of rigid constraints, the objective to minimize the variation in repeater station sizing is of top priority.

(b) After the repeater station sizing objective, profit and communication load balancing are of next highest priority, profit being considered twice as important.

(c) The lowest priority is given to the objective associated with minimizing the message load through link 3–4.

(d) You are to establish your own estimates as to goal aspiration levels.

16.15. A groundwater pumping station, to provide potable water, is to be constructed in a relatively small country town. The site of the station is fixed, because of the availability of well water, and the only questions remaining are:

(a) Which type of monitoring station (differing in terms of automation) should be used?

(b) Which firm should the actual machinery be purchased from?

(c) How many workers need to be hired to run the station? (Three 8-hour shifts are to be used.)

The town wishes to minimize the total initial costs as well as future operating costs (with twice as much importance placed on initial costs). However, since there is a high level of unemployment in the area, they also want to maximize the number of workers gainfully employed (i.e., a worker will not be hired unless there is some amount of actual work to be done). Assuming that the maximization of employment levels has top priority (among the set of non-absolute goals), formulate the baseline and goal programming models for this problem given the data in the accompanying table.

	Monitoring Station Type		*Pumping Machinery Type*		
	A	*B*	*I*	*II*	*III*
(1) Initial costs	\$2,000,000	\$1,500,000	\$5,000,000	\$4,000,000	\$3,500,000
(2) Yearly operating costs	100,000	250,000	200,000	300,000	800,000
(3) Number of personnel required per shift and per hour costs	4 @\$15/hr	6 @\$15/hr	6 @\$10/hr	10 @\$12/hr	15 @\$12/hr

METHODS OF SOLUTION

Introduction

In this chapter we introduce two basic approaches to the solution of the linear goal programming model with an achievement function that is to be *lexicographically minimized*. That is, we shall be working within a preemptive priority structure. Once the lexicographic minimal solution has been achieved, we may use the sensitivity analysis methods of Chapter 19 or the extensions of Chapter 20 to investigate the impact of relaxing the preemptive priority structure or even to extend the notion of the achievement function. Consequently, our methods of solution herein represent just one phase in the entire decision analysis framework.

The earliest method of solution proposed for the lexicographic minimum form of linear goal programming (or for the prioritized vector maximum problem) was, to this author's knowledge, a very simple if not transparent technique that relies upon the solution of a sequence of single-objective linear programming models. Thus, the author terms this technique *sequential linear goal programming* or *SLGP*. The author employed a primitive form of SLGP in actual problem solving during the mid-1960s (for problems involving the deployment of antennas on the Saturn V launch vehicle [1, 3, 4]) and in 1973, Kornbluth [6] published an independent description of the approach.

17

The second approach to the lexicographic linear goal programming problem is obtained through a relatively minor modification of the two-phase simplex method presented in Chapter 7. This method is probably the better known of the two approaches and appears in the goal programming texts by both Lee [7] and Ignizio [2]. It is referred to by various names, including the modified simplex method. However, since it is but an extension of the two-phase method, the author prefers the more descriptive name *multiphase simplex*.

A comparison of the two approaches readily shows that the multiphase algorithm is by far the more efficiently structured. However, because SLGP can use already developed simplex computer codes, it has (as of this date) exhibited an ability to solve larger problems.

Before introducing the SLGP or multiphase algorithms, we first illustrate the graphical approach to lexicographic linear goal programming. This serves three purposes:

1. It indicates the basic philosophy of solution approach.
2. It helps to define, through illustration, some of the terminology of goal programming.
3. It shows some of the fundamental *differences* between goal programming and conventional linear programming.

A Graphical Approach[a]

The fundamental difference between the approach to single-objective linear programming and lexicographic linear goal programming is that the conventional approach seeks a *point* (i.e., an extreme point) *that maximizes a single objective*, whereas goal programming seeks a *region that provides a compromise* to a set of conflicting goals. Otherwise, most of the steps of the graphical approach should be familiar. We list these below:

1. Plot all the goals (including the absolute goals or rigid constraints) in terms of the decision variables (this, of course, restricts the approach to problems with three variables or less).
2. Determine the solution space for the priority 1 goals.
3. Move to the set of goals having the next-highest priority and determine the "best" solution space for this set of goals, where this "best" solution cannot degrade the achievement values already obtained for higher-priority goals.
4. If, at any time in the process, the solution space is reduced to a single point, we may terminate the procedure because no further improvement is possible.
5. Repeat steps 3 and 4 until either we converge to a single point (step 4) or we have evaluated all the priority levels.

The reader should note that, as indicated in step 4, it may not always be necessary to evaluate every priority level since the lexicographic goal programming method seeks a compromise solution space. This space normally shrinks as we include the consideration of each priority and thus, obviously, can shrink no further once the solution space consists of but a single point. We illustrate this concept, as well as others, in the examples that follow.

Example 17-1

For this example we solve, graphically, the problem originally described in Example 16-3. The model is:

Find x_1 and x_2 so as to

$$\text{lexicographically minimize} \quad \mathbf{a} = \{(\rho_1 + \rho_2), (\eta_3), (\rho_4), (\rho_5)\}$$

subject to

$$\begin{aligned} G_1:&\quad 4x_1 + 5x_2 + \eta_1 - \rho_1 = 80 \\ G_2:&\quad 4x_1 + 2x_2 + \eta_2 - \rho_2 = 48 \\ G_3:&\quad 80x_1 + 100x_2 + \eta_3 - \rho_3 = 800 \end{aligned}$$

[a]The reader may wish to review the material in Chapter 5, the graphical approach to single-objective linear programming.

$$
\begin{array}{lllll}
G_4: & x_1 & & + \eta_4 - \rho_4 = & 6 \\
G_5: & x_1 + & x_2 & + \eta_5 - \rho_5 = & 7
\end{array}
$$

$$\mathbf{x}, \boldsymbol{\eta}, \boldsymbol{\rho} \geq 0$$

The designation G_i beside each goal is used to identify the corresponding graphical equivalent.

The five goals are plotted as straight lines in Figure 17-1. Note that only the *decision variables* (i.e., x_1 and x_2) are used in the plot. However, the effect of an increase in any deviation variable (η_i or ρ_i) is reflected by the arrows perpendicular to each goal line. The particular deviation variables to be minimized (i.e., those which appear within the achievement vector) have been circled.

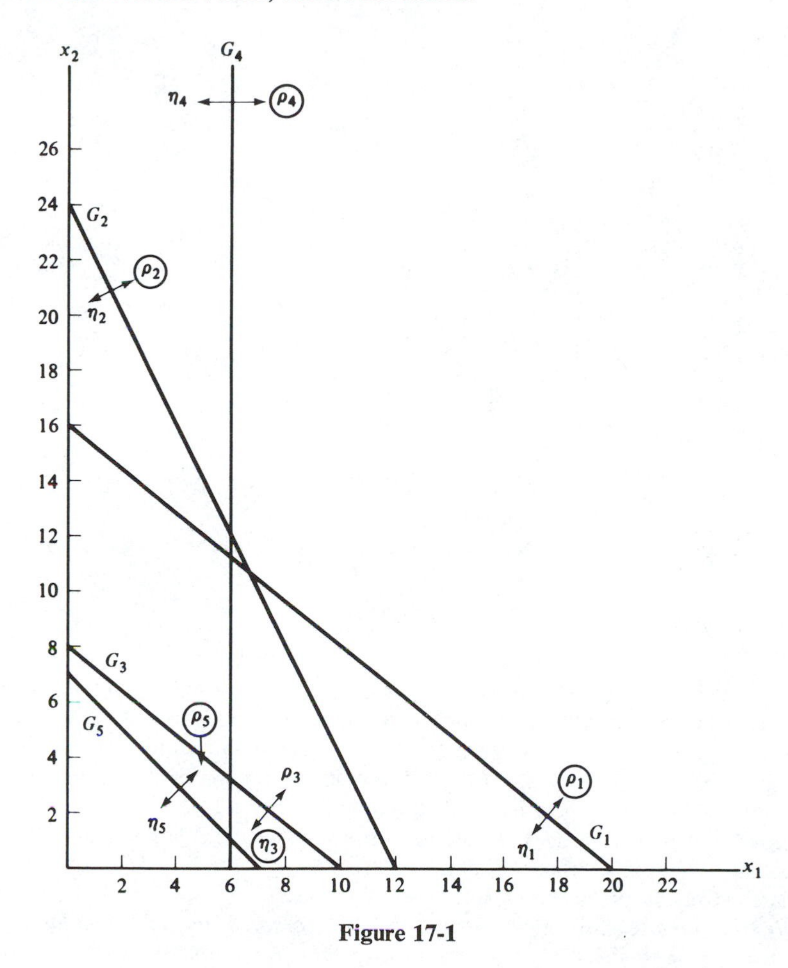

Figure 17-1

The two goals with the highest priority (G_1 and G_2) are considered first. Both goals may be satisfied by simultaneously minimizing ρ_1 and ρ_2 and, in fact, may be completely achieved by setting $\rho_1 = \rho_2 = 0$. The region remaining (i.e., the region for nonnegative x_1 and x_2 wherein $\rho_1 = \rho_2 = 0$) is shown in Figure 17-2 as the cross-hatched area.

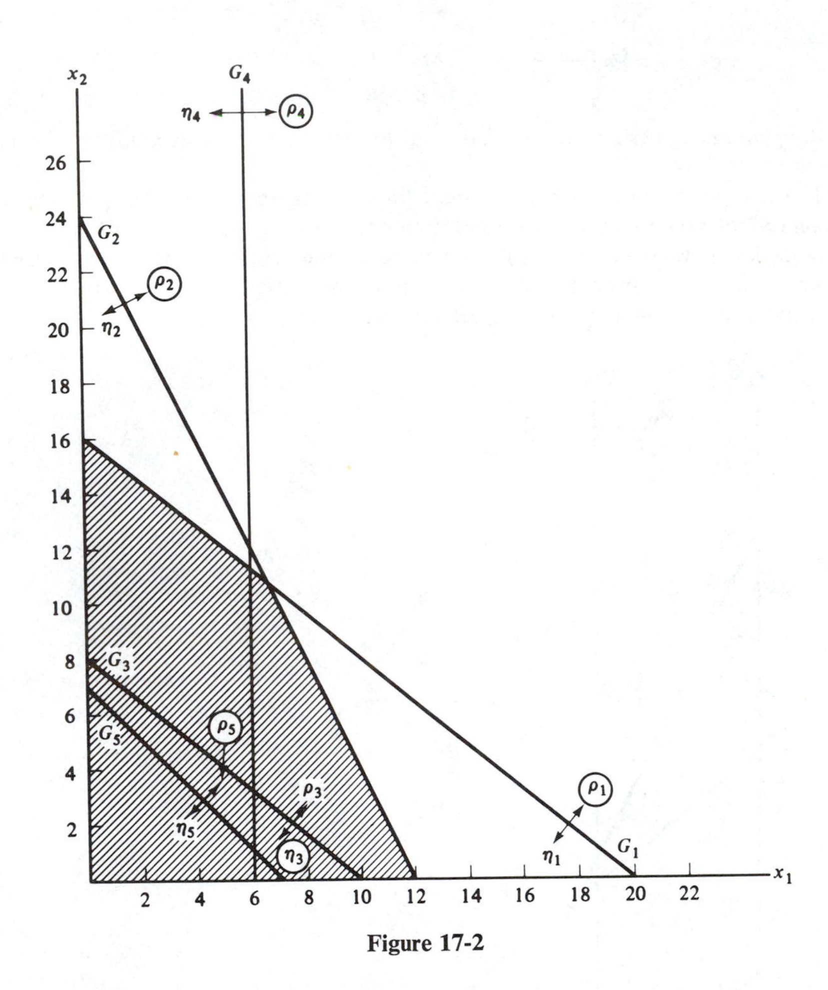

Figure 17-2

We next move to priority level 2, which is achieved through the minimization of η_3. Notice that, in Figure 17-2, η_3 may be set to zero without degrading the solution achieved for priority 1. That is, η_3 may be set to zero without any increase in either ρ_1 or ρ_2 (which combine to form the achievement term for a_1). The new, reduced area is now indicated as the crosshatched region in Figure 17-3.

Moving to priority level 3, we attempt to minimize ρ_4. Again, ρ_4 may be set to zero without increasing either the value of a_1 ($\rho_1 + \rho_2$) or a_2 (η_3) leaving us with the crosshatched region of Figure 17-4.

We have now reached our final priority level, achieved through the minimization of ρ_5. Examining Figure 17-4, it should be clear that ρ_5 cannot be set to zero without degrading the achievement level previously obtained for a higher priority (priority 2 with $a_2 = \eta_3$ to be specific). The smallest value of ρ_5 is a value of 1 at the point $x_1 = 0$, $x_2 = 8$. Thus, our solution space has converged to the point $\mathbf{x} = (0, 8)$, as

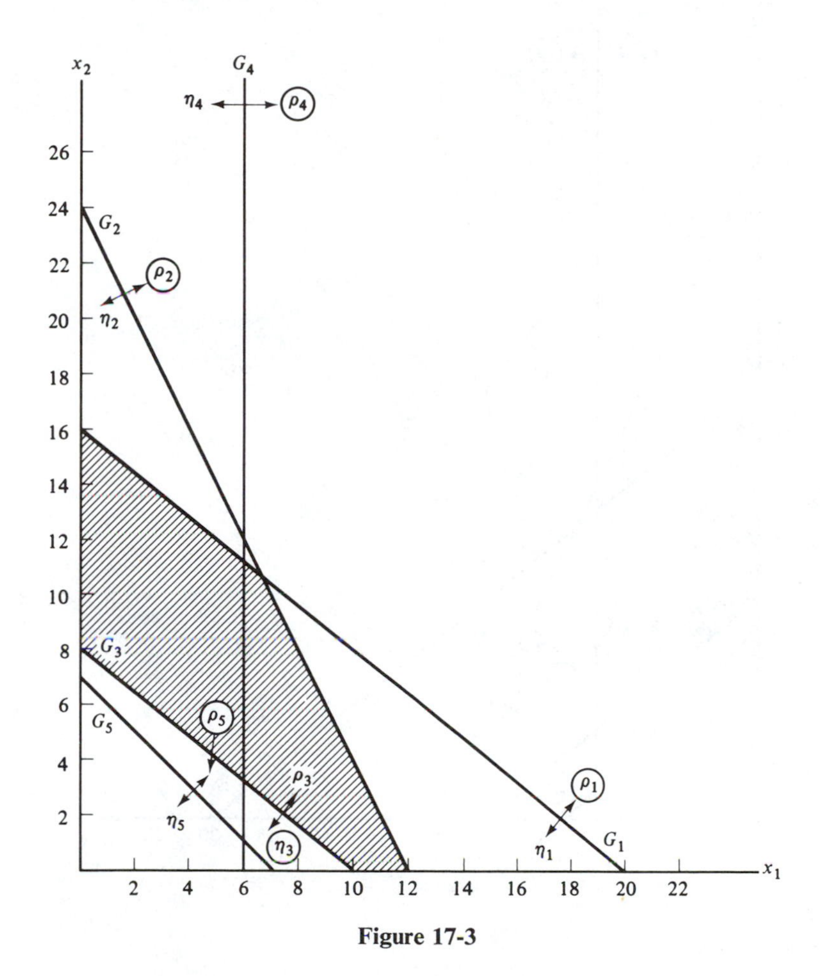

Figure 17-3

shown in Figure 17-5. The corresponding achievement vector is then $\mathbf{a} = (0, 0, 0, 1)$, and we have a solution to our model.

Notice in particular that, for every priority except the last, the corresponding solution that best satisfied the achievement vector was a convex solution *space*. We now consider another illustration in Example 17-2.

Example 17-2

The purpose of this example is to illustrate (1) what may happen if one picks an unreasonably high value for the aspiration level for a relatively high level goal, and (2) the relationship between linear goal programming and single-objective linear programming. To demonstrate, consider the following model:

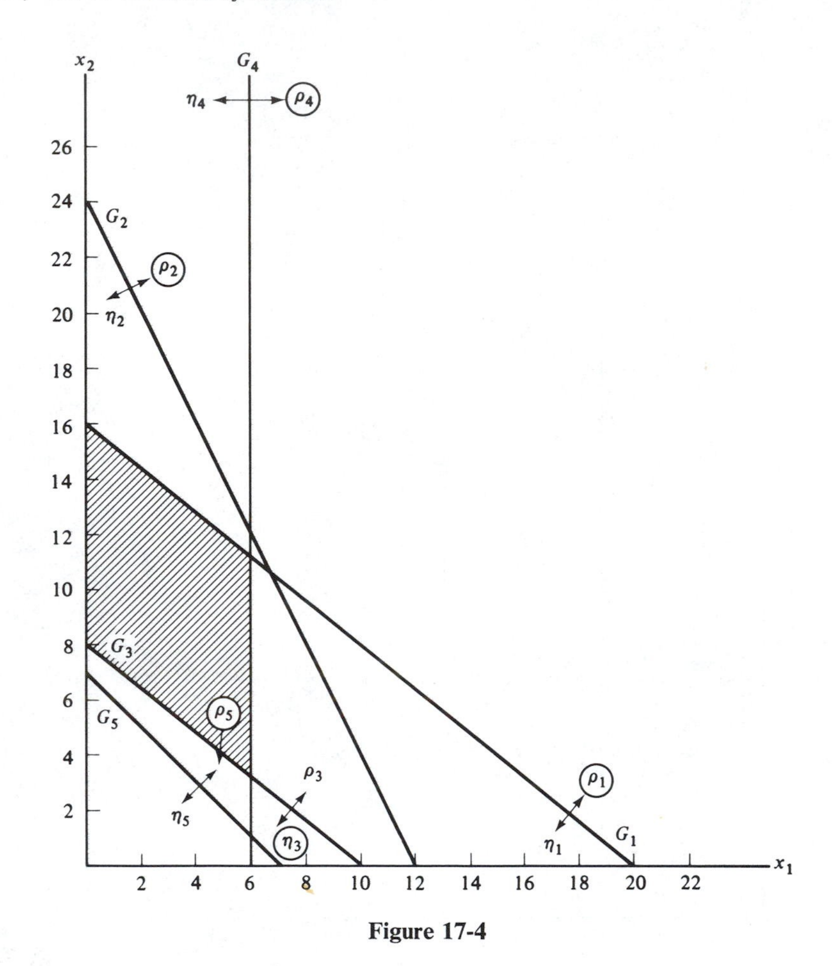

Figure 17-4

Find x_1, x_2 so as to

$$\text{lexicographically minimize} \quad \mathbf{a} = \{(\rho_1), (\eta_2), (\eta_3)\}$$

subject to

$$G_1: \quad x_1 + x_2 + \eta_1 - \rho_1 = 10$$

$$G_2: \quad 2x_1 + x_2 + \eta_2 - \rho_2 = 26$$

$$G_3: \quad -x_1 + 2x_2 + \eta_3 - \rho_3 = 6$$

$$\mathbf{x}, \boldsymbol{\eta}, \boldsymbol{\rho} \geq 0$$

We plot this problem initially in Figure 17-6. The region satisfying priority 1 ($\rho_1 = 0$) is shown crosshatched in Figure 17-6. When we next consider priority 2, we note that the solution space immediately converges to the point $x_1 = 10$, $x_2 = 0$. Consequently, priority level 3 has absolutely no impact on the final solution.

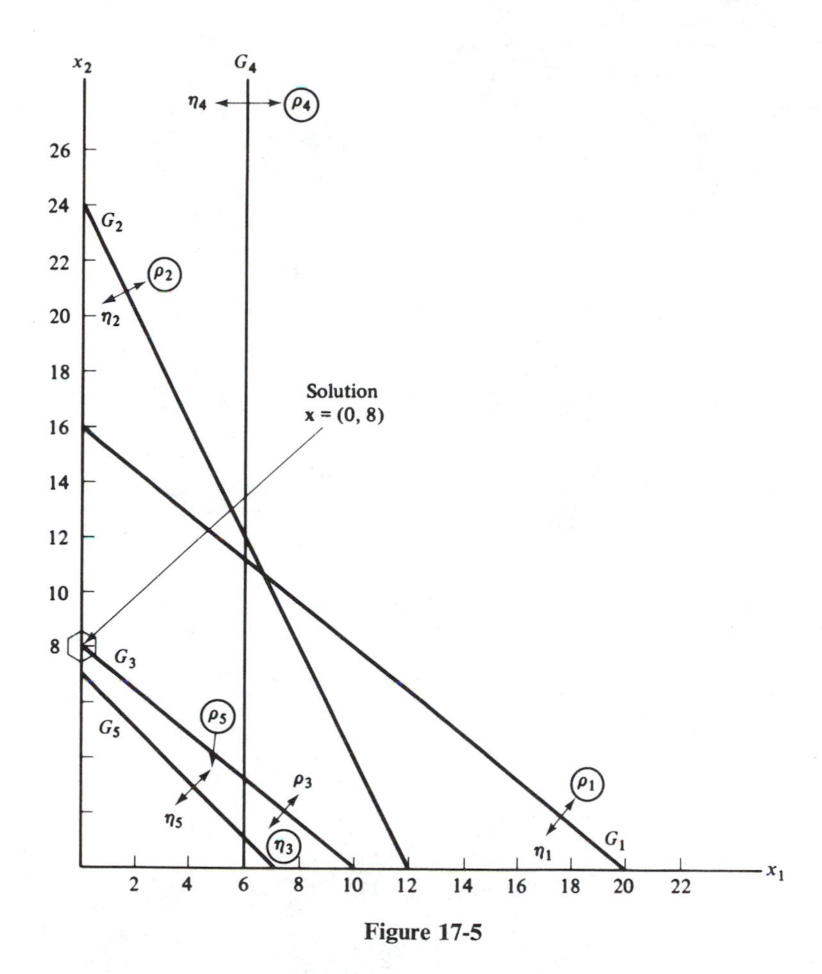

Figure 17-5

What *may* have happened, for such a problem, is that the achievement level for goal 2 (G_2) was set unreasonably high. For example, if the level had been set at, say 14 instead of 26, we would have had the opportunity to at least consider goal 3 (G_3).

The result demonstrated in this example also indicates how one may solve any single-objective linear programming problem within the goal programming framework. All we need to do is:

1. Associate all "constraints" with priority level 1.
2. Establish a very high (for maximizing, very low if minimizing), unreachable level for the single objective and set the achievement of this goal at priority 2.

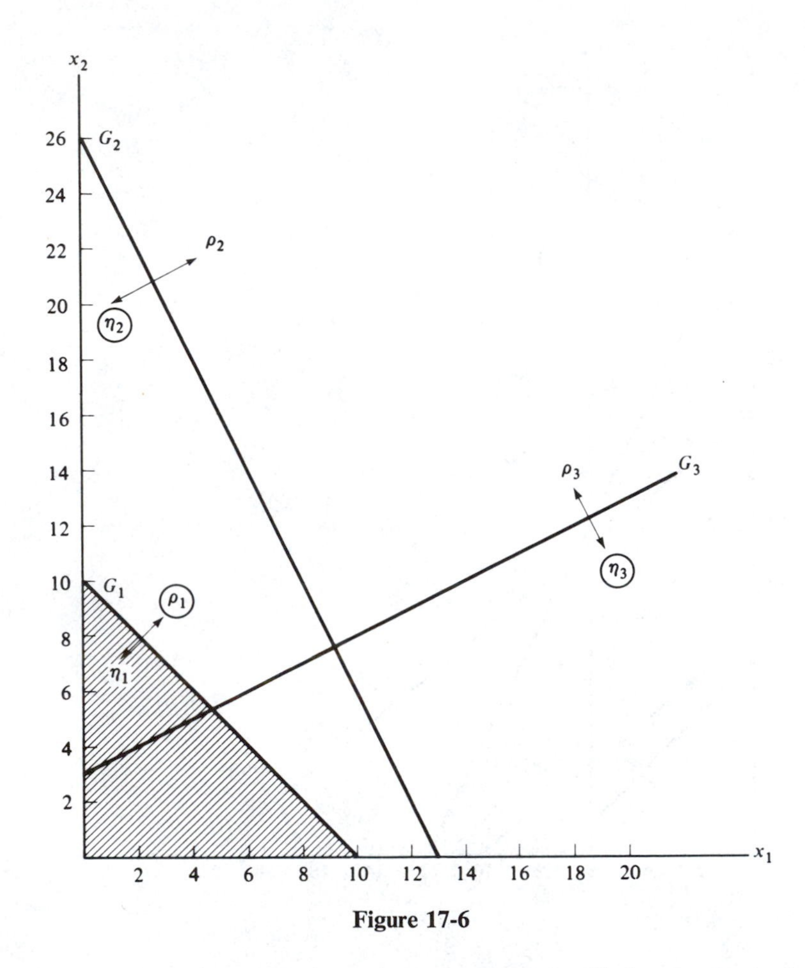

Figure 17-6

The solution to the linear goal programming model will then be equivalent to that reached if the problem had been treated in the conventional, single-objective framework. (A more detailed discussion of the relationship between linear programming and linear goal programming appears in reference [2].)

Some Definitions

We may now summarize (and, in some cases, review) some of the definitions and terminology that should serve to enhance our understanding of the goal programming methodology. We shall relate, where appropriate, some of these ideas back to our previous graphical examples.

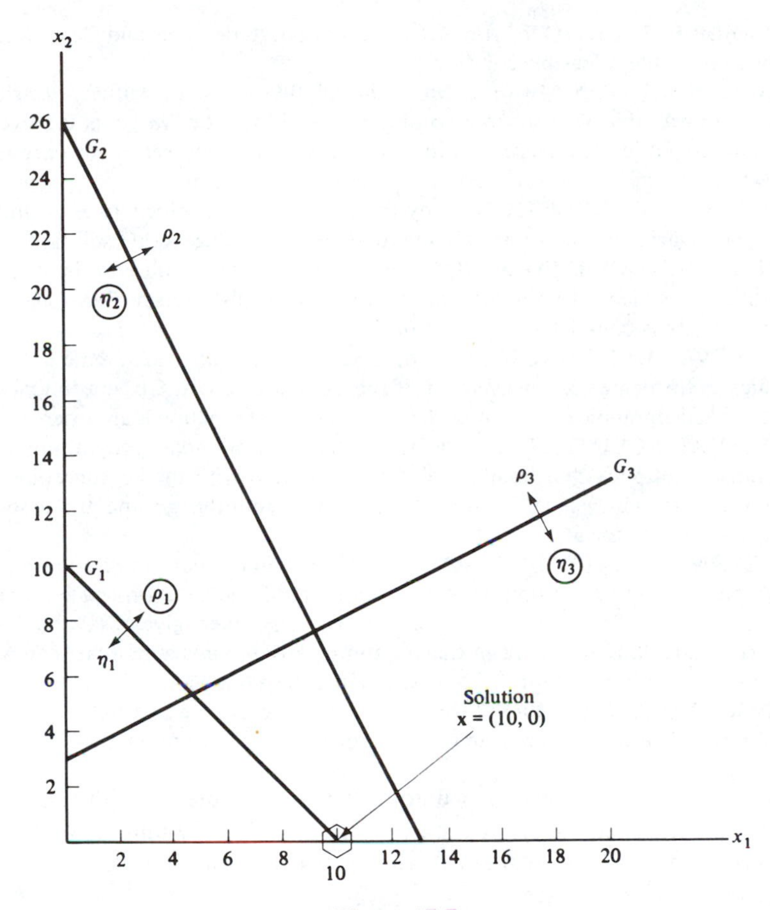

Figure 17-7

DECISION VARIABLE A decision variable, x_j $(j = 1, 2, \ldots, n)$, is a variable that is both under the control of the decision maker and one that can have an impact on the problem solution. Unless otherwise designated, all decision variables will be assumed nonnegative. In some disciplines, particularly engineering, such variables are often denoted as "control variables."

DEVIATION VARIABLES A deviation variable reflects either the underachievement (negative deviation and denoted as η_i) or overachievement (positive deviation and denoted as ρ_i) of an objective i. All deviation variables are assumed nonnegative unless otherwise noted.

LINEAR GOAL PROGRAM A linear goal program, wherein one treats objectives and constraints in a symmetric manner (i.e., we append positive and negative deviation variables in either case), is composed of m linear functions in n decision variables and $2m$ deviation variables. Each of these functions, whether originally an objective or constraint, is denoted as a goal.

FEASIBLE SOLUTION *Any* set of nonnegative decision and deviation variables constitute a feasible solution.

BASIC SOLUTION If $(n + 2m) - m$ of the variables (either decision or deviation variables) are set to zero and the resulting set of m goals is solved, the resultant solution is a basic solution. The m variables not set to zero are termed basic, while the $n + m$ variables set to zero are nonbasic.

DEGENERATE SOLUTION Any basic solution in which one or more of the basic variables takes on a zero value is termed a degenerate solution.

IMPLEMENTABLE SOLUTION An implementable solution is a feasible solution in which all rigid constraints (absolute goals) are satisfied. That is, the first priority is completely achieved ($a_1 = 0$).

ACHIEVEMENT FUNCTION The goal programming achievement function indicates the degree of achievement of the associated goals. Given a function that is to be lexicographically minimized, the achievement function is an ordered vector.

OPTIMAL SOLUTION For lexicographic linear goal programming, the optimal solution is that feasible solution associated with the lexicographic minimum of the achievement vector. We term this solution $\mathbf{x}^*$ and the associated achievement vector $\mathbf{a}^*$.

ALTERNATIVE OPTIMAL SOLUTIONS A linear goal programming problem has an alternative optimal set of solutions if the solution space associated with that problem is other than a single point. Further, given that the solution space is a region, any point *in* that region or any boundary of that region is an alternative optimal solution. That is, it gives the same value of $\mathbf{a}$.

UNBOUNDED SOLUTION Since aspiration levels are associated with every objective, a linear goal programming problem *cannot* be unbounded.

Going back to Figure 17-1, we note that the feasible solution space is *any* point within the first quadrant (i.e., $x_1, x_2 \geq 0$). This is unlike conventional linear programming, which requires that a feasible solution also be one that satisfies the "constraint" set.

Considering Figure 17-2, we note that an implementable solution is any point within the crosshatched area. That is, any solution satisfying the priority 1 goals may be implemented in the actual problem.

Examine next Figure 17-4, which shows the solution space satisfying the first three priorities. Notice that *any* point in the crosshatched region gives the same value for the achievement vector, and thus, unlike conventional linear programming, an alternative optimal solution may exist at the *interior* of a convex set.

Sequential Linear Goal Programming: SLGP

We begin our presentation of algorithms for (lexicographic) linear goal programming with a brief presentation of the earliest, at least to this author's knowledge, approach to that model. The underlying basis for this method, the

SLGP algorithm, is the sequential solution to a series of conventional linear programming models [4]. This is accomplished by partitioning the goal programming model according to priority levels. We first present a narrative summary and then discuss and illustrate the actual algorithm.

Given the linear goal programming model, let us first consider just the portion of the achievement vector and the goals associated with priority level 1. This results in the establishment of a single-objective linear programming model given as:

$$\text{minimize} \quad a_1 = g_1(\boldsymbol{\eta}, \boldsymbol{\rho})$$
$$\text{subject to}$$
$$\sum_{j=1}^{n} c_{i,j}x_j + \eta_i - \rho_i = b_i \qquad \text{for } i \in P_1$$
$$\mathbf{x}, \boldsymbol{\eta}, \boldsymbol{\rho} \geq \mathbf{0}$$

That is, we minimize the first term in the achievement function subject only to those goals in priority level 1 (i.e., $i \in P_1$). Once this is done, we have the best solution to a_1, designated as a_1^*.

We now move to the next priority level. Here we must minimize the second term in the achievement function, a_2. However, we must do so subject to:

1. All goals at priority 1.
2. All goals at priority 2.
3. Plus an extra goal (or rigid constraint) that assures that any solution to priority 2 cannot degrade the achievement level previously obtained in priority 1. That is,

$$g_1(\boldsymbol{\eta}, \boldsymbol{\rho}) = a_1^*$$

We continue this procedure until all priorities have been considered.[b] The solution to the final linear programming model is then also the solution to the equivalent linear goal program. The algorithm to implement the procedure is given below.

Sequential Linear Goal Programming (SLGP) Algorithm

Step 1. Set $k = 1$ (where k is used to represent the priority level under consideration and K is the total of these).

Step 2. Establish the mathematical formulation *for priority level 1 only*: That is,

$$\text{minimize} \quad a_1 = g_1(\boldsymbol{\eta}, \boldsymbol{\rho})$$
$$\text{subject to}$$
$$\sum_{j=1}^{n} c_{i,j}x_j + \eta_i - \rho_i = b_i \qquad \text{for } i \in P_1$$
$$\mathbf{x}, \boldsymbol{\eta}, \boldsymbol{\rho} \geq \mathbf{0}$$

[b]There are ways to shorten the procedure, as discussed later.

The resultant problem is simply a conventional (single-objective) linear programming problem and may be solved by simplex.

Step 3. Solve the single-objective problem associated with priority level k via any appropriate algorithm or computer code. Let the optimal solution to this problem be given as a_k^*, where

$$\mathbf{a}_k^* \text{ is the optimal value of } g_k(\boldsymbol{\eta}, \boldsymbol{\rho})$$

Step 4. Set $k = k + 1$. If $k > K$, go to step 7.

Step 5. Establish the equivalent, single-objective model for the next priority level (level k). This model is given by:

$$\text{minimize} \quad a_k = g_k(\boldsymbol{\eta}, \boldsymbol{\rho})$$
$$\text{subject to}$$
$$f_t(\mathbf{x}) + \eta_t - \rho_t = b_t$$
$$g_s(\boldsymbol{\eta}, \boldsymbol{\rho}) = a_s^*$$
$$\mathbf{x}, \boldsymbol{\eta}, \boldsymbol{\rho} \geq \mathbf{0}$$

where $s = 1, \ldots, k - 1$

t = set of subscripts associated with those goals or constraints included in priority levels $1, 2, \ldots, k$

Step 6. Go to step 3.

Step 7. The solution vector $\mathbf{x}^*$, associated with the last single-objective model solved, is the optimal vector for the original goal programming model.

In practice, there are several shortcuts that may be taken to reduce the computational aspects. These are generally transparent and we shall touch on a few in a moment. First, we demonstrate the basic approach via an example.

Example 17-3

We employ SLGP to solve the following model. The simplex method and condensed tableau of Chapter 7 are employed to solve each resultant single-objective model.

Find $\mathbf{x}$ so as to

$$\text{lexicographically minimize} \quad \mathbf{a} = \{(\rho_1 + \rho_2), (\eta_3), (\rho_4), (\eta_1 + 1.5\eta_2)\}$$
$$\text{subject to}$$
$$x_1 + \eta_1 - \rho_1 = 30$$
$$x_2 + \eta_2 - \rho_2 = 15$$
$$8x_1 + 12x_2 + \eta_3 - \rho_3 = 1{,}000$$
$$x_1 + 2x_2 + \eta_4 - \rho_4 = 40$$
$$\mathbf{x}, \boldsymbol{\eta}, \boldsymbol{\rho} \geq \mathbf{0}$$

We begin by formulating the model associated with priority level 1 only:

$$\text{minimize} \quad a_1 = (\rho_1 + \rho_2)$$
$$\text{subject to}$$
$$\begin{aligned} x_1 \qquad + \eta_1 - \rho_1 &= 30 \\ x_2 + \eta_2 - \rho_2 &= 15 \\ \mathbf{x}, \boldsymbol{\eta}, \boldsymbol{\rho} &\geq \mathbf{0} \end{aligned}$$

The initial tableau for this portion of the problem is given in Table 17-1. This initial tableau is also optimal, and thus the solution to the first level is:

$$\begin{array}{lll} x_1 = 0 & \eta_1 = 30 & \rho_1 = 0 \\ x_2 = 0 & \eta_2 = 15 & \rho_2 = 0 \end{array}$$
$$a_1^* = \rho_1 + \rho_2 = 0$$

Table 17-1. INITIAL TABLEAU, LEVEL 1

	$\mathbf{c}_N \longrightarrow$	*0*	*0*	*1*	*1*	
$\mathbf{c}_B$	V	x_1	x_2	ρ_1	ρ_2	$\mathbf{x}_B$
0	η_1	1	0	−1	0	30
0	η_2	0	1	0	−1	15
		0	0	−1	−1	0

We now move to the formulation for priority level 2, which is given, per our algorithm, as:

$$\text{minimize} \quad a_2 = \eta_3$$
$$\text{subject to}$$
$$\begin{aligned} x_1 \qquad + \eta_1 - \rho_1 &= 30 \\ x_2 + \eta_2 - \rho_2 &= 15 \\ 8x_1 + 12x_2 + \eta_3 - \rho_3 &= 1{,}000 \\ \rho_1 + \rho_2 = a_1^* &= 0 \\ \mathbf{x}, \boldsymbol{\eta}, \boldsymbol{\rho} &\geq \mathbf{0} \end{aligned}$$

The first two constraints correspond to those from priority 1, the third from priority 2, while the fourth constraint is imposed so as to avoid any degradation in the solution already obtained for priority 1. However, this constraint ($\rho_1 + \rho_2 = 0$) is trivial and may be dropped (i.e., it forces ρ_1 and ρ_2 to be zero) if we also drop ρ_1 and ρ_2 from the formulation. Consequently, the simplified formulation for level 2 is:

$$\text{minimize} \quad a_2 = \eta_3$$

$$\text{subject to}$$

$$\begin{aligned} x_1 \qquad\quad + \eta_1 \qquad\qquad &= 30 \\ x_2 + \eta_2 \qquad\qquad &= 15 \\ 8x_1 + 12x_2 + \eta_3 - \rho_3 &= 1{,}000 \\ \mathbf{x}, \boldsymbol{\eta}, \boldsymbol{\rho} \geq \mathbf{0} \end{aligned}$$

The initial tableau for this level is given in Table 17-2. The second and third tableaux are shown in Tables 17-3 and 17-4, with the result in Table 17-4 being optimal. That is, at level 2:

$$x_1 = 30 \qquad \eta_1 = 0 \qquad \rho_3 = 0$$

$$x_2 = 15 \qquad \eta_2 = 0$$

$$\eta_3 = 580$$

$$a_2^* = \eta_3 = 580$$

Table 17-2. INITIAL TABLEAU, LEVEL 2

	$\mathbf{c}_N \longrightarrow$	0	0	0	
$\mathbf{c}_B$	V	x_1	x_2	ρ_3	$\mathbf{x}_B$
0	η_1	1	0	0	30
0	η_2	0	1	0	15
1	η_3	8	12	−1	1,000
		8	12	−1	1,000

Table 17-3. SECOND TABLEAU, LEVEL 2

V	x_1	η_2	ρ_3	$\mathbf{x}_B$
η_1	1	0	0	30
x_2	0	1	0	15
η_3	8	−12	−1	820
	8	−12	−1	820

Table 17-4. FINAL TABLEAU, SECOND LEVEL

V	η_1	η_2	ρ_3	$\mathbf{x}_B$
x_1	1	0	0	30
x_2	0	1	0	15
η_3	−8	−12	−1	580
	−8	−12	−1	580

Before moving to priority level 3, we should examine the final tableau (Table 17-4). Note that the indicator row elements under the nonbasic variables η_1, η_2, and ρ_3 are *all negative in this optimal tableau.* That is, the marginal contributions (or shadow prices) of η_1, η_2, or ρ_3 indicate that if any of these variables are *ever* brought into solution, they would degrade the previous solution obtained for level 2. We may then cite the following auxiliary rule:

COLUMN DROP RULE Any nonbasic variable that has a negative indicator row value *in the optimal tableau* may be dropped (and its corresponding column in the tableau dropped) from the problem, as the introduction of such a variable would degrade the solution.

Thus, η_1, η_2, and ρ_3 may all be dropped, giving rise to the following formulation for level 3:

$$\text{minimize} \quad a_3 = \rho_4$$

$$\text{subject to}$$

$$\begin{aligned}
x_1 \qquad\qquad\qquad\qquad &= 30 \\
x_2 \qquad\qquad\qquad &= 15 \\
8x_1 + 12x_2 + \eta_3 \qquad\quad &= 1{,}000 \\
x_1 + \ 2x_2 + \eta_4 - \rho_4 &= 40 \\
\eta_3 \qquad\quad &= 580
\end{aligned}$$

$$\mathbf{x}, \boldsymbol{\eta}, \boldsymbol{\rho} \geq \mathbf{0}$$

The first four constraints are those associated with priorities 1, 2, and 3, while the final constraint ($\eta_3 = 580$) is added to assure that any solution will not degrade the result obtained for priority 2.

The foregoing problem is trivial, as the values of x_1 and x_2 are fixed by the first two constraints. That is, the solution is

$$x_1 = 30 \qquad \eta_3 = 580 \qquad \rho_4 = 20$$

$$x_2 = 15 \qquad \eta_4 = \quad 0$$

$$a_3^* = \rho_4 = 20$$

Moving to priority level 4, note that

$$a_4 = (\eta_1 + 1.5\eta_2)$$

but η_1 and η_2 have already been dropped (i.e., set to zero) in the problem. Thus, it is unnecessary to formulate or solve the model for level 4 as the solution is fixed, wherein

$$x_1^* = 30$$

$$x_2^* = 15$$

$$\mathbf{a}^* = (0, 580, 20, 0)$$

Thus, we have solved the equivalent linear goal programming problem through our sequence of conventional linear programming problems. A process made even simpler by noting such aspects as the column drop rule. We now consider one additional example.

Example 17-4

Using SLGP, we now solve:
Find $\mathbf{x}$ so as to

$$\text{lexicographically minimize} \quad \mathbf{a} = \{(\rho_1 + \rho_2), (\eta_3), (\rho_4)\}$$

subject to

$$\begin{aligned} 2x_1 + x_2 + \eta_1 - \rho_1 &= 12 \\ x_1 + x_2 + \eta_2 - \rho_2 &= 10 \\ x_1 \qquad + \eta_3 - \rho_3 &= 7 \\ x_1 + 4x_2 + \eta_4 - \rho_4 &= 4 \\ \mathbf{x}, \boldsymbol{\eta}, \boldsymbol{\rho} &\geq \mathbf{0} \end{aligned}$$

The formulation for level 1 is given below and its solution is given in Table 17-5.

Table 17-5. INITIAL TABLEAU, LEVEL 1

	$\mathbf{c}_N \longrightarrow$	*0*	*0*	*1*	*1*	
$\mathbf{c}_B$	V	x_1	x_2	ρ_1	ρ_2	$\mathbf{x}_B$
0	η_1	2	1	−1	0	12
0	η_2	1	1	0	−1	10
		0	0	−1	−1	0
				✓	✓	

$$\text{Minimize} \quad a_1 = (\rho_1 + \rho_2)$$

subject to

$$\begin{aligned} 2x_1 + x_2 + \eta_1 - \rho_1 &= 12 \\ x_1 + x_2 + \eta_2 - \rho_2 &= 10 \\ \mathbf{x}, \boldsymbol{\eta}, \boldsymbol{\rho} &\geq \mathbf{0} \end{aligned}$$

The initial solution in Table 17-5 is also the optimal solution and we note that ρ_1 and ρ_2 may both be dropped (i.e., set to zero) by the column drop rule. That is, at level 1:

$$\begin{aligned} x_1 &= 0 \qquad \eta_1 = 12 \qquad \rho_1 = 0 \text{ (dropped)} \\ x_2 &= 0 \qquad \eta_2 = 10 \qquad \rho_2 = 0 \text{ (dropped)} \end{aligned}$$

$$a_1^* = (\rho_1 + \rho_2) = 0$$

The check marks (✓) under the columns for ρ_1 and ρ_2 in Table 17-5 indicate that these columns and variables have been dropped.

Also, in formulating the model for level 2, we need not add the constraint $a_1^* = (\rho_1 + \rho_2) = 0$, since ρ_1 and ρ_2 are dropped. The model for level 2 is then:

$$\text{minimize} \quad a_2 = (\eta_3)$$

$$\text{subject to}$$

$$\begin{aligned} 2x_1 + x_2 + \eta_1 \qquad\qquad &= 12 \\ x_1 + x_2 + \eta_2 \qquad\qquad &= 10 \\ x_1 \qquad + \eta_3 - \rho_3 &= 7 \\ \mathbf{x}, \boldsymbol{\eta}, \boldsymbol{\rho} \geq 0 & \end{aligned}$$

The initial tableau for this model is given in Table 17-6, and the final solution is shown in Table 17-7.

Table 17-6. INITIAL TABLEAU, LEVEL 2

	$\mathbf{c}_N \longrightarrow$	0	0	0	
$\mathbf{c}_B$	V	x_1	x_2	ρ_3	$\mathbf{x}_B$
0	η_1	2	1	0	12
0	η_2	1	1	0	10
1	η_3	1	0	-1	7
		1	0	-1	7

Table 17-7. FINAL TABLEAU, LEVEL 2

V	η_1	x_2	ρ_3	$\mathbf{x}_B$
x_1	$\frac{1}{2}$	$\frac{1}{2}$	0	6
η_2	$-\frac{1}{2}$	$\frac{1}{2}$	0	4
η_3	$-\frac{1}{2}$	$-\frac{1}{2}$	-1	1
	$-\frac{1}{2}$	$-\frac{1}{2}$	-1	1
	✓	✓	✓	

Examining Table 17-7, we note that variables η_1, x_2, and ρ_3 may all be dropped via the column drop rule. The solution at this level is

$x_1 = 6$ $\qquad$ $\eta_1 = 0$ (dropped) $\qquad$ $\rho_3 = 0$ (dropped)

$x_2 = 0$ (dropped) $\qquad$ $\eta_2 = 4$

$\eta_3 = 1$

$$a_2^* = \eta_3 = 1$$

The linear programming equivalent model for level 3 is then:

$$\text{minimize} \quad a_3 = \rho_4$$

$$\text{subject to}$$

$$\begin{aligned} 2x_1 \qquad\qquad &= 12 \\ x_1 + \eta_2 \qquad &= 10 \\ x_1 + \eta_3 \qquad &= 7 \\ x_1 + \eta_4 - \rho_4 &= 4 \end{aligned}$$

$$a_2^* = \eta_3 = 1 \qquad \text{and} \qquad \mathbf{x}, \boldsymbol{\eta}, \boldsymbol{\rho} \geq \mathbf{0}$$

The solution to this model is obvious. Consider the first constraint. Obviously, x_1 must be 6 and, then, from the fourth constraint, ρ_4 must be two. The solution to level 3 is thus

$$x_1 = 6$$

$$x_2 = 0$$

$$a_3 = \rho_4 = 2$$

Consequently, the solution to the original linear goal program is

$$x_1^* = 6$$

$$x_2^* = 0$$

$$\mathbf{a}^* = (0, 1, 2)$$

Computer Codes for SLGP

Almost any linear goal programming problem for an actual problem would be so large as to require the use of the computer for solution. The author developed a SLGP code in the mid-1960s for use on the IBM 7094. This code was later refined and upgraded and implemented on the IBM 370–3033 using the IBM MPSX package (a simplex routine for large single-objective linear programs) as the procedure to solve each linear programming model in the sequence. A more detailed discussion may be found in reference [4]. This latter code, denoted as SLGP/MPSX, can solve lexicographic linear goal programming models with as many as 16,000 rows (i.e., goals and rigid constraints) and as many variables as one can store. An early version of the code solved a multiple-objective model with 100 variables and 100 rows in 60 seconds (less than 10 seconds of CPU time) and a larger problem, 2,025 variables and 138 rows, in 161 seconds (22 seconds of CPU time).

Some Characteristics of the SLGP Algorithm

One of the most appealing facets of SLGP is that one always deals with the (usually) more familiar single-objective model. Consequently, one may also use any commercial simplex code to implement the procedure. This has led to fairly significant recent interest in the algorithm, which the author believes is somewhat misplaced. The multiphase code, to be presented, is actually more straightforward and generally requires fewer computations. The only advantage of SLGP is temporary: it takes advantage of powerful commercial codes to solve larger problems than have yet to be addressed by the multiphase algorithm.

In addition to the shortcuts demonstrated in the examples and the column drop rule, there are several other techniques that may be implemented so as to reduce computation with SLGP. However, analogous approaches are available for the multiphase algorithm, and thus we will not consider this matter further.

Those familiar with single-objective linear programming should recall that there are certain conditions (e.g., alternative optimal solutions) that may be detected by the examination of the simplex tableau. The SLGP tableaux offer similar information as described below:

> **INFEASIBILITY** Recalling our definition of feasibility, we see that no solution to a linear goal programming model can be infeasible since, with the introduction of the goal deviation variables (for both objectives and rigid constraints), no basic solution may include a negative variable.
>
> **UNIMPLEMENTABLE SOLUTION** If, when applying SLGP, a_1 takes on a positive value, the problem has no solution that satisfies the rigid constraints of priority 1. However, the final program derived will indicate the solution that is nearest to being implementable and, by means of sensitivity analysis, we may determine which rigid constraints must be relaxed if we are to obtain an implementable solution.
>
> **UNBOUNDED SOLUTION** No goal programming problem can be unbounded because aspiration levels are associated with each objective. We either satisfy these levels or come as close as possible to satisfying them.
>
> **ALTERNATIVE OPTIMAL SOLUTION** An alternative optimal solution is any that will produce an identical achievement vector $\mathbf{a}^*$. The existence of alternative optimal solutions is detected in the final tableau for the last priority level, in exactly the same manner as conventional linear programming. That is, if any indicator row element under a nonbasic variable is zero-valued in the final, optimal tableau, then alternative optimal solutions exist.

We shall not go into any other characteristics or conditions since, for SLGP, they will be equivalent to those in single-objective linear programming. We now address the more promising multiphase algorithm.

The Multiphase Linear Goal Programming Algorithm

As mentioned before, the multiphase (or modified simplex) algorithm is simply a refinement of the well-known two-phase method (see Chapter 7). Before discussing the algorithm, let us first examine the special tableau that is used in the procedure, as it differs somewhat from those employed for the single-objective model.

The Condensed Tableau for the Multiphase Algorithm

The multiphase tableau, at first glance, is somewhat formidable. However, after one examines a few examples, the convenience offered by the tableau becomes obvious, and working with it should be about as easy as with the traditional simplex tableaux. Table 17-8 depicts the general, initial multiphase tableau in its condensed form (i.e., only nonbasic columns are included).

Table 17-8. THE INITIAL MULTIPHASE TABLEAU

	Left Stub										
			P_K	$w_{K,1}$	$\cdots$	$w_{K,n}$	$w_{K,n+1}$	$\cdots$	$w_{K,n+m}$	*Top Stub*	
			$\vdots$	$\vdots$			$\vdots$				
			P_1	$w_{1,1}$	$\cdots$	$w_{1,n}$	$w_{1,n+1}$	$\cdots$	$w_{1,n+m}$		
P_K	$\cdots$	P_1	V	x_1	$\cdots$	x_n	ρ_1	$\cdots$	ρ_m	$\mathbf{x}_B$	
$u_{1,K}$	$\cdots$	$u_{1,1}$	η_1	$y_{1,1}$	$\cdots$	$y_{1,n}$	$y_{1,n+1}$	$\cdots$	$y_{1,n+m}$	b_1	
	$\vdots$		$\vdots$		$\vdots$			$\vdots$		$\vdots$	
$u_{m,K}$	$\cdots$	$u_{m,1}$	η_m	$y_{m,1}$	$\cdots$	$y_{m,n}$	$y_{m,n+1}$	$\cdots$	$y_{m,n+m}$	b_m	
Indicator Rows			P_1	$R_{1,1}$	$\cdots$	$R_{1,n}$	$R_{1,n+1}$	$\cdots$	$R_{1,n+m}$	a_1	
			$\vdots$		$\vdots$			$\vdots$		$\vdots$	
			P_K	$R_{K,1}$	$\cdots$	$R_{K,n}$	$R_{K,n+1}$	$\cdots$	$R_{K,n+m}$	a_K	

The headings and elements within this tableau may be defined as follows:

Headings:

P_k = kth priority level, $k = 1, \ldots, K$

V = problem variables—both decision and deviation; the variables to the right of V (x_j and ρ_i) are the initial set of nonbasic variables, the variables below $V(\eta_i)$ are the initial set of basic variables

$\mathbf{x}_B$ = elements below $\mathbf{x}_B$—the initial *values* of the basic variables; since the initial basis (associated with $\eta_1, \ldots, \eta_m$) is an identity matrix, these initial values are simply the original right-hand side values (b_i's) of the model

Elements:

$j = 1, 2, \ldots, n$

$i = 1, 2, \ldots, m$

$s = 1, 2, \ldots, S$

$k = 1, 2, \ldots, K$

$y_{i,s}$ = interior tableau element in the ith row under the sth nonbasic variable; initially, $y_{i,s}$ is simply the coefficient of the sth nonbasic variable in the ith goal

$w_{k,s}$ = weighting factor for the nonbasic variable in column s at priority level $k(P_k)$

$u_{i,k}$ = weighting factor for the basic variable in row i at the kth priority level

$R_{k,s}$ = indicator row element for priority level k under the sth nonbasic variable, that is, the "shadow price" or "marginal utility" for the sth nonbasic variable at the kth priority level

a_k = level of achievement of the goals in priority k, where $\mathbf{a} = (a_1, \ldots a_k, \ldots a_K)$

All the elements in the initial tableau, except for $R_{k,s}$ and a_k, are simply obtained from the mathematical model. However, $R_{k,s}$ and a_k must be computed as follows[c]:

$$R_{k,s} = \mathbf{u}_k^T \mathbf{y}_s - w_{k,s} \tag{17.1a}$$

or

$$R_{k,s} = \sum_{i=1}^{m} (y_{i,s} u_{i,k}) - w_{k,s} \tag{17.1b}$$

and

$$a_k = \mathbf{u}_k^T \mathbf{x}_B \tag{17.2a}$$

or

$$a_k = \sum_{i=1}^{m} (x_{B,i} u_{i,k}) \tag{17.2b}$$

The establishment of the initial multiphase tableau is clarified by means of an illustration.

Example 17-5

We shall now establish the initial multiphase tableau for the following model. Find $\mathbf{x}$ so as to

$$\text{lexicographically minimize} \quad \mathbf{a} = \{(\rho_1 + \rho_2), (\eta_3 + 2\eta_4), (\eta_1)\}$$

subject to

$$\begin{aligned} x_1 \qquad\quad + \eta_1 - \rho_1 &= 20 \\ x_2 + \eta_2 - \rho_2 &= 35 \\ 5x_1 + 3x_2 + \eta_3 - \rho_3 &= 220 \\ x_1 + x_2 + \eta_4 - \rho_4 &= 60 \\ \mathbf{x}, \boldsymbol{\eta}, \boldsymbol{\rho} &\geq \mathbf{0} \end{aligned}$$

The initial multiphase tableau is shown in Table 17-9. For clarity, the zero elements of the tableau have been omitted. *Some* of the computations that were necessary

---

[c]The reader should note the analogy between (17.1) and (17.2) and the single-objective equivalents of (7.2) and (7.1), respectively.

Table 17-9. INITIAL TABLEAU, EXAMPLE 17-5

			P_3							
			P_2							
			P_1			1	1			
P_3	P_2	P_1	V	x_1	x_2	ρ_1	ρ_2	ρ_3	ρ_4	$\mathbf{x}_B$
1			η_1	1		−1				20
			η_2		1		−1			35
	1		η_3	5	3			−1		220
	2		η_4	1	1				−1	60
			P_1			−1	−1			0
			P_2	7	5			−1	−2	340
			P_3	1		−1				20

to compute the $R_{k,s}$ and a_k elements are listed below.

$$R_{1,1} = (0 \quad 0 \quad 0 \quad 0)\begin{pmatrix}1\\0\\5\\1\end{pmatrix} - 0 = 0$$

$$R_{1,2} = (0 \quad 0 \quad 0 \quad 0)\begin{pmatrix}0\\1\\3\\1\end{pmatrix} - 0 = 0$$

$$R_{1,3} = (0 \quad 0 \quad 0 \quad 0)\begin{pmatrix}-1\\0\\0\\0\end{pmatrix} - 1 = -1$$

$$R_{2,1} = (0 \quad 0 \quad 1 \quad 2)\begin{pmatrix}1\\0\\5\\1\end{pmatrix} - 0 = 7$$

$$a_1 = (0 \quad 0 \quad 0 \quad 0)\begin{pmatrix}20\\35\\220\\60\end{pmatrix} = 0$$

$$a_2 = (0 \quad 0 \quad 1 \quad 2)\begin{pmatrix}20\\35\\220\\60\end{pmatrix} = 340$$

The interpretation of the initial tableau of Table 17-9 is as follows:

1. The initial solution is:

$$\left.\begin{aligned} \eta_1 &= 20 \\ \eta_2 &= 35 \\ \eta_3 &= 220 \\ \eta_4 &= 60 \end{aligned}\right\} \text{basic variables}$$

$$\left.\begin{aligned} x_1 &= 0 \\ x_2 &= 0 \\ \rho_1 &= 0 \\ \rho_2 &= 0 \\ \rho_3 &= 0 \\ \rho_4 &= 0 \end{aligned}\right\} \text{nonbasic variables}$$

2. The achievement vector is

$$\mathbf{a} = (0, 340, 20)$$

   Thus, only priority level 1 is completely achieved.

3. The present solution, via an examination of the $R_{k,s}$ elements, is seen to not yet be optimal. We shall explain how this conclusion is reached in the algorithm presented next.

The Multiphase Simplex Algorithm

By following the steps given below, the optimal solution to the linear, lexicographic goal programming model may be derived. As with most any algorithm, numerous means are available to improve the computational process, and we address a few of the more obvious of these in the discussion to follow.

Step 1. *Initialization.* Establish the initial multiphase tableau and the indicator row for priority level 1 *only* (the $R_{1,s}$ elements). Set $k = 1$ and proceed to step 2.

Step 2. *Check for optimality.* Examine each *positive*-valued indicator row element ($R_{k,s}$) in indicator row k. Select the largest positive $R_{k,s}$ for which there are no negative-valued *indicator* numbers, at a higher priority, in the same column. Designate this column as s'. In the event of ties, the selection of $R_{k,s}$ may be made arbitrarily. If no such $R_{k,s}$ may be found in the kth row, go to step 6. Otherwise, go to step 3.

Step 3. *Determining the entering variable.* The nonbasic variable associated with column s' is the new entering variable. (Ties are broken arbitrarily.)

Step 4. *Determining the departing variable.* Determine the row associated with the minimum *nonnegative* value of

$$\frac{x_{B,i}}{y_{i,s'}}$$

In the event of ties, select the row having the basic variable with the higher-priority level. Designate this row as i'. The basic variable associated with row i' is the departing variable.

Step 5. *Establishment of the new tableau*

(a) Set up a new tableau with all $y_{i,s}$, $x_{B,i}$, $R_{k,s}$, and a_k elements empty. Exchange the positions of the basic variable heading in row i' (of the preceding tableau) with the nonbasic variable heading in column s' (of the preceding tableau).

(b) Row i' of the new tableau (except for $y_{i',s'}$) is obtained by dividing row i' of the preceding tableau by $y_{i',s'}$.

(c) Column s' of the new tableau (except for $y_{i',s'}$) is obtained by dividing column s' of the preceding tableau by the *negative* of $y_{i',s'}$ (i.e., by $-y_{i',s'}$).

(d) The new element at position $y_{i',s'}$ is given by the reciprocal of $y_{i',s'}$ (from the preceding tableau). The remaining tableau elements are computed as follows. Let any element with a caret over it (i.e., $\hat{x}_{B,i}$, $\hat{y}_{i,s}$, etc.) represent the *new* set of elements, while those without the caret denote the values of these elements from the preceding tableau. Then, for those elements *not* in either row i' or column s':

$$\hat{x}_{B,i} = x_{B,i} - \frac{(x_{B,i'})(y_{i,s'})}{y_{i',s'}} \tag{17.3}$$

$$\hat{y}_{i,s} = y_{i,s} - \frac{(y_{i',s})(y_{i,s'})}{y_{i',s'}} \tag{17.4}$$

$$\hat{R}_{k,s} = R_{k,s} - \frac{(y_{i',s})(R_{k,s'})}{y_{i',s'}} \tag{17.5}$$

$$\hat{a}_k = a_k - \frac{(x_{B,i'})(R_{k,s'})}{y_{i',s'}} \tag{17.6}$$

An alternative approach to computing the new $R_{k,s}$ and a_k values is to employ (17.1) and (17.2). Note that (17.3) through (17.6) all have the following form:

$$\text{new value} = \text{old value} - \frac{(\text{APRV})(\text{APCV})}{\text{PNV}} \tag{17.7}$$

where APRV = associated pivot row value

APCV = associated pivot column value

PNV = pivot number value (i.e., $y_{i',s'}$)

i' = pivot row

s' = pivot column

(e) Return to step 2.

Step 6. *Convergence check*. Examine each column vector of indicator elements ($\mathbf{R}_s$) in the present tableau. At least one of these column vectors must consist solely of zeros if the present solution is to be improved. If so, go to step 7. Otherwise, we have reached the optimal solution and may stop.

Step 7. *Evaluate the next-lower priority level*. Set $k = k + 1$. If k now exceeds K (the total number of priorities), then stop, as the present solution is optimal. If

$k \leq K$, establish the indicator row for priority k (P_k) from (17.1) and (17.2) and go to step 2.

We may now demonstrate the implementation of the algorithm with a few example problems.

Example 17-6

We shall use the multiphase algorithm to solve the problem previously addressed by SLGP in Example 17-4. That is,

Find $\mathbf{x}$ so as to

$$\text{lexicographically minimize} \quad \mathbf{a} = \{(\rho_1 + \rho_2), (\eta_3), (\rho_4)\}$$

subject to

$$\begin{aligned} 2x_1 + \ x_2 + \eta_1 - \rho_1 &= 12 \\ x_1 + \ x_2 + \eta_2 - \rho_2 &= 10 \\ x_1 \qquad + \eta_3 - \rho_3 &= 7 \\ x_1 + 4x_2 + \eta_4 - \rho_4 &= 4 \\ \mathbf{x}, \boldsymbol{\eta}, \boldsymbol{\rho} &\geq \mathbf{0} \end{aligned}$$

The initial tableau for the model is shown in Table 17-10. Note that only the indicator row for P_1 is calculated at this stage.

Table 17-10. INITIAL TABLEAU, EXAMPLE 17-6

			P_3						1	
			P_2							
			P_1			1	1			
P_3	P_2	P_1	V	x_1	x_2	ρ_1	ρ_2	ρ_3	ρ_4	$\mathbf{x}_B$
			η_1	2	1	−1				12
			η_2	1	1		−1			10
	1		η_3	1				−1		7
			η_4	1	4				−1	4
			P_1			−1	−1			0

Step 1. The initial tableau is given in Table 17-10. Set $k = 1$ and go to step 2.
Step 2. Since there are no positive elements in the indicator row, we go to step 6.
Step 6. Since at least one column vector in the indicator row consists solely of zeros (actually, four, under x_1, x_2, ρ_3, and ρ_4, consist solely of zeros), we go to step 7.
Step 7. $k = k + 1 = 2$. Since $k \leq K$ (i.e., $2 \leq 3$), we compute the indicator row for priority 2 to form the new tableau shown in Table 17-11. Go to step 2.

Table 17-11. SECOND TABLEAU, EXAMPLE 17-6

			P_3						1	
			P_2							
			P_1			1	1			
P_3	P_2	P_1	V	x_1	x_2	ρ_1	ρ_2	ρ_3	ρ_4	$\mathbf{x}_B$
			η_1	2	1	−1				12
			η_2	1	1		−1			10
	1		η_3	1				−1		7
			η_4	1	4				−1	4
			P_1			−1	−1			0
			P_2	1				−1		7

Step 2. The largest positive (actually, the only positive) $R_{1,s}$ value without any negative values above it is associated with column 1. Thus, $s' = 1$. We go to step 3.

Step 3. Since $s' = 1$, x_1 is the entering variable.

Step 4. Computing all nonnegative $x_{B,i}/y_{i,s'}$ ratios, we obtain

$$\frac{x_{B,1}}{y_{1,1}} = \frac{12}{2} = 6$$

$$\frac{x_{B,2}}{y_{2,1}} = \frac{10}{1} = 10$$

$$\frac{x_{B,3}}{y_{3,1}} = \frac{7}{1} = 7$$

$$\frac{x_{B,4}}{y_{4,1}} = \frac{4}{1} = 4 \quad \text{(minimum value)}$$

Thus, row 4 has the smallest nonnegative value and $i' = 4$. Consequently, η_4 is the departing variable.

Step 5. (a) The new tableau, with x_1 and η_4 interchanged in the column and row headings, is shown in Table 17-12.

Table 17-12

			P_3						1	
			P_2							
			P_1			1	1			
P_3	P_2	P_1	V	η_4	x_2	ρ_1	ρ_2	ρ_3	ρ_4	$\mathbf{x}_B$
			η_1							
			η_2							
	1		η_3							
			x_1							
			P_1							
			P_2							

(b) Row $i' = 4$ of the new tableau (except for $y_{i', s'}$) is obtained by dividing row 4 of the preceding tableau (in Table 17-11) by $y_{i', s'} = y_{4, 1} = 1$.

(c) Column $s' = 1$ of the new tableau, except for $y_{4, 1}$, is obtained by dividing column 1 of the preceding tableau by $-y_{4, 1} = -1$.

(d) The new element at $y_{4, 1}$ is the reciprocal of the element in row 4, column 1 of the preceding tableau. Thus, $y_{4, 1} = \frac{1}{1} = 1$. The remaining elements in the matrix are computed using (17.3) through (17.6), producing the tableau shown in Table 17-13.

Table 17-13

P_3	P_2	P_1	V	η_4	x_2	ρ_1	ρ_2	ρ_3	ρ_4	$\mathbf{x}_B$
			P_3						1	
			P_2							
			P_1			1	1			
			η_1	−2	−7	−1			2	4
			η_2	−1	−3		−1		1	6
	1		η_3	−1	−4			−1	1	3
			x_1	1	4				−1	4
			P_1			−1	−1			0
			P_2	−1	−4			−1	1	3

(e) Return to step 2.

Step 2. The largest (and only) positive $R_{2, s}$ value without any negative values above it is associated with column 6 (i.e., under ρ_4). Thus, $s' = 6$ and we go to step 3.

Step 3. Since $s' = 6$, the new entering variable is ρ_4.

Step 4. Computing all nonnegative $x_{B, i}/y_{i, s'}$ $(x_{B, i}/y_{i, 4})$ ratios we find that the smallest is associated with row 1 (i.e., $\frac{4}{2}$), and thus $i' = 1$. Consequently, η_1 is the departing variable.

Step 5. Using the rules in step 5, we obtain the new tableau, as shown in Table 17-14. Go to step 2.

Table 17-14

P_3	P_2	P_1	V	η_4	x_2	ρ_1	ρ_2	ρ_3	η_1	$\mathbf{x}_B$
			P_3							
			P_2							
			P_1			1	1			
1			ρ_4	−1	$-\frac{7}{2}$	$-\frac{1}{2}$			$\frac{1}{2}$	2
			η_2		$\frac{1}{2}$	$\frac{1}{2}$	−1		$-\frac{1}{2}$	4
	1		η_3		$-\frac{1}{2}$	$\frac{1}{2}$		−1	$-\frac{1}{2}$	1
			x_1		$\frac{1}{2}$	$-\frac{1}{2}$			$\frac{1}{2}$	6
			P_1			−1	−1			0
			P_2		$-\frac{1}{2}$	$\frac{1}{2}$		−1	$-\frac{1}{2}$	1

Step 2. Since there are no positive $R_{2,s}$ elements without negative elements above (i.e., in indicator row 1), we go to step 6.
Step 6. Since $\mathbf{R}_1$ consists solely of zeros, we go to step 7.
Step 7. $k = k + 1 = 2 + 1 = 3$. We set up the indicator row for priority 3 as shown in Table 17-15. Go to step 2.

Table 17-15

			P_3							
			P_2							
			P_1			1	1			
P_3	P_2	P_1	V	η_4	x_2	ρ_1	ρ_2	ρ_3	η_1	$\mathbf{x}_B$
1			ρ_4	-1	$-\frac{7}{2}$	$-\frac{1}{2}$			$\frac{1}{2}$	2
			η_2		$\frac{1}{2}$	$\frac{1}{2}$	-1		$-\frac{1}{2}$	4
	1		η_3		$-\frac{1}{2}$	$\frac{1}{2}$		-1	$-\frac{1}{2}$	1
			x_1		$\frac{1}{2}$	$-\frac{1}{2}$			$\frac{1}{2}$	6
			P_1			-1	-1			0
			P_2		$-\frac{1}{2}$	$\frac{1}{2}$		-1	$-\frac{1}{2}$	1
			P_3	-1	$-\frac{7}{2}$	$-\frac{1}{2}$			$\frac{1}{2}$	2

Step 2. Since there is no positive $R_{3,s}$ without a negative value above it, we go to step 6.
Step 6. Since there are no column vectors ($\mathbf{R}_s$) consisting solely of zeros, we stop with the optimal solution given as:

$$x_1^* = 6 \quad \eta_1^* = 0 \quad \rho_1^* = 0$$
$$x_2^* = 0 \quad \eta_2^* = 4 \quad \rho_2^* = 0$$
$$\eta_3^* = 1 \quad \rho_3^* = 0$$
$$\eta_4^* = 0 \quad \rho_4^* = 2$$
$$\mathbf{a}^* = (0, 1, 2)$$

Comparing results, we observe that SLGP and the multiphase algorithm produced identical results, as they should, for this example. However, note that we did not take advantage of any column drop rule as was done in SLGP. We could, if we so wish, drop columns from the multiphase tableaux by observing the following rule:

> **COLUMN DROP RULE** (Multiphase Tableaux) Given any multiphase tableau *that is optimal for the priorities under consideration* (i.e., there are no positive $R_{k,s}$ values without negative elements above them), any nonbasic variable and associated column may be dropped (the variables may be set permanently to zero and the columns removed from the tableau) if its associated $\mathbf{R}_s$ vector has, as its first nonzero element, a negative value.

Using this rule, we note that:

1. In Table 17-10, the columns associated with ρ_1 and ρ_2 could have been dropped.

2. In Table 17-14, the columns associated with x_2, ρ_3, and η_1 could have been dropped.
3. In Table 17-15, *all* the columns could have been dropped.

The column drop rule can provide a significant reduction in time when solving problems by hand. However, its sole disadvantage appears when we wish to perform a sensitivity analysis on the final solution obtained. If columns have been dropped, we must first recompute their values and insert these columns before we can conduct the subsequent analysis.

As yet another note, it should be obvious that both the top stub and left stub need not actually be included in the tableaux as long as we save the original formulation for use in computing each new indicator row. We now solve another example, in which both the column drop rule and the dropping of the stubs is used.

Example 17-7

We now solve the problem previously solved by SLGP in Example 17-3 by the multiphase algorithm.

Find $\mathbf{x}$ so as to

$$\text{lexicographically minimize} \quad \mathbf{a} = \{(\rho_1 + \rho_2), (\eta_3), (\rho_4), (\eta_1 + 1.5\eta_2)\}$$

subject to

$$\begin{aligned}
x_1 \qquad\qquad + \eta_1 - \rho_1 &= 30 \\
x_2 + \eta_2 - \rho_2 &= 15 \\
8x_1 + 12x_2 + \eta_3 - \rho_3 &= 1{,}000 \\
x_1 + 2x_2 + \eta_4 - \rho_4 &= 40 \\
\mathbf{x}, \boldsymbol{\eta}, \boldsymbol{\rho} &\geq \mathbf{0}
\end{aligned}$$

The tableaux that document the solution process are given in Tables 17-16 through 17-21. Pivot rows and columns have been circled and a check mark (✓) appears beneath any column that is to be dropped.

Table 17-16 is our initial tableau, which is optimal for priority 1, and both ρ_1 and

Table 17-16. INITIAL TABLEAU, EXAMPLE 17-7

				P_4							
				P_3						1	
				P_2							
				P_1			1	1			
P_4	P_3	P_2	P_1	V	x_1	x_2	ρ_1	ρ_2	ρ_3	ρ_4	$\mathbf{x}_B$
1				η_1	1		−1				30
1.5				η_2		1		−1			15
		1		η_3	8	12			−1		1,000
				η_4	1	2				−1	40
				P_1			−1	−1			0
							✓	✓			

Table 17-17

V	x_1	x_2	ρ_3	ρ_4	$\mathbf{x}_B$
η_1	1				30
η_2		1			15
η_3	8	12	−1		1,000
η_4	1	2		−1	40
P_1					0
P_2	8	12	−1		1,000

Table 17-18

V	x_1	η_2	ρ_3	ρ_4	$\mathbf{x}_B$
η_1	1				30
x_2		1			15
η_3	8	−12	−1		820
η_4	1	−2		−1	10
P_1					0
P_2	8	−12	−1		820

Table 17-19

V	η_4	η_2	ρ_3	ρ_4	$\mathbf{x}_B$
η_1	−1	2		1	20
x_2		1			15
η_3	−8	4	−1	8	740
x_1	1	−2		−1	10
P_1					0
P_2	−8	4	−1	8	740

Table 17-20

V	η_4	η_2	ρ_3	η_1	$\mathbf{x}_B$
ρ_4	−1	2		1	20
x_2		1			15
η_3		−12	−1	−8	580
x_1				1	30
P_1					0
P_2		−12	−1	−8	580
		✓	✓	✓	

ρ_2 may be dropped. Moving to priority level 2, we develop Table 17-17, which is not yet optimal, and we pivot on x_2 and η_2 to arrive at Table 17-18.

The tableau in Table 17-18 is still not optimal and we pivot on x_1 and η_4 to obtain

Table 17-21. FINAL TABLEAU, EXAMPLE 17-7

V	η_4	$\mathbf{x}_B$
ρ_4	-1	20
x_2		15
η_3		580
x_1		30
P_1		0
P_2		580
P_3	-1	20

Table 17-19, which is, again, not yet optimal. Pivoting on ρ_4 and η_1, we arrive at Table 17-20, which is optimal for both priorities 1 and 2. Using the column drop rule we may drop variables η_2, ρ_3, and η_1 at this stage. Our next tableau, shown in Table 17-21, now considers priority level 3. Not only is it optimal, but the single remaining column (η_4) may be dropped. Thus, we need not go to priority level 4. Our final solution is then:

$$
\begin{array}{lll}
x_1^* = 30 & \eta_1^* = 0 & \rho_1^* = 0 \\
x_2^* = 15 & \eta_2^* = 0 & \rho_2^* = 0 \\
 & \eta_3^* = 580 & \rho_3^* = 0 \\
 & \eta_4^* = 0 & \rho_4^* = 20
\end{array}
$$

$$\mathbf{a}^* = (0, 580, 20, 0)$$

Whether or not we take advantage of the shortcuts, the solution of problems involving more than about four or five variables and rows is best accomplished by the computer [2–5]. We now address our attention to some additional facets of problem formulation and tableau interpretation when employing the multiphase algorithm.

Some Other Considerations

Next, we discuss some of the more common problems and/or questions that arise in the employment of the multiphase algorithm. These include:

—Negative right-hand side values in the goals
—Alternative optimal solutions
—Infeasible and unbounded problems
—Unimplementable solutions
—Reconstruction of the final tableau when column drop is employed

Negative Right-Hand Side

Consider the following goal:

$$G_1: -x_1 + 8x_2 + \eta_1 - \rho_1 = -20$$

There is basically nothing wrong with such a goal (or rigid constraint) from a mathematical or physical point of view. However, the form of the multiphase simplex algorithm that we have presented requires that all right-hand side values be nonnegative. This requirement is easily satisfied.

First, one should place the goal in its form *without* the deviation variables. That is, either

$$f(\mathbf{x}) \geq -b$$

or

$$f(\mathbf{x}) \leq -b$$

or

$$f(\mathbf{x}) = -b$$

We then simply multiply the form above by -1, which changes the right-hand side to a positive value (and reverses the direction of any inequalities). Finally, we add the deviation variables to re-form the goal.

For example, if, in G_1, we had wished to minimize η_1, the form of the goal without deviation variables is

$$G_1: -x_1 + 8x_2 \geq -20$$

Thus,

$$-G_1 = G_1' = x_1 - 8x_2 \leq 20$$

and

$$G_1': x_1 - 8x_2 + \eta_1 - \rho_1 = 20$$

where ρ_1 must now be minimized.

Alternative Optimal Solutions

Alternative optimal solutions occur when two or more programs (i.e., values of the decision variables) attain the same optimal achievement vector. The existence of alternative optimal solutions is indicated by an entire column of zero valued $R_{k,s}$ elements (i.e., $\mathbf{R}_s = \mathbf{0}$) in the final tableau (where all indicator rows appear).

If we examine the final tableau from Table 17-15, we see that there are *no* alternative optimal solutions. However, the tableau in Table 17-22 does indicate the existence of alternative optimal solutions since $\mathbf{R}_2 = \mathbf{0}$ (i.e., the second indicator column has all zero elements). If the reader brings x_2 into solution (η_3 will depart), he or she will discover that an alternative optimal program is $x_1^* = 1, x_2^* = 3$.

Infeasible and Unbounded Problems

As mentioned in the discussion of SLGP, we do not encounter either infeasibility or unboundedness in our goal programming models.[d] Consequently, these phenomena, although associated with single-objective linear programming, are not of interest in our approach.

Table 17-22. FINAL TABLEAU WITH ALTERNATIVE OPTIMAL SOLUTION

		P_2						
		P_1				1	1	
P_2	P_1	V	η_2	x_2	ρ_1	ρ_2	ρ_3	$\mathbf{x}_B$
1		η_1	−1		−1	1		6
		x_1	1	1		−1		4
		η_3		1			−1	3
		P_1				−1	−1	0
		P_2	−1		−1	1		6

Unimplementable Solutions

Although a goal programming solution will never be infeasible,[d] it may well be unimplementable. That is, one or more rigid constraints are not achieved, leading to a positive value for a_1, the first term in the achievement vector. When such a condition is encountered, it usually means that we must relax one or more of the rigid constraints.

Reconstruction of the Final Tableau

When the column drop rule is employed, the resulting final tableau is reduced and, if sensitivity analysis is to be employed, must be reconstructed. In theory, this is a straightforward task. In practice, with large problems, it can present some difficulty.

Before addressing the reconstruction process, let us first list the basic formulas associated with linear goal programming. The reader should notice their equivalence with those for the single-objective model. These are:

$$\mathbf{x}_B = \mathbf{B}^{-1}\mathbf{b} \tag{17.8}$$

$$\mathbf{y}_s = \mathbf{B}^{-1}\mathbf{c}_s \tag{17.9}$$

$$R_{k,s} = \mathbf{u}_k^T\mathbf{y}_s - w_{k,s} \tag{17.10}$$

$$a_k = \mathbf{u}_k^T\mathbf{x}_B \tag{17.11}$$

where $\mathbf{x}_B$ = present basic solution

$\mathbf{B}$ = basis matrix

$\mathbf{B}^{-1}$ = inverse of the basis matrix

$\mathbf{y}_s$ = tableau interior column under the sth nonbasic variable

$\mathbf{c}_s$ = *original* set of problem coefficients for the sth nonbasic variable

$R_{k,s}$, $\mathbf{u}_k^T$, $w_{k,s}$, and a_k are as defined previously

---

[d]That is, if we begin with a feasible solution and utilize either the SLGP or multiphase simplex algorithm, we shall not encounter an infeasible solution. However, in the chapters to follow, we may encounter infeasible solutions and deal with such situations by means of other approaches.

Consequently, if one wishes to reconstruct the tableau, the crucial operation is to derive the inverse of the corresponding basis matrix. With $\mathbf{B}^{-1}$ derived, all the other elements in the tableau may be derived by use of (17.8) through (17.11). Consequently, we center our attention on the development of $\mathbf{B}^{-1}$.

Our first step is to identify **B**, the basis matrix corresponding to the present solution, $\mathbf{x}_B$. As you recall from Chapter 3 (or Chapter 6),

$$\mathbf{B} = (\mathbf{b}_1 \quad \mathbf{b}_2 \quad \cdots \quad \mathbf{b}_m)$$

where $\mathbf{b}_i$ is a column vector in **B** associated with the basic variable, $x_{B,i}$. In fact, $\mathbf{b}_i$ is simply the vector of coefficients ($\mathbf{c}_j$) from the original formulation as associated with the basic variable $x_{B,i}$. To illustrate, consider Example 17-7, where, from Table 17-21, the final basic solution consists of

$$x_{B,1} \longrightarrow p_4$$
$$x_{B,2} \longrightarrow x_2$$
$$x_{B,3} \longrightarrow \eta_3$$
$$x_{B,4} \longrightarrow x_1$$

The columns ($\mathbf{c}_j$) under each of these variables, in the original formulation, are then:

p_4	x_2	η_3	x_1
0	0	0	1
0	1	0	0
0	12	1	8
−1	2	0	1

Thus, **B** is given as

$$\mathbf{B} = \begin{pmatrix} 0 & 0 & 0 & 1 \\ 0 & 1 & 0 & 0 \\ 0 & 12 & 1 & 8 \\ -1 & 2 & 0 & 1 \end{pmatrix}$$

and the inverse of **B** is

$$\mathbf{B}^{-1} = \begin{pmatrix} 1 & 2 & 0 & -1 \\ 0 & 1 & 0 & 0 \\ -8 & -12 & 1 & 0 \\ 1 & 0 & 0 & 0 \end{pmatrix}$$

The reader may then use (17.9) to compute all the interior column vectors ($\mathbf{y}_s$) for each nonbasic variable and (17.10) to find the indicator row elements.

Computer Codes for Multiphase Simplex

The amount of effort, time, and expenditure of funds for the development of computer codes for the multiphase algorithm is not yet anywhere near that expended for the conventional simplex algorithm. As a result, *and only because of this*, computer codes for the multiphase algorithm do not at this time solve very large problems. For example, an early code provided by the author [2, 5] has been redimensioned and refined to solve problems with a few hundred variables and a like number of rows. However, the elementary form of the algorithm leads to round-off errors that may become pronounced. Typical performance of such a code, for a 100 variable-by-100 row example, would be about 3 minutes on the IBM 370–3033.

Summary

In this chapter we have dealt solely with the solution of a linear goal programming model with an achievement function that is to be *lexicographically* minimized. That is, our assumption is that the goals within each priority level are preemptive. In Chapter 19 we investigate one way in which the preemptive structure may be relaxed and then, in Chapter 20, several alternative approaches to the multiple-objective model.

As we saw, problems with up to three variables may be approached graphically. However, the typical problem requires a more general approach and two of these were discussed: the sequential method (SLGP) and the multiphase method. In general, the multiphase method is preferred.

A fundamental aspect of linear programming is the concept of duality. Consequently, before addressing any other methods of solution or analysis, in Chapter 19 we turn our attention to the multidimensional dual for linear goal programming together with two of its adjuncts: the multidimensional dual simplex algorithm and the primal–dual algorithm.

REFERENCES

1. IGNIZIO, J. P. "S-II Trajectory Study and Optimum Antenna Placement," Downey, Calif.: North American Aviation, SID-63, 1963.
2. IGNIZIO, J. P. *Goal Programming and Extensions*. Lexington, Mass.: Heath (Lexington Books), 1976.
3. IGNIZIO, J. P. "Goal Programming: A Tool for Multiobjective Analysis," *Journal of Operational Research*, Vol. 29, II, 1978, pp. 1109–1119.
4. IGNIZIO, J. P., AND PERLIS, J. H. "Sequential Linear Goal Programming: Implementation via MPSX," *International Journal of Computers and Operations Research*, Vol. 5, 1979, pp. 141–145.
5. IGNIZIO, J. P. "An Introduction to Goal Programming: With Applications to Urban Systems," *Journal of Computers, Environment and Urban Systems*, Vol. 5, 1980, pp. 15–33.
6. KORNBLUTH, J. S. H. "A Survey of Goal Programming," *Omega*, Vol. 1, No. 2, 1973, pp. 193–205.
7. LEE, S. M. *Goal Programming for Decision Analysis*. Philadelphia, Auerbach, 1972.

PROBLEMS

17.1. Develop the graphical model and solution to Problem 16.10.

17.2. Given the single-objective linear programming models shown below, formulate as a goal programming model and solve graphically. Discuss and explain the differences, if any, between the goal programming solution and that obtained via conventional linear programming.

(a) Maximize $z = 4x_1 + 3x_2$ subject to

$$x_1 + x_2 \leq 10$$
$$3x_1 + 4x_2 \leq 12$$
$$\mathbf{x} \geq \mathbf{0}$$

(b) Maximize $z = 2x_1 + x_2$ subject to

$$-x_1 + x_2 \leq 10$$
$$3x_1 + 6x_2 \geq 9$$
$$\mathbf{x} \geq \mathbf{0}$$

17.3. Solve Example 17-1 graphically if, for the third goal (G_3) the right-hand side value is changed from 800 to 2,000.

17.4. Discuss the difference between a "feasible solution" in linear goal programming and in linear programming. Illustrate graphically.

17.5. Discuss why a linear goal programming problem cannot be unbounded. Illustrate your discussion via Problem 17.2(b).

17.6. Why are the existence of alternative optimal solutions so important to goal programming?

17.7. Show that a "feasible solution" to a linear goal programming model need not be implementable.

17.8. Illustrate, graphically, how an optimal solution to the lexicographic linear goal programming model may not be an efficient (i.e., nondominated) solution.

Solve Problems 17.9 through 17.13 via both the SLGP and multiphase simplex algorithm. Use column drop and, in the case of the multiphase algorithm, early stop[e] wherever possible.

17.9. Lexicographically minimize $\mathbf{a} = \{(\eta_1), (\eta_3), (\eta_2), (\rho_1 + \rho_2)\}$ subject to

$$2x_1 + x_2 + \eta_1 - \rho_1 = 20$$
$$x_1 + \eta_2 - \rho_2 = 12$$
$$x_2 + \eta_3 - \rho_3 = 10$$
$$\mathbf{x}, \boldsymbol{\eta}, \boldsymbol{\rho} \geq \mathbf{0}$$

17.10. Lexicographically minimize $\mathbf{a} = \{(\eta_1), (\rho_2), (8\eta_3 + 5\eta_4), (\rho_1)\}$ subject to

[e]That is, stop whenever the solution has converged to a point (no alternative optimal solutions exist) as indicated in step 6 of the multiphase algorithm.

$$\begin{aligned} x_1 + x_2 + \eta_1 - \rho_1 &= 100 \\ x_1 + x_2 + \eta_2 - \rho_2 &= 90 \\ x_1 \qquad + \eta_3 - \rho_3 &= 80 \\ x_2 + \eta_4 - \rho_4 &= 55 \\ \mathbf{x}, \boldsymbol{\eta}, \boldsymbol{\rho} &\geq \mathbf{0} \end{aligned}$$

17.11. Lexicographically minimize $\mathbf{a} = \{(\eta_1 + \rho_1), (2\rho_2 + \rho_3)\}$ subject to

$$\begin{aligned} x_1 - 10x_2 + \eta_1 - \rho_1 &= 50 \\ 3x_1 + 5x_2 + \eta_2 - \rho_2 &= 20 \\ 8x_1 + 6x_2 + \eta_3 - \rho_3 &= 100 \\ \mathbf{x}, \boldsymbol{\eta}, \boldsymbol{\rho} &\geq \mathbf{0} \end{aligned}$$

17.12. Lexicographically minimize $\mathbf{a} = \{(\rho_1 + \rho_2), (\eta_3), (\rho_4)\}$ subject to

$$\begin{aligned} x_1 + 2x_2 + \eta_1 - \rho_1 &= 4 \\ 4x_1 + 3x_2 + \eta_2 - \rho_2 &= 12 \\ x_1 + x_2 + \eta_3 - \rho_3 &= 8 \\ x_1 \qquad + \eta_4 - \rho_4 &= 2 \\ \mathbf{x}, \boldsymbol{\eta}, \boldsymbol{\rho} &\geq \mathbf{0} \end{aligned}$$

17.13. Lexicographically minimize $\mathbf{a} = \{(\rho_1), (\eta_2), (3\eta_1 + \rho_3)\}$ subject to

$$\begin{aligned} -x_1 + x_2 + \eta_1 - \rho_1 &= -20 \\ 5x_1 + 6x_2 + \eta_2 - \rho_2 &= 60 \\ x_2 + \eta_3 - \rho_3 &= 10 \\ \mathbf{x}, \boldsymbol{\eta}, \boldsymbol{\rho} &\geq \mathbf{0} \end{aligned}$$

17.14. In Example 17-17, change the priority structure to that shown below and resolve the problem via either SLGP or the multiphase algorithm:

$$\text{lexicographically minimize} \quad \mathbf{a} = \{(\rho_1 + \rho_2), (\rho_4), (\eta_1 + 1.5\eta_2), (\eta_3)\}$$

17.15. Solve the following graphically, via SLGP, and finally, by the multiphase simplex algorithm.

$$\text{Lexicographically minimize} \quad \mathbf{a} = \{(\rho_1 + \rho_2), (\eta_3), (\eta_4)\}$$

subject to

$$\begin{aligned} x_1 + x_2 + \eta_1 - \rho_1 &= 400 \\ 2x_1 + x_2 + \eta_2 - \rho_2 &= 500 \\ x_1 \qquad + \eta_3 - \rho_3 &= 300 \\ 0.4x_1 + 0.3x_2 + \eta_4 - \rho_4 &= 240 \\ \mathbf{x}, \boldsymbol{\eta}, \boldsymbol{\rho} &\geq \mathbf{0} \end{aligned}$$

DUALITY IN LINEAR GOAL PROGRAMMING

Introduction

Much of the power of single-objective linear programming stems from the existence and exploitation of the property known as duality (see Chapters 8 through 10). Obviously, it would be highly desirable to isolate an analogous property in linear goal programming and, fortunately, this can be done. Consequently, in this chapter we devote our attention to duality in linear goal programming as well as its use in the development of two new algorithms: the multidimensional dual simplex algorithm and the primal–dual algorithm for linear goal programming. In Chapter 19 we address the extremely important area of sensitivity analysis and find that the results of this chapter play a major part in such analysis.

Although duality in linear goal programming is not a difficult topic, it does *appear* complex, owing, for the most part, to the notation that must be used. Those who wish to learn about this property must take this into consideration. However, for those readers who merely wish to exploit duality in problem solving and analysis, it should be noted that much of the material in this chapter may be skipped (or skimmed), and we shall draw attention to this in the footnotes.

The discussion of duality in linear goal programming is dependent upon the approach used for modeling and solution (i.e., whether one employs the SLGP or the multiphase algorithm). If one selects SLGP, then, since each problem in the sequence is simply a conventional linear program, there is a corresponding sequence of conventional linear programming duals. However, if the

18

multiphase algorithm is used, the corresponding dual is a linear programming problem with *multiple, prioritized right-hand sides*. The author developed this latter form in the early 1970s [1, 3] and has termed it the *multidimensional dual*.

It should also be noted that the sequence of conventional duals in SLGP may be related to the multidimensional dual [1, 5] and, further, if one *relaxes the priority structure and accounts for the introduction of deviation variables*, the multidimensional dual is directly equivalent to the dual of the multiobjective linear programming (MOLP) model as discussed in the references [1, 4, 5]. Unfortunately, a thorough and complete discussion of duality in linear goal programming would require an extensive amount of space; consequently, we provide only an overview of the topic in this chapter. Our primary purpose herein is to develop certain aspects of the multidimensional dual which will provide assistance in problem solution and sensitivity analysis.

Duality in SLGP[a,b]

As mentioned, probably the most difficult part of duality in goal programming lies in the relative complexity of the notation. Consequently, to help simplify the discussion, we use a numerical example to assist in the illustration

---

[a]Those readers wishing to merely exploit duality in problem solving and analysis may skip (or simply skim) this section.

[b]Throughout Chapter 18, we shall omit the matrix or vector transpose designation (e.g., $\mathbf{uy}$ rather than $\mathbf{u}^T\mathbf{y}$) simply so as to reduce the complexity of notation.

and development of the dual for any model within the SLGP sequence. This example is:

Find $\mathbf{x}$ so as to

$$\text{lexicographically minimize} \quad a = \{(\rho_1 + \rho_2), (2\eta_3 + 3\eta_4)\} \tag{18.1}$$

subject to

$$x_1 + x_2 + \eta_1 - \rho_1 = 12 \tag{18.2}$$

$$2x_1 + x_2 + \eta_2 - \rho_2 = 20 \tag{18.3}$$

$$16x_1 + 10x_2 + \eta_3 - \rho_3 = 160 \tag{18.4}$$

$$3x_1 + 5x_2 + \eta_4 - \rho_4 = 60 \tag{18.5}$$

$$\mathbf{x}, \boldsymbol{\eta}, \boldsymbol{\rho} \geq \mathbf{0} \tag{18.6}$$

The primal model for the first problem in the sequence (i.e., $k = 1$) may be written:

$$\text{minimize} \quad a_1 = 0\eta_1 + 0\eta_2 + \rho_1 + \rho_2 \tag{18.7}$$

subject to

$$\left.\begin{aligned} x_1 + x_2 + \eta_1 - \rho_1 &= 12 \\ 2x_1 + x_2 + \eta_2 - \rho_2 &= 20 \end{aligned}\right\} \tag{18.8}$$

$$\mathbf{x}, \boldsymbol{\eta}, \boldsymbol{\rho} \geq \mathbf{0} \tag{18.9}$$

For $k = 2$, the second priority, the primal is:

$$\text{minimize} \quad a_2 = 2\eta_3 + 3\eta_4 + 0\rho_3 + 0\rho_4 \tag{18.10}$$

subject to

$$\left.\begin{aligned} x_1 + x_2 + \eta_1 - \rho_1 &= 12 \\ 2x_1 + x_2 + \eta_2 - \rho_2 &= 20 \end{aligned}\right\} \tag{18.11}$$

$$\left.\begin{aligned} 16x_1 + 10x_2 + \eta_3 - \rho_3 &= 160 \\ 3x_1 + 5x_2 + \eta_4 - \rho_4 &= 60 \end{aligned}\right\} \tag{18.12}$$

$$0\eta_1 + 0\eta_2 + \rho_1 + \rho_2 = a_1^* \tag{18.13}$$

$$\mathbf{x}, \boldsymbol{\eta}, \boldsymbol{\rho} \geq \mathbf{0} \tag{18.14}$$

Equation (18.10) is the linear programming objective for priority 2, while the constraints in (18.11) and (18.12) are those associated with priorities 1 and 2, respectively. The constraint in (18.13) is added to ensure that we do not degrade the optimal solution previously obtained for the first priority model.

Since the negative deviation variables (η's) are basic, we do not want them to appear in the objective function. We thus simply solve (18.11) and (18.12) for the η values and substitute into (18.10) and (18.13). The result, in matrix form, is:

$$\text{minimize} \quad a_2 = [-\mathbf{u}^{(2)}\mathbf{C}^{(2)} \,|\, \mathbf{0} \,|\, \{\mathbf{u}^{(2)} + \mathbf{w}^{(2)}\}] \begin{pmatrix} \mathbf{x} \\ \boldsymbol{\eta}^{(2)} \\ \boldsymbol{\rho}^{(2)} \end{pmatrix} + \mathbf{u}^{(2)}\mathbf{b}^{(2)} \tag{18.15}$$

subject to

$$[-\mathbf{C}^{(1,2)} \,|\, -\mathbf{I}^{(1,2)} \,|\, \mathbf{I}^{(1,2)}] \begin{pmatrix} \mathbf{x} \\ \boldsymbol{\eta}^{(1,2)} \\ \boldsymbol{\rho}^{(1,2)} \end{pmatrix} = -\mathbf{b}^{(1,2)} \tag{18.16}$$

$$[-\mathbf{u}^{(1)}\mathbf{C}^{(1)} \,|\, \mathbf{0} \,|\, \{\mathbf{u}^{(1)} + \mathbf{w}^{(1)}\}] \begin{pmatrix} \mathbf{x} \\ \boldsymbol{\eta}^{(1)} \\ \boldsymbol{\rho}^{(1)} \end{pmatrix} = a_1^* - \mathbf{u}^{(1)}\mathbf{b}^{(1)} \tag{18.17}$$

$$\mathbf{x}, \boldsymbol{\eta}, \boldsymbol{\rho} \geq \mathbf{0} \tag{18.18}$$

where

$\mathbf{u}^{(k)}$ = weight vector associated with the negative deviation variables at priority k (a row vector)

$\mathbf{w}^{(k)}$ = weight vector associated with the positive deviation variables at priority k (a row vector)

$\mathbf{b}^{(k)}$ = right-hand side vector for the goals (constraints) associated with priority level k only (a column vector)

$\mathbf{b}^{(1,k)}$ = right-hand side vector for the goals (constraints) associated with priority levels 1 through k (a column vector)

$\mathbf{C}^{(k)}$ = coefficient matrix for the goals (constraints) associated with priority level k only

$\mathbf{C}^{(1,k)}$ = coefficient matrix for the goals (constraints) associated with priority levels 1 through k

$\mathbf{I}$ = identity matrix, where any superscripts are defined similar to those for $\mathbf{C}$

a_k^* = optimal value for a_k

$\boldsymbol{\eta}^{(k)}, \boldsymbol{\rho}^{(k)}$ = deviation variable vectors (negative and positive, respectively) for those deviation variables at priority level k

$\boldsymbol{\eta}^{(1,k)}, \boldsymbol{\rho}^{(1,k)}$ = deviation variable vectors (negative and positive, respectively) for those deviation variables at priority levels 1 through k

For our example, we may relate the foregoing definitions to their actual vectors and matrices as follows:

$$\mathbf{u}^{(1)} = (0 \quad 0) \qquad \mathbf{w}^{(1)} = (1 \quad 1)$$
$$\mathbf{u}^{(2)} = (2 \quad 3) \qquad \mathbf{w}^{(2)} = (0 \quad 0)$$
$$\mathbf{b}^{(1)} = \begin{pmatrix} 12 \\ 20 \end{pmatrix} \qquad \mathbf{b}^{(2)} = \begin{pmatrix} 160 \\ 60 \end{pmatrix}$$
$$\mathbf{b}^{(1,2)} = \begin{pmatrix} 12 \\ 20 \\ 160 \\ 60 \end{pmatrix}$$
$$\mathbf{C}^{(1)} = \begin{pmatrix} 1 & 1 \\ 2 & 1 \end{pmatrix} \qquad \mathbf{C}^{(2)} = \begin{pmatrix} 16 & 10 \\ 3 & 5 \end{pmatrix}$$
$$\mathbf{C}^{(1,2)} = \begin{pmatrix} 1 & 1 \\ 2 & 1 \\ 16 & 10 \\ 3 & 5 \end{pmatrix}$$
$$\boldsymbol{\eta}^{(1)} = \begin{pmatrix} \eta_1 \\ \eta_2 \end{pmatrix} \qquad \boldsymbol{\eta}^{(2)} = \begin{pmatrix} \eta_3 \\ \eta_4 \end{pmatrix} \qquad \boldsymbol{\eta}^{(1,2)} = \begin{pmatrix} \eta_1 \\ \eta_2 \\ \eta_3 \\ \eta_4 \end{pmatrix}$$
$$\boldsymbol{\rho}^{(1)} = \begin{pmatrix} \rho_1 \\ \rho_2 \end{pmatrix} \qquad \boldsymbol{\rho}^{(2)} = \begin{pmatrix} \rho_3 \\ \rho_4 \end{pmatrix} \qquad \boldsymbol{\rho}^{(1,2)} = \begin{pmatrix} \rho_1 \\ \rho_2 \\ \rho_3 \\ \rho_4 \end{pmatrix}$$

THE GENERAL FORM OF THE SLGP PRIMAL [2]

The primal model for the SLGP problem at priority level k may be expressed, in general terms, as:

$$\text{minimize} \quad a_k = [-\mathbf{u}^{(k)}\mathbf{C}^{(k)} \,|\, \mathbf{0} \,|\, \{\mathbf{u}^{(k)} + \mathbf{w}^{(k)}\}] \begin{pmatrix} \mathbf{x} \\ \boldsymbol{\eta}^{(k)} \\ \boldsymbol{\rho}^{(k)} \end{pmatrix} + \mathbf{u}^{(k)}\mathbf{b}^{(k)} \tag{18.19}$$

subject to

$$[-\mathbf{C}^{(1,k)} \,|\, -\mathbf{I}^{(1,k)} \,|\, \mathbf{I}^{(1,k)}] \begin{pmatrix} \mathbf{x} \\ \boldsymbol{\eta}^{(1,k)} \\ \boldsymbol{\rho}^{(1,k)} \end{pmatrix} = -\mathbf{b}^{(1,k)} \tag{18.20}$$

$$\left.\begin{array}{c}[\mathbf{u}^{(1)}\mathbf{C}^{(1)}\,|\,\mathbf{0}\,|\,\{\mathbf{u}^{(1)}+\mathbf{w}^{(1)}\}]\begin{pmatrix}\mathbf{x}\\ \boldsymbol{\eta}^{(1)}\\ \boldsymbol{\rho}^{(1)}\end{pmatrix}=a_1^*-\mathbf{u}^{(1)}\mathbf{b}^{(1)}\\ \vdots \qquad\qquad\qquad \vdots\\ [-\mathbf{u}^{(k-1)}\mathbf{C}^{(k-1)}\,|\,\mathbf{0}\,|\,\{\mathbf{u}^{(k-1)}+\mathbf{w}^{(k-1)}\}]\begin{pmatrix}\mathbf{x}\\ \boldsymbol{\eta}^{(k-1)}\\ \boldsymbol{\rho}^{(k-1)}\end{pmatrix}=a_{k-1}^*-\mathbf{u}^{(k-1)}\mathbf{b}^{(k-1)}\end{array}\right\} \tag{18.21}$$

$$\mathbf{x},\ \boldsymbol{\eta},\ \boldsymbol{\rho} \geq \mathbf{0} \tag{18.22}$$

Consequently, from (18.19) through (18.22), we may write the SLGP primal model at $k = 2$ for our example as:

$$\text{minimize} \quad a_2 = \left[-(2 \quad 3)\begin{pmatrix}16 & 10\\ 3 & 5\end{pmatrix} \,|\, 0 \quad 0 \,|\, 2 \quad 3\right]\begin{pmatrix}x_1\\ x_2\\ \eta_3\\ \eta_4\\ \rho_3\\ \rho_4\end{pmatrix} + (2 \quad 3)\begin{pmatrix}160\\ 60\end{pmatrix}$$

subject to

$$\begin{pmatrix}-1 & -1 & -1 & 0 & 0 & 0 & 1 & 0 & 0 & 0\\ -2 & -1 & 0 & -1 & 0 & 0 & 0 & 1 & 0 & 0\\ -16 & -10 & 0 & 0 & -1 & 0 & 0 & 0 & 1 & 0\\ -3 & -5 & 0 & 0 & 0 & -1 & 0 & 0 & 0 & 1\end{pmatrix}\begin{pmatrix}x_1\\ x_2\\ \eta_1\\ \eta_2\\ \eta_3\\ \eta_4\\ \rho_1\\ \rho_2\\ \rho_3\\ \rho_4\end{pmatrix} = \begin{pmatrix}-12\\ -20\\ -160\\ -60\end{pmatrix}$$

$$\left[-(0 \quad 0)\begin{pmatrix}1 & 1\\ 2 & 1\end{pmatrix} \,|\, 0 \quad 0 \,|\, 1 \quad 1\right]\begin{pmatrix}x_1\\ x_2\\ \eta_1\\ \eta_2\\ \rho_1\\ \rho_2\end{pmatrix} = a_1^* - (0 \quad 0)\begin{pmatrix}12\\ 20\end{pmatrix}$$

$$\mathbf{x},\ \boldsymbol{\eta},\ \boldsymbol{\rho} \geq \mathbf{0}$$

which may be simplified as:

$$\text{minimize} \quad a_2 = [-41 \quad -35 \quad 0 \quad 0 \quad 2 \quad 3] \begin{pmatrix} x_1 \\ x_2 \\ \eta_3 \\ \eta_4 \\ \rho_3 \\ \rho_4 \end{pmatrix} + 500$$

subject to

$$\begin{pmatrix} -1 & -1 & -1 & 0 & 0 & 0 & 1 & 0 & 0 & 0 \\ -2 & -1 & 0 & -1 & 0 & 0 & 0 & 1 & 0 & 0 \\ -16 & -10 & 0 & 0 & -1 & 0 & 0 & 0 & 1 & 0 \\ -3 & -5 & 0 & 0 & 0 & -1 & 0 & 0 & 0 & 1 \\ 0 & 0 & 0 & 0 & 0 & 0 & 1 & 1 & 0 & 0 \end{pmatrix} \begin{pmatrix} x_1 \\ x_2 \\ \eta_1 \\ \eta_2 \\ \eta_3 \\ \eta_4 \\ \rho_1 \\ \rho_2 \\ \rho_3 \\ \rho_4 \end{pmatrix} = \begin{pmatrix} -12 \\ -20 \\ -160 \\ -60 \\ a_1^* \end{pmatrix}$$

$$\mathbf{x}, \boldsymbol{\eta}, \boldsymbol{\rho} \geq \mathbf{0}$$

THE GENERAL FORM OF THE SLGP DUAL [2]

From the general form of the primal, as given earlier, the general form of the dual of the SLGP problem at priority level k is:

$$\text{maximize} \quad Z_k = -\mathbf{b}^{(1,k)}\mathbf{v}^{(1,k)} + \sum_{t=1}^{k-1} \lambda_t\{a_t^* - \mathbf{u}^{(t)}\mathbf{b}^{(t)}\} + \mathbf{u}^{(k)}\mathbf{b}^{(k)}$$

subject to

$$\begin{bmatrix} -\mathbf{C}^{(1,k)} & -\mathbf{I}^{(1,k)} & \mathbf{I}^{(1,k)} \\ -\mathbf{u}^{(1)}\mathbf{C}^{(1)} & \mathbf{0} & \{\mathbf{u}^{(1)} + \mathbf{w}^{(1)}\} \\ & \vdots & \\ -\mathbf{u}^{(k-1)}\mathbf{C}^{(k-1)} & \mathbf{0} & \{\mathbf{u}^{(k-1)} + \mathbf{w}^{(k-1)}\} \end{bmatrix}^T \begin{pmatrix} \mathbf{v}^{(1,k)} \\ \lambda_1 \\ \vdots \\ \lambda_{k-1} \end{pmatrix} \leq [-\mathbf{u}^{(k)}\mathbf{C}^{(k)} \,|\, \mathbf{0} \,|\, \{\mathbf{u}^{(k)} + \mathbf{w}^{(k)}\}]^T$$

with $\mathbf{v}$ and $\boldsymbol{\lambda}$ as *unrestricted* dual variables.

From this rather imposing-looking general form, one may develop the dual problem for level 2 ($k = 2$) in our example as:

$$\text{maximize} \quad Z_2 = -12v_1 - 20v_2 - 160v_3 - 60v_4 + \lambda_1 a_1^* + 500$$

subject to

$$\begin{pmatrix} -1 & -2 & -16 & -3 & 0 \\ -1 & -1 & -10 & -5 & 0 \\ -1 & 0 & 0 & 0 & 0 \\ 0 & -1 & 0 & 0 & 0 \\ 0 & 0 & -1 & 0 & 0 \\ 0 & 0 & 0 & -1 & 0 \\ 1 & 0 & 0 & 0 & 1 \\ 0 & 1 & 0 & 0 & 1 \\ 0 & 0 & 1 & 0 & 0 \\ 0 & 0 & 0 & 1 & 0 \end{pmatrix} \begin{pmatrix} v_1 \\ v_2 \\ v_3 \\ v_4 \\ \lambda_1 \end{pmatrix} \leq \begin{pmatrix} -41 \\ -35 \\ 0 \\ 0 \\ 0 \\ 0 \\ 0 \\ 0 \\ 2 \\ 3 \end{pmatrix}$$

with $\mathbf{v}$, λ_1 unrestricted in sign and a_1^* dependent upon the previous priority level.

The Multidimensional Dual[c] [1, 3]

The general form of the primal for a linear goal programming problem may be expressed as:

$$\text{lexicographically minimize} \quad \mathbf{a} = \{g_1(\boldsymbol{\eta}, \boldsymbol{\rho}), \ldots, g_K(\boldsymbol{\eta}, \boldsymbol{\rho})\} \tag{18.23}$$

subject to

$$\sum_{j=1}^{n} c_{i,j}x_j + \eta_i - \rho_i = b_i \qquad \text{for all } i \tag{18.24}$$

$$\mathbf{x}, \boldsymbol{\eta}, \boldsymbol{\rho} \geq \mathbf{0} \tag{18.25}$$

We may rewrite this in matrix form as follows:

$$\text{lexicographically minimize} \quad \mathbf{a} = \left[(\mathbf{0} \quad \mathbf{u}^{(1)} \quad \mathbf{w}^{(1)}) \begin{pmatrix} \mathbf{x} \\ \boldsymbol{\eta}^{(1)} \\ \boldsymbol{\rho}^{(1)} \end{pmatrix}, \ldots, (\mathbf{0} \quad \mathbf{u}^{(K)} \quad \mathbf{w}^{(K)}) \begin{pmatrix} \mathbf{x} \\ \boldsymbol{\eta}^{(K)} \\ \boldsymbol{\rho}^{(K)} \end{pmatrix} \right] \tag{18.26}$$

[c] Those readers wishing to merely exploit duality in problem solving and analysis may either skim or skip this section.

subject to

$$[\mathbf{C}^{(1,K)} \quad \mathbf{I}^{(1,K)} \quad -\mathbf{I}^{(1,K)}]\begin{pmatrix}\mathbf{x}\\ \boldsymbol{\eta}^{(1,K)}\\ \boldsymbol{\rho}^{(1,K)}\end{pmatrix} = \mathbf{b} \tag{18.27}$$

$$\mathbf{x}, \boldsymbol{\eta}, \boldsymbol{\rho} \geq \mathbf{0} \tag{18.28}$$

where the vectors and matrices are defined as in the preceding section.

Again (as in the preceding section), we solve for $\boldsymbol{\eta}$ to remove the basic variables from the achievement vector. From (18.27), $\boldsymbol{\eta}$ is given as:

$$\boldsymbol{\eta} = -\mathbf{C}\mathbf{x} + \boldsymbol{\rho} + \mathbf{b}$$

and substituting this into (18.26), we obtain:

$$\text{lexicographically minimize} \quad \mathbf{a} = \left[(\mathbf{0} \quad \mathbf{u}^{(1)} \quad \mathbf{w}^{(1)})\begin{pmatrix}\mathbf{x}\\ -\mathbf{C}^{(1)}\mathbf{x} + \boldsymbol{\rho}^{(1)} + \mathbf{b}^{(1)}\\ \boldsymbol{\rho}^{(1)}\end{pmatrix}, \ldots, (\mathbf{0} \quad \mathbf{u}^{(K)} \quad \mathbf{w}^{(K)})\begin{pmatrix}\mathbf{x}\\ -\mathbf{C}^{(K)}\mathbf{x} + \boldsymbol{\rho}^{(K)} + \mathbf{b}^{(K)}\\ \boldsymbol{\rho}^{(K)}\end{pmatrix}\right]$$

subject to

$$[-\mathbf{C}^{(1,K)} \quad -\mathbf{I}^{(1,K)} \quad \mathbf{I}^{(1,K)}]\begin{pmatrix}\mathbf{x}\\ \boldsymbol{\eta}^{(1,K)}\\ \boldsymbol{\rho}^{(1,K)}\end{pmatrix} = -\mathbf{b}^{(1,K)}$$

$$\mathbf{x}, \boldsymbol{\eta}, \boldsymbol{\rho} \geq \mathbf{0}$$

And this may be rewritten as the general form of the following linear goal programming primal:

lexicographically minimize

$$\mathbf{a} = \left[\left\{(-\mathbf{u}^{(1)}\mathbf{C}^{(1)} \,|\, \mathbf{0} \,|\, \{\mathbf{u}^{(1)} + \mathbf{w}^{(1)}\})\begin{pmatrix}\mathbf{x}\\ \boldsymbol{\eta}^{(1)}\\ \boldsymbol{\rho}^{(1)}\end{pmatrix} + \mathbf{u}^{(1)}\mathbf{b}^{(1)}\right\}, \ldots, \left\{(-\mathbf{u}^{(K)}\mathbf{C}^{(K)} \,|\, \mathbf{0} \,|\, \{\mathbf{u}^{(K)} + \mathbf{w}^{(K)}\})\begin{pmatrix}\mathbf{x}\\ \boldsymbol{\eta}^{(K)}\\ \boldsymbol{\rho}^{(K)}\end{pmatrix} + \mathbf{u}^{(K)}\mathbf{b}^{(K)}\right\}\right] \tag{18.29}$$

subject to

$$[-\mathbf{C}^{(1,K)} \quad -\mathbf{I}^{(1,K)} \quad \mathbf{I}^{(1,K)}]\begin{pmatrix}\mathbf{x}\\ \boldsymbol{\eta}^{(1,K)}\\ \boldsymbol{\rho}^{(1,K)}\end{pmatrix} = -\mathbf{b}^{(1,K)} \tag{18.30}$$

$$\mathbf{x}, \boldsymbol{\eta}, \boldsymbol{\rho} \geq \mathbf{0} \tag{18.31}$$

THE GENERAL FORM OF THE MULTIDIMENSIONAL DUAL [1, 3]

The general form of the dual (i.e., the *multidimensional dual*) for the primal given in (18.29) through (18.31) may be written:

$$\text{maximize} \quad Z = -\mathbf{b}^{(1,K)T}\mathbf{v} + \{\mathbf{u}^{(1)}\mathbf{b}^{(1)}, \ldots, \mathbf{u}^{(K)}\mathbf{b}^{(K)}\} \tag{18.32}$$

subject to

$$\begin{bmatrix} (-\mathbf{C}^{(1,K)})^T \\ (-\mathbf{I}^{(1,K)})^T \\ (\mathbf{I}^{(1,K)})^T \end{bmatrix} \mathbf{v} \leq \begin{bmatrix} (-\mathbf{u}^{(1)}\mathbf{C}^{(1)})^T \\ (\mathbf{0})^T \\ (\mathbf{u}^{(1)} + \mathbf{w}^{(1)})^T \end{bmatrix}, \ldots, \begin{bmatrix} (-\mathbf{u}^{(K)}\mathbf{C}^{(K)})^T \\ (\mathbf{0})^T \\ (\mathbf{u}^{(K)} + \mathbf{w}^{(K)})^T \end{bmatrix} \tag{18.33}$$

where the dual variables $\mathbf{v}$ are both *unrestricted and multidimensional.*

Example 18-1

We illustrate the formulation of the multidimensional dual using the same example given originally in (18.1) through (18.6) and repeated below:

$$\text{lexicographically minimize} \quad \mathbf{a} = \{(\rho_1 + \rho_2), (2\eta_3 + 3\eta_4)\}$$

subject to

$$\begin{aligned} x_1 + x_2 + \eta_1 - \rho_1 &= 12 \\ 2x_1 + x_2 + \eta_2 - \rho_2 &= 20 \\ 16x_1 + 10x_2 + \eta_3 - \rho_3 &= 160 \\ 3x_1 + 5x_2 + \eta_4 - \rho_4 &= 60 \\ \mathbf{x}, \boldsymbol{\eta}, \boldsymbol{\rho} &\geq \mathbf{0} \end{aligned}$$

First we place the problem in the general form of the primal as described earlier. That is,

lexicographically minimize

$$\mathbf{a} = \left\{ \left[-(0 \quad 0)\begin{pmatrix} 1 & 1 \\ 2 & 1 \end{pmatrix} \middle| 0 \quad 0 \middle| \{(0 \quad 0) + (1 \quad 1)\} \right] \begin{pmatrix} x_1 \\ x_2 \\ \eta_1 \\ \eta_2 \\ \rho_1 \\ \rho_2 \end{pmatrix} + (0 \quad 0)\begin{pmatrix} 12 \\ 20 \end{pmatrix} \right\},$$

$$\left\{ \left[-(2 \quad 3)\begin{pmatrix} 16 & 10 \\ 3 & 5 \end{pmatrix} \middle| 0 \quad 0 \middle| \{(2 \quad 3) + (0 \quad 0)\} \right] \begin{pmatrix} x_1 \\ x_2 \\ \eta_3 \\ \eta_4 \\ \rho_3 \\ \rho_4 \end{pmatrix} + (2 \quad 3)\begin{pmatrix} 160 \\ 60 \end{pmatrix} \right\}$$

subject to

$$\begin{pmatrix} -1 & -1 & -1 & 0 & 0 & 0 & 1 & 0 & 0 & 0 \\ -2 & -1 & 0 & -1 & 0 & 0 & 0 & 1 & 0 & 0 \\ -16 & -10 & 0 & 0 & -1 & 0 & 0 & 0 & 1 & 0 \\ -3 & -5 & 0 & 0 & 0 & -1 & 0 & 0 & 0 & 1 \end{pmatrix} \begin{pmatrix} x_1 \\ x_2 \\ \eta_1 \\ \eta_2 \\ \eta_3 \\ \eta_4 \\ \rho_1 \\ \rho_2 \\ \rho_3 \\ \rho_4 \end{pmatrix} = \begin{pmatrix} -12 \\ -20 \\ -160 \\ -60 \end{pmatrix}$$

$$\mathbf{x}, \boldsymbol{\eta}, \boldsymbol{\rho} \geq \mathbf{0}$$

The primal may then be rewritten as:

lexicographically minimize

$$\mathbf{a} = \left\{ [0 \quad 0 \quad 0 \quad 0 \quad 1 \quad 1] \begin{pmatrix} x_1 \\ x_2 \\ \eta_1 \\ \eta_2 \\ \rho_1 \\ \rho_2 \end{pmatrix} + 0 \right\}, \left\{ [-41 \quad -35 \quad 0 \quad 0 \quad 2 \quad 3] \begin{pmatrix} x_1 \\ x_2 \\ \eta_3 \\ \eta_4 \\ \rho_3 \\ \rho_4 \end{pmatrix} + 500 \right\}$$

subject to

$$\begin{pmatrix} -1 & -1 & -1 & 0 & 0 & 0 & 1 & 0 & 0 & 0 \\ -2 & -1 & 0 & -1 & 0 & 0 & 0 & 1 & 0 & 0 \\ -16 & -10 & 0 & 0 & -1 & 0 & 0 & 0 & 1 & 0 \\ -3 & -5 & 0 & 0 & 0 & -1 & 0 & 0 & 0 & 1 \end{pmatrix} \begin{pmatrix} x_1 \\ x_2 \\ \eta_1 \\ \eta_2 \\ \eta_3 \\ \eta_4 \\ \rho_1 \\ \rho_2 \\ \rho_3 \\ \rho_4 \end{pmatrix} = \begin{pmatrix} -12 \\ -20 \\ -160 \\ -60 \end{pmatrix}$$

$$\mathbf{x}, \boldsymbol{\eta}, \boldsymbol{\rho} \geq \mathbf{0}$$

Now, using (18.32) and (18.33), we write the multidimensional dual of our example as:

$$\text{maximize} \quad Z = -\mathbf{b}^{(1,2)T}\mathbf{v} + \{\mathbf{u}^{(1)}\mathbf{b}^{(1)}, \mathbf{u}^{(2)}\mathbf{b}^{(2)}\}$$

subject to

$$\begin{bmatrix} (-\mathbf{C}^{(1,2)})^T \\ (-\mathbf{I}^{(1,2)})^T \\ (\mathbf{I}^{(1,2)})^T \end{bmatrix} \mathbf{v} \leq \begin{bmatrix} (-\mathbf{u}^{(1)}\mathbf{C}^{(1)})^T \\ \mathbf{0}^T \\ (\mathbf{u}^{(1)} + \mathbf{w}^{(1)})^T \end{bmatrix}, \begin{bmatrix} (-\mathbf{u}^{(2)}\mathbf{C}^{(2)})^T \\ \mathbf{0}^T \\ (\mathbf{u}^{(2)} + \mathbf{w}^{(2)})^T \end{bmatrix}$$

v unrestricted and multidimensional

For this example, the multidimensional dual is given as:

$$\text{maximize} \quad Z = (-12 \quad -20 \quad -160 \quad -60)\mathbf{v} + \{0, 500\}$$

subject to

$$\begin{pmatrix} -1 & -2 & -16 & -3 \\ -1 & -1 & -10 & -5 \\ -1 & 0 & 0 & 0 \\ 0 & -1 & 0 & 0 \\ 0 & 0 & -1 & 0 \\ 0 & 0 & 0 & -1 \\ 1 & 0 & 0 & 0 \\ 0 & 1 & 0 & 0 \\ 0 & 0 & 1 & 0 \\ 0 & 0 & 0 & 1 \end{pmatrix} \mathbf{v} \leq \begin{pmatrix} 0 \\ 0 \\ 0 \\ 0 \\ 0 \\ 0 \\ 1 \\ 1 \\ 0 \\ 0 \end{pmatrix}, \begin{pmatrix} -41 \\ -35 \\ 0 \\ 0 \\ 0 \\ 0 \\ 0 \\ 0 \\ 2 \\ 3 \end{pmatrix}$$

v unrestricted and multidimensional

and, for this example, **v** may be expressed as

$$\mathbf{v} = \begin{pmatrix} v_{1,1} \\ v_{2,1} \\ v_{3,1} \\ v_{4,1} \end{pmatrix}, \begin{pmatrix} v_{1,2} \\ v_{2,2} \\ v_{3,2} \\ v_{4,2} \end{pmatrix}$$

where $v_{i,k}$ is the value of the ith dual variable for the problem corresponding to the kth right-hand side.

Both the similarities and the differences between the SLGP dual and the multidimensional dual should be evident. SLGP dual is a conventional linear programming dual that includes two terms, λ_k and a_k^*, that are used to take into consideration the previous duals (and primals) in the sequence. In the multidimensional dual we notice two aspects (which are actually related to λ_k and a_k^* indirectly) in particular:

1. The multidimensional dual is a conventional linear programming dual *except* that it has multiple and prioritized right-hand sides.
2. The multidimensional dual variables are themselves multidimensional.

As such, there is an intriguing similarity between the multidimensional dual and the so-called "minimal adjustment" problem. In the (linear) minimal adjustment problem we seek to find a sequence of solutions to a series of linear programming problems that are identical in form except for the right-hand side values. As an example, consider a production facility whose resources (right-hand side values for the constraints) change from year to year. It may well be that if we optimize the problem for year 1, we may have to substantially change or readjust the production lines for the next year (and so on). What would be desirable is to find a solution that *tends* to *both* optimize profit *and* minimize year-to-year changes, or some optimal compromise.

The Multidimensional and the MOLP Duals

In Chapter 16 we mentioned an alternative approach to linear multiple-objective optimization which has, unfortunately, been given the very general and ambiguous name of MOLP (for multiple-objective linear programming). MOLP deals with the vector-maximum (or vector-minimum) problem as follows:

$$\text{maximize} \quad \mathbf{z} = \mathbf{Cx}$$

$$\text{subject to}$$

$$\mathbf{Ax} \leq \mathbf{b}$$

$$\mathbf{x} \geq \mathbf{0}$$

where $\mathbf{z}$ indicates that we have a vector of objectives to be considered. The dual of the MOLP, which is discussed in detail in the references (see [4] in particular) may be written in general as:

$$\text{minimize} \quad Z = \mathbf{b}^T\mathbf{v}$$

$$\text{subject to}$$

$$\mathbf{A}^T\mathbf{v} \geq \mathbf{C}^T$$

$$\mathbf{v} \geq \mathbf{0}$$

That is, we have a linear programming model with *multiple, unprioritized* right-hand sides. It should be intuitively obvious (see references [1, 4] for details) that if we relax the preemptive priorities in the multidimensional dual and account for the use of negative and positive deviation variables in the primal, the multidimensional dual encompasses, and may be reduced to, the MOLP dual.

The Multidimensional Dual Simplex Algorithm [1, 3]

The reader may recall (see Chapter 9) that duality in single-objective linear programming may be exploited so as to develop an algorithm, known as dual simplex, that is useful in solving certain problems and in sensitivity analysis. The author has developed an analogous tool in linear goal programming that is termed the multidimensional dual simplex algorithm [1, 3]. We introduce and illustrate this algorithm in this section.

Before proceeding to our discussion of the multidimensional dual simplex algorithm, let us briefly comment on how one would solve the multidimensional dual formulation given earlier. An algorithm for solution and several illustrations are given in reference [3] and we shall, herein, present only a narrative overview.

Examining the general multidimensional dual formulation of (18.32) and (18.33) or the specific formulation given in Example 18-1, it should be obvious (if one draws an analogy to the techniques of conventional linear programming) that a solution to the multidimensional dual is arrived at by forcing (if possible) the right-hand side values to nonnegative values. However, if forcing say $v_{3,2}$ nonnegative would drive a higher-priority dual variable negative, we do not consider such a pivot. In essence, this approach to solving the multidimensional dual is analogous to solving a transposition of the primal. Along the same lines, we can develop a multidimensional dual simplex algorithm for the solution to the primal linear goal programming model, which is:

1. Optimal according to the indicator rows, *and*
2. At least one element in $\mathbf{x}_B$ is negative.

The algorithm then attempts to drive all right-hand side values nonnegative *without* degrading the optimality of the tableau (i.e., according to the indicator rows). The steps of this algorithm are given below.

Algorithm

Step 1. To employ this algorithm, the problem must be optimal according to the indicator row values (i.e., the topmost nonzero element in each indicator column, $\mathbf{R}_s$, cannot be positive) *and* at least one element in $\mathbf{x}_B$ must be negative. If these conditions are satisfied, go to step 2.

Step 2. Select the row associated with the most negative $x_{B,i}$ element. The basic variable associated with this row is the departing variable. Denote this row as i'.

Step 3. Determine the column associated with the lexicographically minimal "column ratio."[d] Determine this column ratio set for only those columns having a *negative* element in row i'. The column ratio vector for column s is computed as follows:

[d]Note that we have extended here our definition of the lexicographic minimum from Chapter 16, so as to include negative elements.

(a) Compute $R_{k,s}/y_{i',s}$ for each priority level k. Order these ratios from highest to lowest priority level. That is,

$$\mathbf{r}_s = (R_{1,s}/y_{i',s}, \ldots, R_{K,s}/y_{i',s}) \qquad \text{where } y_{i',s} < 0 \tag{18.34}$$

(b) Select the lexicographically minimal column ratio set. For example, if

$$\mathbf{r}_q = (2, -3, 0)$$

and

$$\mathbf{r}_t = (2, -5, 7)$$

then $\mathbf{r}_t$ is preferred to $\mathbf{r}_q$.

(c) Designate the column associated with the most lexicographically minimal $\mathbf{r}_s$ (in the case of ties, break these arbitrarily) as column s'. The nonbasic variable associated with column s' is the entering variable.

Step 4. Using the procedure of our multiphase algorithm of Chapter 17 (step 5 of the algorithm), exchange the departing variable for the entering variable, and compute the new tableau.

Step 5. Repeat steps 2 through 4 until all $x_{B,i}$ are nonnegative.

Example 18-2

Consider the tableau shown in Table 18-1. The indicator rows satisfy the optimality conditions, but we note that one right-hand side value ($x_{B,1} = -2$) is negative. Thus, we may apply the multidimensional dual simplex algorithm to obtain an optimal *and feasible* solution.

Applying the algorithm, we first note that there is only one negative right-hand side value (in the first row), and thus the departing variable is x_2 and $i' = 1$. Taking the

Table 18-1. OPTIMAL, INFEASIBLE TABLEAU

			P_3						1	
			P_2							
			P_1			1	1			
P_3	P_2	P_1	V	η_2	η_1	ρ_1	ρ_2	ρ_3	ρ_4	$\mathbf{x}_B$
			x_2	−1	1	−1	1			−2
			x_1	1			−1			12
	1		η_3	−2	−3	3	2	−1		2
			η_4		−1	1			−1	2
			P_1			−1	−1			
			P_2	−2	−3	3	2	−1		2
			P_3						−1	

$i' = 1, s' = 1$

column ratios as in step 3, we have

$$\mathbf{r}_1 = (0/-1, -2/-1, 0/-1) = (0, 2, 0)$$
$$\mathbf{r}_3 = (-1/-1, 3/-1, 0/-1) = (1, -3, 0)$$

and thus $\mathbf{r}_1$ is preferred. Consequently, $s' = 1$ and η_2 should enter. The next tableau, shown in Table 18-2, is optimal and feasible and we are finished.

Table 18-2. FINAL TABLEAU FOR EXAMPLE 18-2

			P_3						1	
			P_2							
			P_1			1	1			
P_3	P_2	P_1	V	x_2	η_1	ρ_1	ρ_2	ρ_3	ρ_4	$\mathbf{x}_B$
			η_2	−1	−1	1	−1			2
			x_1	1	1	−1				10
	1		η_3	−2	−5	5		−1		6
			η_4		−1	1			−1	2
			P_1			−1	−1			0
			P_2	−2	−5	5		−1		6
			P_3						−1	0

Problems leading directly to a tableau that is both optimal *and* infeasible would rarely be encountered in practice, and thus the multidimensional dual simplex algorithm (like the dual simplex algorithm of single-objective linear programming) is not a general problem-solving technique. Its primary use, in our presentation, is in the sensitivity analysis presentation of Chapter 19.

The Primal-Dual LGP Algorithm [2]

We recall from Chapter 9 that a very useful and robust algorithm, denoted as the primal–dual algorithm, could be developed for conventional linear programming by utilizing a (heuristic) combination of the primal simplex and dual simplex algorithms. If we consider a slightly modified form of the linear goal programming model, an analogous algorithm may be developed in linear goal programming. We call this the primal–dual algorithm for linear goal programming (LGP).

Problem Formulation

The formulation that we shall employ is identical to that previously used in linear goal programming except that:

1. We shall not add *positive* deviation variables to any *rigid constraints*.
2. There will be (as a result of the above) no term in the achievement vector corresponding to the achievement of the rigid constraints.

The general form is then:

$$\text{lexicographically minimize} \quad \mathbf{a} = \{a_1, \ldots, a_K\} \tag{18.35}$$

subject to

$$\sum_{j=1}^{n} c_{i,j} x_j \leq b_i \qquad \text{for all } i \text{ associated with rigid constraints} \tag{18.36}$$

$$\left.\begin{aligned} &\sum_{j=1}^{n} c_{t(1),j} x_j + \eta_{t(1)} - \rho_{t(1)} = b_{t(1)}, \qquad t(1) \in P_1 \\ &\qquad \vdots \\ &\sum_{j=1}^{n} c_{t(K),j} x_j + \eta_{t(K)} - \rho_{t(K)} = b_{t(K)}, \qquad t(K) \in P_K \end{aligned}\right\} \tag{18.37}$$

$$\mathbf{x}, \boldsymbol{\eta}, \boldsymbol{\rho} \geq \mathbf{0} \tag{18.38}$$

Notice that the rigid constraints are reflected in (18.36) and, as indicated, are all formulated as equal to or less than ($\leq$) inequalities. Also, realize that the first term in **a** (i.e., a_1) is now associated with the highest-ranked set of goals rather than with rigid constraints.

Algorithm

The algorithm for solution operates as follows. We first determine the best multidimensional dual simplex pivot (if one is possible) and the best primal pivot (if one is possible). If none are possible, the problem is infeasible (analogous to unimplementable with our regular multiphase algorithm). Otherwise, we then select the pivot that produces the most preferred **a** and establish a new tableau. This is repeated until the tableau is either both feasible and optimal or we determine that the model is infeasible.

Notice in particular that a *multidimensional dual simplex* pivot is possible only if:

1. One or more $x_{B,i} < 0$.
2. Letting the pivot row be denoted as i', the pivot column, s', must have

   $y_{i',s'} < 0$

   $\mathbf{R}_{s'}$ must not have, as its first nonzero element, a positive value

Further, a *primal pivot* is possible only if:

1. One or more $\mathbf{R}_s$ has, as its first nonzero element, a positive value.
2. Letting the pivot column be denoted as s', the pivot row, i', must have

$$y_{i',s'} > 0$$
$$x_{B,i'} \geq 0$$

This is best demonstrated by means of an example.

Example 18-3

Consider the following problem:

$$\left.\begin{array}{l} 6x_1 + 10x_2 \geq 60 \\ 3x_1 + x_2 \geq 18 \end{array}\right\} \text{rigid constraints}$$
$$x_1 + x_2 \geq 12 \quad \text{priority 1 goal}$$
$$x_1 + x_2 \leq 4 \quad \text{priority 2 goal}$$

The resultant model is then:

$$\text{lexicographically minimize} \quad \mathbf{a} = \{(\eta_3), (\rho_4)\}$$

subject to

$$\begin{aligned} -6x_1 - 10x_2 + \eta_1 \phantom{{}+\eta_3 - \rho_3} &= -60 \\ -3x_1 - x_2 + \eta_2 \phantom{{}+\eta_3 - \rho_3} &= -18 \\ x_1 + x_2 + \eta_3 - \rho_3 &= 12 \\ x_1 + x_2 + \eta_4 - \rho_4 &= 4 \\ \mathbf{x}, \boldsymbol{\eta}, \boldsymbol{\rho} &\geq \mathbf{0} \end{aligned}$$

The solution process is shown in Tables 18-3 through 18-5. Notice carefully that, in Table 18-3, there is no multidimensional dual simplex pivot despite the fact that some of the $x_{B,i}$ are negative. This is because both $\mathbf{R}_1$ and $\mathbf{R}_2$ are headed by $+1$. In Table 18-4 there is a dual pivot (row 1, column 2) producing an $\mathbf{a} = (8, 0)$, but the primal pivot gives $\mathbf{a} = (0, 8)$ and is thus selected. Table 18-5 indicates both an optimal and feasible solution, and thus we are finished. Since the tableau is optimal and feasible, the interior elements need not be computed. The solution is

$$\mathbf{x}^* = (12, 0)$$
$$\mathbf{a}^* = (0, 8)$$

Table 18-3. PRIMAL PIVOT

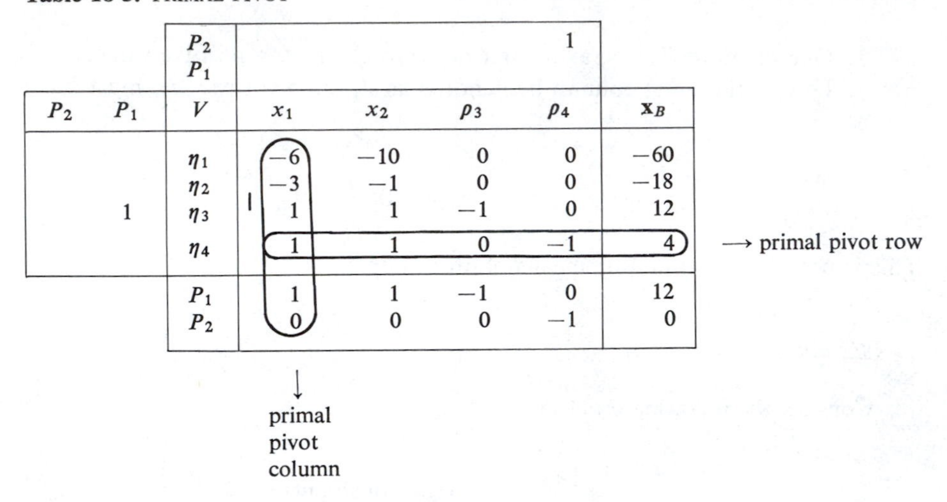

P_2	P_1	V	x_1	x_2	ρ_3	ρ_4	$\mathbf{x}_B$
		P_2				1	
		P_1					
		η_1	−6	−10	0	0	−60
		η_2	−3	−1	0	0	−18
	1	η_3	1	1	−1	0	12
		η_4	1	1	0	−1	4
		P_1	1	1	−1	0	12
		P_2	0	0	0	−1	0

Table 18-4. SECOND PRIMAL PIVOT

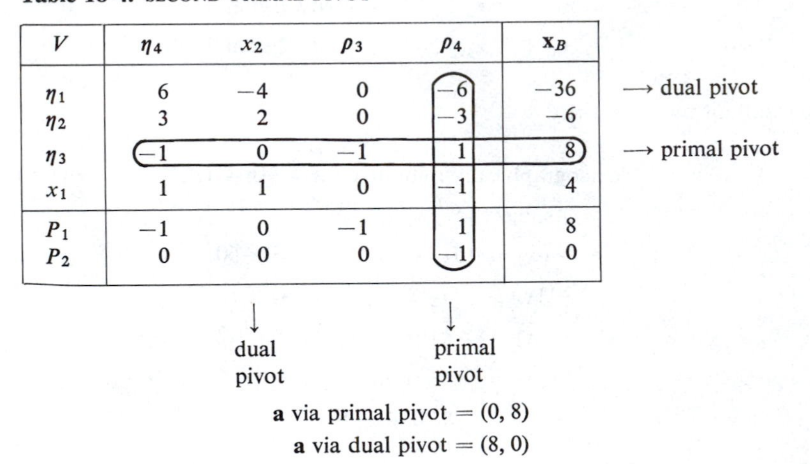

V	η_4	x_2	ρ_3	ρ_4	$\mathbf{x}_B$
η_1	6	−4	0	−6	−36
η_2	3	2	0	−3	−6
η_3	−1	0	−1	1	8
x_1	1	1	0	−1	4
P_1	−1	0	−1	1	8
P_2	0	0	0	−1	0

a via primal pivot = (0, 8)

a via dual pivot = (8, 0)

Table 18-5. FINAL TABLEAU

V	η_4	x_2	ρ_3	η_3	$\mathbf{x}_B$
η_1					12
η_2					18
ρ_4					8
x_1					12
P_1	0	0	0	−1	0
P_2	−1	0	−1	1	8

EVALUATION OF THE PRIMAL–DUAL ALGORITHM IN LGP

The obvious advantages of the primal–dual algorithm are:

1. The number of variables are reduced (no ρ_i's in rigid constraints).
2. The dimensionality of **a**, the top stub, the left stub, and the indicator rows is reduced by one.

However, no extensive comparison of computational efficiency of the primal–dual versus the multiphase algorithm has yet been made and thus, at this point, a choice between the two must be made on a mainly subjective basis.

Summary

In this chapter we have examined the dual forms for both SLGP and the multiphase algorithms. Two immediate extensions of the multidimensional dual were the multidimensional dual simplex algorithm and the primal–dual algorithm for LGP. In Chapter 19 we actually use the multidimensional dual simplex algorithm to aid our computations in sensitivity analysis. It should be obvious that we have only begun to exploit the possibilities offered by duality in linear goal programming, and those readers wishing further details should consult the list of references [1, 2, 3, 5].

REFERENCES

1. IGNIZIO, J. P. "The Development of the Multidimensional Dual in Linear Goal Programming," Working Paper, Pennsylvania State University, September 19, 1974.
2. IGNIZIO, J. P. "A Primal–Dual Algorithm for Linear Goal Programming," Working Paper, Pennsylvania State University, September 20, 1974.
3. IGNIZIO, J. P. *Goal Programming and Extensions*. Lexington, Mass.: Heath (Lexington Books), 1976.
4. KORNBLUTH, J. S. H. "Duality, Indifference and Sensitivity Analysis in Multiple Objective Linear Programming," *Operational Research Quarterly*, Vol. 25, 1974, pp. 599–614.
5. MARKOWSKI, C. "Duality in Linear Goal Programming," unpublished Ph.D. dissertation, Pennsylvania State University, 1980.

PROBLEMS

Note: Problems 18.1 through 18.4 may be omitted if the sections on "Duality in SLGP" and "The Multidimensional Dual" were not covered.

18.1. Develop the SLGP dual formulations, for each priority level, for the problems listed below.

(a) Lexicographically minimize $\mathbf{a} = \{(\eta_1), (\eta_3), (\eta_2), (\rho_1 + \rho_2)\}$ subject to

$$\begin{aligned} 2x_1 + x_2 + \eta_1 - \rho_1 &= 20 \\ x_1 \qquad + \eta_2 - \rho_2 &= 12 \\ x_2 + \eta_3 - \rho_3 &= 10 \\ \mathbf{x}, \boldsymbol{\eta}, \boldsymbol{\rho} &\geq \mathbf{0} \end{aligned}$$

(b) Lexicographically minimize $\mathbf{a} = \{(\eta_1), (\rho_2), (2\eta_3 + \eta_4), (\rho_1)\}$ subject to

$$\begin{aligned} x_1 + x_2 - \eta_1 - \rho_1 &= 100 \\ x_1 + x_2 + \eta_2 - \rho_2 &= 90 \\ x_1 \qquad + \eta_3 - \rho_3 &= 80 \\ x_2 + \eta_4 - \rho_4 &= 55 \\ \mathbf{x}, \boldsymbol{\eta}, \boldsymbol{\rho} &\geq \mathbf{0} \end{aligned}$$

18.2. Develop the multidimensional dual formulations for the two examples given in Problem 18.1.

18.3. Develop and list the steps of an algorithm that may be used to solve the multidimensional dual.

18.4. Given the multidimensional dual variable, $v_{i,k}$, discuss the physical interpretation of the values of such a variable in the "minimal adjustment" problem.

18.5. Given the tableau shown, first explain why the multidimensional dual simplex algorithm may be employed and then use this algorithm to obtain an optimal and feasible solution.

		P_2						1	
		P_1				1	1		
P_2	P_1	V	x_1	x_2	x_3	ρ_1	ρ_2	ρ_3	$\mathbf{x}_B$
		η_1	−1	−2	1	−1			−5
		η_2	−2	−1	−1		−1		−4
		η_3	5	2	3			−1	0
		P_1	0	0	0	−1	−1	0	0
		P_2	0	0	0	0	0	−1	0

18.6. Constrast the results obtained in Problem 18.5 with those obtained in Example 9-1.

18.7. Apply the multidimensional dual simplex algorithm to the problem given in the accompanying tableau. Discuss your results.

		P_2					1	
		P_1			1	1		
P_2	P_1	V	x_1	x_2	ρ_1	ρ_2	ρ_3	$\mathbf{x}_B$
		η_1	1	1	-1			-10
		η_2	1			-1		50
		η_3	2	1			-1	100
		P_1	0	0	-1	-1	0	0
		P_2	0	0	0	0	-1	0

18.8. Solve Problem 18.5 via the primal–dual algorithm for linear goal programming (first mathematically formulate the problem).

18.9. Solve Problem 17.10 via the primal–dual algorithm.

18.10. Solve Example 18-3 by the primal–dual algorithm wherein the priority 1 goal, $x_1 + x_2 \geq 12$, is replaced by $x_1 + x_2 \leq 12$.

SENSITIVITY ANALYSIS IN LINEAR GOAL PROGRAMMING

The Importance of Sensitivity Analysis

We must always realize that the solution we obtain in solving a linear goal programming (or the single objective) model is the solution that is optimal only with respect to the model and not necessarily with respect to the actual problem. In constructing such a model we have employed certain simplifying assumptions that may or may not affect the validity of the derived solution. For example, we form a deterministic model wherein it is likely that the actual values of model coefficients are not constant but stochastic. Further, we assume a static situation where, in reality, most processes and organizations vary and change with time. It is also possible that we may not be totally confident in either the priority structure or intrapriority weights.

Another reason to look beyond the initial solution is that, at least in most real-life situations, the solution is only part (and often but a minor part) of the information that is *really needed or desired.* Quite often, more important than model solution is any information that will enable us to improve the system itself, such as information with regard to:

1. Which facilities and/or products may be discontinued.
2. What may be gained from, and how much we should pay for, additional resources.
3. What the impact will be of increases or shortages in resources and increases or decreases in inflation and/or interest rates.

19

Sensitivity analysis provides us with a systematic procedure for analyzing all of the aspects listed above and, as such, can well be the most important phase in the total decision-making framework. Consequently, in this chapter we introduce the basics of sensitivity analysis in linear goal programming [1]. As we shall see, *any of the analyses* that we performed in single-objective linear programming (see Chapter 10) may also be performed in linear goal programming.

Types of Analysis

As in Chapter 10, we investigate two particular types of sensitivity analysis: (1) the analysis of *discrete changes*, and (2) the analysis of *variations across a continuous range*. The types of discrete changes that we investigate are:

1. $w_{k,s}$: a change in the weighting factor at priority k for the sth *nonbasic* variable.
2. $u_{i,k}$: a change in the weighting factor at priority k for the ith *basic* variable.
3. b_i: a change in the original right-hand side value of goal (or rigid constraint) i.
4. $c_{i,s}$: a change in the original coefficient associated with the ith goal (or rigid constraint) and the sth *nonbasic* variable.

5. The addition of a new goal.
6. The addition of a new decision variable.
7. A reordering and/or permutation of the original priority levels.

In our analysis of continuous variations over a range, known as parametric linear goal programming, we address only two types of changes:

1. Range variations of b_i.
2. Range variations of $w_{k,s}$ or $u_{i,k}$.

In all cases, the presentation of the methodology shall be illustrated through example. The particular example that will be employed throughout the chapter is as follows:

Find $\mathbf{x}$ so as to

$$\text{lexicographically minimize} \quad \mathbf{a} = \{(\rho_1 + \rho_2), (\eta_3 + 2\eta_4), (\eta_1)\} \tag{19.1}$$

subject to

$$G_1: \quad x_1 + \eta_1 - \rho_1 = 20 \tag{19.2}$$

$$G_2: \quad x_2 + \eta_2 - \rho_2 = 35 \tag{19.3}$$

$$G_3: \quad -5x_1 + 3x_2 + \eta_3 - \rho_3 = 220 \tag{19.4}$$

$$G_4: \quad x_1 - x_2 + \eta_4 - \rho_4 = 60 \tag{19.5}$$

$$\mathbf{x}, \boldsymbol{\eta}, \boldsymbol{\rho} \geq \mathbf{0} \tag{19.6}$$

The final tableau for this example is given in Table 19-1.

Table 19-1. ILLUSTRATIVE FINAL TABLEAU

			P_3							
			P_2							
			P_1			1	1			
P_3	P_2	P_1	V	x_1	η_2	ρ_1	ρ_2	ρ_3	ρ_4	$\mathbf{x}_B$
1			η_1	1		−1				20
			x_2		1		−1			35
	1		η_3	−5	−3		3	−1		115
	2		η_4	1	1		−1		−1	95
			P_1			−1	−1			0
			P_2	−3	−1		1	−1	−2	305
			P_3	1		−1				20

It should be stressed that the example has been selected simply to demonstrate the sensitivity analyses with a minimum of computational effort and is not, therefore, representative of the computational burden that will normally be encountered. We now proceed to some preliminary remarks and then to a demonstration of the various analyses.

Some Limitations on Changes

In investigating any changes, we must always recall that only commensurable goals may be included at the same priority level—other than priority level 1, which is maintained for rigid constraints (i.e., absolute goals and thus independent of weights). This means, for example, that if we investigate a new goal, say

$$2x_1 + x_2 + \eta_5 - \rho_5 = 30$$

where ρ_5 is to be minimized, this goal may be included in priority level 2 (for example) only if the new goal is commensurable (or may be made commensurable by weighting) with the other goals in priority level 2.

Finding the Inverse of the Basis Matrix

The inverse of the basis matrix, $\mathbf{B}^{-1}$, plays an important role in sensitivity analysis. Consequently, we now indicate how it may be found simply from the inspection of a tableau. Recall from Chapter 10 that

$$\mathbf{B}^{-1} = (\mathbf{e}_1 \ldots \mathbf{e}_r \ldots \mathbf{e}_m)$$

where $\mathbf{e}_r$ is the $\mathbf{y}_s$ column, from the tableau under consideration, under the rth *original* basic (negative deviation) variable. Thus, to find $\mathbf{B}^{-1}$ for the final tableau of Table 19-1, we simply look under the columns associated with η_1, η_2, η_3, and η_4 (in that order), where we realize that we are using a *condensed* form of the tableau.

In Table 19-1, the $\mathbf{y}$ column associated with η_1 does not appear explicitly as η_1 is basic in the first row. Thus, the associated column is part of the identity matrix associated with the present basis (but not shown in a condensed tableau) and is

$$\begin{pmatrix} 1 \\ 0 \\ 0 \\ 0 \end{pmatrix}$$

The $\mathbf{y}_s$ column under η_2 may be read directly from the tableau as

$$\begin{pmatrix} 0 \\ 1 \\ -3 \\ 1 \end{pmatrix}$$

Finally, since η_3 and η_4 are basic in the third and fourth rows, respectively, their columns are:

$$\begin{pmatrix} 0 \\ 0 \\ 1 \\ 0 \end{pmatrix} \quad \text{and} \quad \begin{pmatrix} 0 \\ 0 \\ 0 \\ 1 \end{pmatrix}$$

Putting these all together, in the proper order, we obtain

$$\mathbf{B}^{-1} = \begin{pmatrix} 1 & 0 & 0 & 0 \\ 0 & 1 & 0 & 0 \\ 0 & -3 & 1 & 0 \\ 0 & 1 & 0 & 1 \end{pmatrix}$$

To check, recall that $\mathbf{x}_B = \mathbf{B}^{-1}\mathbf{b}$, or

$$\mathbf{x}_B = \begin{pmatrix} 1 & 0 & 0 & 0 \\ 0 & 1 & 0 & 0 \\ 0 & -3 & 1 & 0 \\ 0 & 1 & 0 & 1 \end{pmatrix} \begin{pmatrix} 20 \\ 35 \\ 220 \\ 60 \end{pmatrix} = \begin{pmatrix} 20 \\ 35 \\ 115 \\ 95 \end{pmatrix}$$

which is the same solution as shown in Table 19-1.

We may now proceed to the illustrations of the manner by which certain discrete changes in the original problem may be evaluated.

A Discrete Change in $w_{k,s}$ or $u_{i,k}$

Although both $w_{k,s}$ and $u_{i,k}$ represent a weighting factor for a term in the achievement vector, the *impact* of any discrete changes will differ. We let any term with a caret over it represent the new (changed) value of a parameter, while terms without the caret indicate the original values. Consequently, given a change in $w_{k,s}$, we note that only a *single* element in the final tableau will be affected, the indicator row element $R_{k,s}$. That is,

$$\hat{R}_{k,s} = \mathbf{u}_k^T \mathbf{y}_s - \hat{w}_{k,s} \tag{19.7}$$

As a result of a change in the value of $R_{k,s}$, there *may* be an impact on the optimality of the new tableau. For example, if $R_{k,s}$ was negative and is changed to a positive value where there are no negative-valued indicator numbers above it, the optimal solution will change.

A change in $u_{i,k}$ may affect *an entire indicator row* as well as *the value of* a_k. Thus,

$$\hat{R}_{k,s} = \hat{\mathbf{u}}_k^T \mathbf{y}_s - w_{k,s} \qquad \text{for all } s \tag{19.8}$$

$$\hat{a}_k = \hat{\mathbf{u}}_k^T \mathbf{x}_B \tag{19.9}$$

Since $R_{k,s}$ values may change, the optimality of the solution may be affected. A change in a_k affects the achievement of a priority level and, if this is level 1, may affect the implementability of a solution.

Example 19-1: A Change in $w_{k,s}$

Consider the example given in (19.1) through (19.6). Let us assume that we wish to evaluate the effect of changing the weight on ρ_2 at priority 1 from a value of 1 to a value of zero (this is actually equivalent, for this problem, to removing the entire second goal from consideration). That is, $\hat{w}_{1,4} = 0$. The only impact, then, is on $R_{1,4}$, as given by

$$\hat{R}_{1,4} = \mathbf{u}_1^T \mathbf{y}_4 - \hat{w}_{1,4}$$

$$= (0 \quad 0 \quad 0 \quad 0)\begin{pmatrix} 0 \\ -1 \\ 3 \\ -1 \end{pmatrix} - 0 = 0$$

With a change in $R_{1,4}$ from -1 (as in Table 19-1) to a value of zero, the tableau is no longer optimal. To regain optimality, we must pivot on ρ_2 and η_3 (we leave the completion of this to the reader).

Example 19-2: A Change in $u_{i,k}$

Returning to the illustrative problem and Table 19-1, let us now assume that the weighting factor for η_4 at priority 2 ($u_{4,2}$) is changed from its present value of 2 to a new value of 6. Thus, $\hat{u}_{4,2} = 6$ and the resulting impact is on an entire indicator row and a_2 as follows:

$$\hat{R}_{2,s} = \hat{\mathbf{u}}_2^T \mathbf{y}_s - w_{2,s} \qquad \text{for all } s$$

$$\hat{a}_2 = \hat{\mathbf{u}}_2^T \mathbf{x}_B$$

where

$$\hat{\mathbf{u}}_2^T = (0 \quad 0 \quad 1 \quad 6)$$

Thus,

$$\begin{aligned}\hat{R}_{2,1} &= 1\\ \hat{R}_{2,2} &= 3\\ \hat{R}_{2,3} &= 0\\ \hat{R}_{2,4} &= -3\\ \hat{R}_{2,5} &= -1\\ \hat{R}_{2,6} &= -6\\ \hat{a}_2 &= 685\end{aligned}$$

As a result, the solution is no longer optimal and we must employ the multiphase algorithm to regain optimality (first pivoting on η_2 and x_2).

The reader is now invited to analyze the effect of a change on $w_{3,2}$ (the weight on η_3 at priority level 2). Let the new value of $w_{3,2}$ be 2 and note that this does *not* change the optimality of the present program, although it does change the degree of achievement.

A Discrete Change in b_i

Since the b_i values often represent estimates of resource availabilities or aspiration levels and are, in addition, likely to change with time, an investigation of the impact of any such changes is usually an important aspect of real-world problem solution. The impact of a discrete change in b_i is evident in both $\mathbf{x}_B$ and $\mathbf{a}$. That is,

$$\hat{\mathbf{x}}_B = \mathbf{B}^{-1}\hat{\mathbf{b}} \tag{19.10}$$

and

$$\hat{a}_k = \mathbf{u}^T\hat{\mathbf{x}}_B \tag{19.11}$$

Example 19-3: A Change in b_i

Going back to (19.3), let us examine the effect of a change in b_2 from its original value of 35 to a new value of 75. Thus, $\hat{b}_2 = 75$ and

$$\hat{\mathbf{b}} = \begin{pmatrix} 20 \\ 75 \\ 220 \\ 60 \end{pmatrix}$$

From Table 19-1 we may obtain $\mathbf{B}^{-1}$ and substitute this value into (19.10), to obtain

$$\hat{\mathbf{x}}_B = \mathbf{B}^{-1}\hat{\mathbf{b}} = \begin{pmatrix} 1 & 0 & 0 & 0 \\ 0 & 1 & 0 & 0 \\ 0 & -3 & 1 & 0 \\ 0 & 1 & 0 & 1 \end{pmatrix}\begin{pmatrix} 20 \\ 75 \\ 220 \\ 60 \end{pmatrix} = \begin{pmatrix} 20 \\ 75 \\ -5 \\ 135 \end{pmatrix}$$

From (19.11) we find $\mathbf{a}$:

$$\hat{a}_1 = \mathbf{u}_1^T\hat{\mathbf{x}}_B = (0 \quad 0 \quad 0 \quad 0)\begin{pmatrix} 20 \\ 75 \\ -5 \\ 135 \end{pmatrix} = 0$$

$$\hat{a}_2 = \mathbf{u}_2^T\hat{\mathbf{x}}_B = (0 \quad 0 \quad 1 \quad 2)\begin{pmatrix} 20 \\ 75 \\ -5 \\ 135 \end{pmatrix} = 265$$

$$\hat{a}_3 = \mathbf{u}_3^T\hat{\mathbf{x}}_B = (1 \quad 0 \quad 0 \quad 0)\begin{pmatrix} 20 \\ 75 \\ -5 \\ 135 \end{pmatrix} = 20$$

Notice that the change in b_2 has led to an *infeasible* tableau (since $\eta_3 = -5$). Since the tableau is also optimal (according to the indicator rows), we may employ the multidimensional dual simplex algorithm to regain feasibility. This is shown in Tables 19-2 and 19-3.

The effect, therefore, of changing b_2 from 35 to 75 may be seen by comparing Table 19-1 with Table 19-3. That is, the major effects are:

1. x_2 changes from 35 to $\frac{220}{3}$.
2. The achievement vector changes.

Table 19-2. TABLEAU AFTER A CHANGE IN b_2

			P_3							
			P_2							
			P_1			1	1			
P_3	P_2	P_1	V	x_1	η_2	ρ_1	ρ_2	ρ_3	ρ_4	$\mathbf{x}_B$
1			η_1	1		−1				20
			x_2		1		−1			75
	1		η_3	−5	−3		3	−1		−5
	2		η_4	1	1		−1		−1	135
			P_1			−1	−1			0
			P_2	−3	−1		1	−1	−2	265
			P_3	1		−1				20

Table 19-3. FINAL TABLEAU

			P_3							
			P_2		1					
			P_1			1	1			
P_3	P_2	P_1	V	x_1	η_3	ρ_1	ρ_2	ρ_3	ρ_4	$\mathbf{x}_B$
1			η_1	1	0	-1	0	0	0	20
			x_2	$-\frac{5}{3}$	$\frac{1}{3}$	0	0	$-\frac{1}{3}$	0	$\frac{220}{3}$
			η_2	$\frac{5}{3}$	$-\frac{1}{3}$	0	-1	$\frac{1}{3}$	0	$\frac{5}{3}$
	2		η_4	$-\frac{2}{3}$	$\frac{1}{3}$	0	0	$-\frac{1}{3}$	-1	$\frac{410}{3}$
			P_1	0	0	-1	0	0	0	0
			P_2	$-\frac{4}{3}$	$-\frac{1}{3}$	0	0	$-\frac{2}{3}$	-2	$\frac{800}{3}$
			P_3	1	0	-1	0	0	0	20

A Discrete Change in $c_{i,s}$

The only changes that we shall consider with regard to the $c_{i,j}$ values are those associated with *nonbasic* variables (in the final tableau under consideration), that is, with only the $c_{i,s}$ coefficients. Given a change in the coefficient $c_{i,s}$, we have an impact on the associated $\mathbf{y}_s$ column in the tableau and that, in turn, will then affect the indicator elements in $\mathbf{R}_s$. This is given by

$$\hat{\mathbf{y}}_s = \mathbf{B}^{-1}\hat{\mathbf{c}}_s \tag{19.12}$$

and

$$\hat{R}_{k,s} = \mathbf{u}_k^T\hat{\mathbf{y}}_s - w_{k,s} \qquad \text{for all } k \tag{19.13}$$

Example 19-4: A Change in $c_{i,s}$

Returning to the example given in (19.1) through (19.6), let us assume that we wish to determine the impact of a change in $c_{3,2}$ (presently a value of 3). Since $c_{3,2}$ is associated with x_2 and x_2, in turn, is *basic* in Table 19-1, such a change does not fall within the scope of our method.

Let us, however, consider a change in $c_{3,1}$ from -5 to 2. Thus, $\hat{c}_{3,1} = 2$ and we may now employ (19.12) and (19.13), to obtain

$$\hat{\mathbf{y}}_1 = \mathbf{B}^{-1}\hat{\mathbf{c}}_1$$

$$= \begin{pmatrix} 1 & 0 & 0 & 0 \\ 0 & 1 & 0 & 0 \\ 0 & -3 & 1 & 0 \\ 0 & 1 & 0 & 1 \end{pmatrix} \begin{pmatrix} 1 \\ 0 \\ 2 \\ 1 \end{pmatrix} = \begin{pmatrix} 1 \\ 0 \\ 2 \\ 1 \end{pmatrix}$$

and

$$\hat{R}_{k,1} = \mathbf{u}_k^T\hat{\mathbf{y}}_1 - w_{k,1}$$

or

$$\hat{R}_{1,1} = (0 \quad 0 \quad 0 \quad 0)\begin{pmatrix}1\\0\\2\\1\end{pmatrix} - 0 = 0$$

$$\hat{R}_{2,1} = (0 \quad 0 \quad 1 \quad 2)\begin{pmatrix}1\\0\\2\\1\end{pmatrix} - 0 = 4$$

$$R_{3,1} = (1 \quad 0 \quad 0 \quad 0)\begin{pmatrix}1\\0\\2\\1\end{pmatrix} - 0 = 1$$

The impact of the change in $c_{3,1}$ is then that the problem is no longer optimal (we must next pivot on x_1 and η_1).

Addition of a New Goal

Particular care must be given to the addition of a new goal. First, as mentioned before, this goal must be commensurable (unless it represents a rigid constraint) with any other goals at the same priority level. Second, we must "clear" the coefficients in the new goal that are associated with variables that are presently basic. This is best explained by example.

Example 19-5: Addition of a New Goal

Returning again to our illustrative example, let us evaluate the impact of adding the following new goal, G_5:

$$G_5: \quad -4x_1 + x_2 + \eta_5 - \rho_5 = 8$$

where ρ_5 is to be minimized at priority level 1 (i.e., a rigid constraint).

Examining Table 19-1, we see that η_1, x_2, η_3, and η_4 are basic and thus cannot appear in G_5. We need then to "get rid" of the coefficient of x_2 in G_5 (i.e., drive it to zero). This can be accomplished as follows. The equation for x_2, *from Table 19-1*, is

$$x_2 + 0x_1 + \eta_2 + 0\rho_1 - \rho_2 + 0\rho_3 + 0\rho_4 = 35$$

and is found in the second row of the table. Subtracting this equation from G_5, we obtain

$$G_5': \quad -4x_1 + 0x_2 - \eta_2 - 0\rho_1 + \rho_2 - 0\rho_3 - 0\rho_4 + \eta_5 - \rho_5 = -27$$

We now have an expression for G_5 that does not include any nonzero coefficients for any basic variables.

At this point we are ready to construct the new tableau by "simply" adding a new column for ρ_5 and a new row for η_5, as shown in Table 19-4. Note that recalculation of the first indicator row and a_1 was also necessary to develop this tableau.

Table 19-4. TABLEAU WITH NEW GOAL

			P_3								
			P_2								
			P_1			1	1			1	
P_3	P_2	P_1	V	x_1	η_2	ρ_1	ρ_2	ρ_3	ρ_4	ρ_5	x_B
1			η_1	1	0	−1	0	0	0	0	20
			x_2	0	1	0	−1	0	0	0	35
	1		η_3	−5	−3	0	3	−1	0	0	115
	2		η_4	1	1	0	−1	0	−1	0	95
			η_5	−4	−1	0	1	0	0	−1	−27
			P_1	0	0	−1	−1	0	0	−1	0
			P_2	−3	−1	0	1	−1	−2	0	305
			P_3	1	0	−1	0	0	0	0	20

The impact of the addition of the new goal is seen by examining Table 19-4; the previous solution is no longer feasible ($\eta_5 = -27$) and we must employ multidimensional dual simplex to recover feasibility (which leads to the solution $\mathbf{x} = (\frac{27}{4}, 35)$, as may be verified by the reader). In some cases, the addition of a new goal can cause the problem to become *both* nonoptimal and infeasible, leading to the need for the employment of the primal–dual algorithm discussed in Chapter 18. (To illustrate, try the preceding example, wherein η_5, rather than ρ_5, is to be minimized at priority 1.)

Addition of a New Variable

The question of the impact of the inclusion of a new variable is fairly common. For example, given that a firm is presently manufacturing N items, should they consider the production of a new product, and if so, what will be the effect on the present program?

The addition of a new variable requires a new column in the tableau. This column is associated with the new, nonbasic variable and, actually, we have already discussed how to handle such a change when we presented the approach for dealing with a change in $\mathbf{c}_s$. That is, originally, the new variable did not exist and thus all its $c_{i,j}$ coefficients were zero. Once the new variable has been added to the problem, we have, in effect, changed the $c_{i,j}$ coefficients of a nonbasic variable from all zero to some new values. Thus, we may find the new $\mathbf{y}_s$ column under the new variable by means of (19.12). This, of course, also requires computation of a new $\mathbf{R}_s$ column and may affect the optimality. Consider the following example.

Example 19-6: Addition of a New Variable

We assume that our previous model of (19.1) through (19.6) is to be modified through the incorporation of a new variable, x_3. The new model is as follows:

$$\text{lexicographically minimize} \quad \mathbf{a} = \{(\rho_1 + \rho_2), (\eta_3 + 2\eta_4), (\eta_1)\}$$

subject to

$$\begin{aligned}
x_1 \qquad\qquad\qquad + \eta_1 - \rho_1 &= 20 \\
x_2 + x_3 + \eta_2 - \rho_2 &= 35 \\
-5x_1 + 3x_2 + x_3 + \eta_3 - \rho_3 &= 220 \\
x_1 - x_2 - x_3 + \eta_4 - \rho_4 &= 60 \\
\mathbf{x}, \boldsymbol{\eta}, \boldsymbol{\rho} &\geq \mathbf{0}
\end{aligned}$$

We solve for $\mathbf{y}_s$, under x_3, as:

$$\hat{\mathbf{y}}_s = \mathbf{B}^{-1}\hat{\mathbf{c}}_s = \begin{pmatrix} 1 & 0 & 0 & 0 \\ 0 & 1 & 0 & 0 \\ 0 & -3 & 1 & 0 \\ 0 & 1 & 0 & 1 \end{pmatrix} \begin{pmatrix} 0 \\ 1 \\ 1 \\ -1 \end{pmatrix} = \begin{pmatrix} 0 \\ 1 \\ -2 \\ 0 \end{pmatrix}$$

The new tableau is shown in Table 19-5. The new tableau is no longer optimal and we must bring in the new variable, x_3 (pivot on x_3 and x_2).

Table 19-5

			P_3								
			P_2								
			P_1			1	1				
P_3	P_2	P_1	V	x_1	η_2	ρ_1	ρ_2	ρ_3	ρ_4	x_3	$\mathbf{x}_B$
1			η_1	1		−1					20
			x_2		1		−1			1	35
	1		η_3	−5	−3		3	−1		−2	115
	2		η_4	1	1		−1		−1		95
			P_1			−1	−1				0
			P_2	−3	−1		1	−1	−2	2	305
			P_3	1		−1					20

A Reordering of Priority Levels

As in the previous cases, it is not necessary to solve a problem from the beginning if one wishes to determine the impact of a reordering of the initial priorities. Any reasonable (i.e., all goals with a priority, other than priority level 1, must always be commensurable or be made commensurable via weights) reordering of priorities may be analyzed by simply employing the rules used to analyze a change in $w_{k,s}$ and $u_{i,k}$. However, the easiest approach is to simply:

1. Reorder the priorities (and associated weighting factors) in the top and left stubs.
2. Compute the new values for all $R_{k,s}$ and a_k.
3. The only two major impacts are that:
   a. The present solution may no longer be optimal, and/or
   b. The present solution may no longer be implementable.

Parametric Linear Goal Programming

We now move on to the analysis of changes across continuous ranges, a particularly useful tool. Unfortunately, the computational burden of such an analysis does not make solution by hand very attractive. Consequently, we examine only a few simple examples.

Example 19-7: A Parameter in the Achievement Vector

Consider a modified form of our example from (19.1) through (19.6) as follows:

$$\text{lexicographically minimize} \quad \mathbf{a} = \{(\rho_1 + \rho_2), [\eta_3 + \eta_4(1 + d)], (\eta_1)\}$$

subject to

$$\begin{aligned} x_1 \qquad\quad + \eta_1 - \rho_1 &= 20 \\ x_2 + \eta_2 - \rho_2 &= 35 \\ -5x_1 + 3x_2 + \eta_3 - \rho_3 &= 220 \\ x_1 - \quad x_2 + \eta_4 - \rho_4 &= 60 \\ \mathbf{x}, \boldsymbol{\eta}, \boldsymbol{\rho} &\geq \mathbf{0} \end{aligned}$$

In the original model, the weight associated with η_4 at priority level 2 was 2 units. In the foregoing model we have replaced this *fixed* value with the term $(1 + d)$, where we wish to explore the effect of a range of variation in d. That is, will the optimal program be affected by a variation in the weight on η_4 and, if so:

1. What will these effects be?
2. Where will they occur?

The approach to be used to conduct the parametric analysis is given in the following steps:

Step 1. Formulate the model and include all the parameters that we wish to investigate.
Step 2. Set the parameters (d in our example) to zero and solve for the optimal solution.
Step 3. Introduce the parameter(s) back into the tableau (via the discrete change processes discussed earlier).
Step 4. Investigate the ranges of the parameter(s) over which the tableau in step 3 is still optimal.
Step 5. Set one of the parameters to its *finite* borderline value(s) and develop the new, optimal tableaux.
Step 6. Repeat steps 3 and 5 until all the appropriate ranges for all the parameters have been investigated.

For our example, the final optimal tableau with d set to zero is shown in Table 19-6. Reintroducing d is simply a change in a weighting factor for a basic variable, and this results in Table 19-7.

Table 19-6. FINAL TABLEAU WITH $d = 0$

			P_3							
			P_2							
			P_1			1	1			
P_3	P_2	P_1	V	x_1	η_2	ρ_1	ρ_2	ρ_3	ρ_4	$\mathbf{x}_B$
1			η_1	1		−1				20
			x_2		1		−1			35
	1		η_3	−5	−3		3	−1		115
	1		η_4	1	1		−1		−1	95
			P_1			−1	−1			0
			P_2	−4	−2		2	−1	−1	210
			P_3	1		−1				20

Table 19-7. FINAL TABLEAU WITH d INTRODUCED

			P_3							
			P_2							
			P_1			1	1			
P_3	P_2	P_1	V	x_1	η_2	ρ_1	ρ_2	ρ_3	ρ_4	$\mathbf{x}_B$
1			η_1	1		−1				20
			x_2		1		−1			35
	1		η_3	−5	−3		3	−1		115
	$(1+d)$		η_4	1	1		−1		−1	95
			P_1			−1	−1			0
			P_2	$(-4+d)$	$(-2+d)$		$(2-d)$	−1	$(-1-d)$	$210 + 95d$
			P_3	1		−1				20

The solution shown in Table 19-7 is feasible (since all right-hand-side values are nonnegative). However, it will only be optimal if

$$R_{2,1} \leq 0$$
$$R_{2,2} \leq 0$$
$$R_{2,6} \leq 0$$

That is,

$$(-4 + d) \leq 0 \quad \text{or} \quad d \leq 4$$
$$(-2 + d) \leq 0 \quad \text{or} \quad d \leq 2$$
$$(-1 - d) \leq 0 \quad \text{or} \quad d \geq -1$$

Thus, for this solution to be optimal, d must satisfy all three of the relationships above, and this is accomplished whenever

$$-1 \leq d \leq 2$$

Consequently, the program shown in Table 19-7 is optimal only for those ranges of d between -1 and $+2$. This corresponds to a weighting factor for η_4 of 0 to 3 (i.e., $1 + d$).

We next investigate the new optimal solutions for the *finite* borderline, or extreme, values of d. These values are at -1 and $+2$. However, values less than -1 are not of interest because we do not deal with negative weighting factors. Thus, we investigate the optimal solutions that occur in Table 19-7 when $d = 2$. Notice that when $d = 2$, there is a column of zeros, in the indicator rows, under η_2 and thus an alternative optimal solution. We thus pivot on η_2 and x_2 to produce the new (alternative optimal) tableau shown in Table 19-8.

The tableau in Table 19-8 is optimal only if

$$-4 + d \leq 0 \quad \text{or} \quad d \leq 4$$
$$2 - d \leq 0 \quad \text{or} \quad d \geq 2$$
$$-1 - d \leq 0 \quad \text{or} \quad d \geq -1$$

Table 19-8. TABLEAU GENERATED BY SETTING $d = 2$

			P_3							
			P_2							
			P_1			1	1			
P_3	P_2	P_1	V	x_1	x_2	ρ_1	ρ_2	ρ_3	ρ_4	$\mathbf{x}_B$
1			η_1	1		-1				20
			η_2		1		-1			35
	1		η_3	-5	3			-1		220
	$(1 + d)$		η_4	1	-1				-1	60
			P_1			-1	-1			0
			P_2	$(-4 + d)$	$(2 - d)$			-1	$(-1 - d)$	$280 + 60d$
			P_3	1		-1				20

which is satisfied for the range

$$2 \leq d \leq 4$$

Looking at Table 19-8, we see that as long as d is between 2 and 4 (a weight of between 3 and 5 for η_4), our optimal solution is $\mathbf{x}^* = (0, 0)$. That is, we should "do nothing."

We may now evaluate the new optimal tableau at the next borderline value of d (i.e., at $d = 4$). With $d = 4$, the tableau in Table 19-8 is no longer optimal and we may pivot on x_1 and η_1 to produce Table 19-9.

Table 19-9. TABLEAU GENERATED BY SETTING $d = 4$

			P_3	1						
			P_2							
			P_1			1	1			
P_3	P_2	P_1	V	η_1	x_2	ρ_1	ρ_2	ρ_3	ρ_4	$\mathbf{x}_B$
			x_1	1		-1				20
			η_2		1		-1			35
	1		η_3	5	3	-5		-1		320
	$(1 + d)$		η_4	-1	-1	1			-1	40
			P_1			-1	-1			0
			P_2	$(4 - d)$	$(2 - d)$	$(-4 + d)$		-1	$(-1 - d)$	$360 + 40d$
			P_3	-1						0

The tableau in Table 19-9 is optimal only if

$$4 - d \leq 0 \quad \text{or} \quad d \geq 4$$
$$2 - d \leq 0 \quad \text{or} \quad d \geq 2$$
$$-1 - d \leq 0 \quad \text{or} \quad d \geq -1$$

which is satisfied whenever $d \geq 4$. That is, for any values of d from 4 up to infinity, the optimal program is as shown in Table 19-9 [i.e., $\mathbf{x}^* = (20, 0)$].

There are no further *finite* borderline values for d to examine, so we have now analyzed the problem for values of d ranging continuously from -1 to ∞ as shown in Table 19-10. Examining Table 19-10, we see that our optimal program (i.e., values

Table 19-10. SUMMARY OF SOLUTIONS

Range of d	*Optimal Program*	*Optimal Achievement Vector*
$-1 \leq d \leq 2$ (see Table 19-7)	$\mathbf{x}^* = (0, 35)$	$\mathbf{a}^* = (0, 210 + 95d, 20)$
$2 \leq d \leq 4$ (see Table 19-8)	$\mathbf{x}^* = (0, 0)$	$\mathbf{a}^* = (0, 280 + 60d, 20)$
$d \geq 4$ (see Table 19-9)	$\mathbf{x}^* = (20, 0)$	$\mathbf{a}^* = (0, 360 + 40d, 0)$

for x_1 and x_2) is indeed dependent on the weighting factor for η_4 at priority level 2. For weights $(1 + d)$ between 0 and 3, $x_1^* = 0$, $x_2^* = 35$. For weights between 3 and 5, $x_1^* = x_2^* = 0$. Finally, for weights from 5 up, $x_1^* = 20$, $x_2^* = 0$.

Example 19-8: A Parameter in the Right-Hand Side Column

The six steps given in Example 19-7 for the parametric analysis of the achievement vector weights may be modified so as to accommodate the analysis of parameters on the right-hand side. We simply change the wording in steps 4 and 5, replacing the word "optimal" by the word "feasible." Consider the modified form of our example, as follows:

$$\text{lexicographically minimize} \quad \mathbf{a} = \{(\rho_1 + \rho_2), (\eta_3 + 2\eta_4), (\eta_1)\}$$

subject to

$$\begin{aligned}
G_1:& \quad x_1 + + \eta_1 - \rho_1 = 20 - d \\
G_2:& \quad x_2 + \eta_2 - \rho_2 = 35 + d \\
G_3:& \quad -5x_1 + 3x_2 + \eta_3 - \rho_3 = 220 \\
G_4:& \quad x_1 - x_2 + \eta_4 - \rho_4 = 60 + 2d \\
& \quad \mathbf{x}, \boldsymbol{\eta}, \boldsymbol{\rho} \geq \mathbf{0}
\end{aligned}$$

Notice the inclusion of the parameter d in the right-hand side for goals 1, 2, and 4. That is, the right-hand side of G_1 *decreases* with the increases in d, while that of G_2 *increases* with increases in d. The right-hand side of G_4 indicates that the increase is proportional to twice the increase in d. Physically, this could correspond to (linear) increases or decreases in resources over time (d would then correspond to a measure of time) or due to inflation (where d would represent the rate of inflation).

As before, we first set $d = 0$ and find the optimal solution. This is simply the same solution, and tableau, as shown in Table 19-1. Next, we reintroduce the parameter d that corresponds to a discrete change in the original right-hand side vector. We find the new right hand side via (19.10) and the new achievement vector via (19.11). That is,

$$\begin{aligned}
\hat{\mathbf{x}}_B &= \mathbf{B}^{-1}\hat{\mathbf{b}} \\
&= \begin{pmatrix} 1 & 0 & 0 & 0 \\ 0 & 1 & 0 & 0 \\ 0 & -3 & 1 & 0 \\ 0 & 1 & 0 & 1 \end{pmatrix} \begin{pmatrix} 20 - d \\ 35 + d \\ 220 + 0d \\ 60 + 2d \end{pmatrix} = \begin{pmatrix} 20 - d \\ 35 + d \\ 115 - 3d \\ 95 + 3d \end{pmatrix}
\end{aligned}$$

and

$$\hat{a}_k = \mathbf{u}_k \hat{\mathbf{x}}_B$$

Table 19-11 represents the optimal tableau with d reintroduced.

Table 19-11. TABLEAU FOR EXAMPLE 19-8

				x_1	η_2	ρ_1	ρ_2	ρ_3	ρ_4	
			P_3							
			P_2							
			P_1			1	1			
P_3	P_2	P_1	V	x_1	η_2	ρ_1	ρ_2	ρ_3	ρ_4	$\mathbf{x}_B$
1			η_1	1		-1				$20 - d$
			x_2		1		-1			$35 + d$
	1		η_3	-5	-3		3	-1		$115 - 3d$
	2		η_4	1	1		-1		-1	$95 + 3d$
			P_1			-1	-1			0
			P_2	-3	-1		1	-1	-2	$305 + 3d$
			P_3	1		-1				$20 - d$

The tableau in Table 19-11 is optimal regardless of the value of d. However, it is feasible only if $\mathbf{x}_B$ includes no negative elements. That is, the tableau is feasible only if

$$\begin{aligned} 20 - d &\geq 0 \quad &\text{or} \quad d &\leq 20 \\ 35 + d &\geq 0 \quad &\text{or} \quad d &\geq -35 \\ 115 - 3d &\geq 0 \quad &\text{or} \quad d &\leq \tfrac{115}{3} \\ 95 + 3d &\geq 0 \quad &\text{or} \quad d &\geq -\tfrac{95}{3} \end{aligned}$$

and the range of d that satisfies all four of the constraints above is

$$-\tfrac{95}{3} \leq d \leq 20$$

Over this range in d, the optimal program is $\mathbf{x}^* = (0, 35 + d)$.

We next generate new, *feasible* tableaux when d is set to its (finite) borderline values. Notice that if $d > 20$, the right-hand side of G_1 will go negative. Assuming that this is physically possible, we thus restrict our interest to $d = -\frac{95}{3}$. Setting d to $-\frac{95}{3}$, we note that η_4 becomes zero in Table 19-11, while η_1, x_2, and η_3 are all positive values. We may thus use the multidimensional dual simplex algorithm to pivot on η_4 and ρ_4 to obtain Table 19-12.

Table 19-12. TABLEAU GENERATED BY SETTING $d = -\frac{95}{3}$

				x_1	η_2	ρ_1	ρ_2	ρ_3	η_4	
			P_3							
			P_2						2	
			P_1			1	1			
P_3	P_2	P_1	V	x_1	η_2	ρ_1	ρ_2	ρ_3	η_4	$\mathbf{x}_B$
1			η_1	1		-1				$20 - d$
			x_2		1		-1			$35 + d$
		1	η_3	-5	-3		3	-1		$115 - 3d$
			ρ_4	-1	-1		1		-1	$-95 - 3d$
			P_1			-1	-1			0
			P_2	-5	-3		3	-1	-2	$115 - 3d$
			P_3	1		-1				$20 - d$

Examining the tableau in Table 19-12, we see that it is feasible only if

$$\begin{aligned} 20 - d &\geq 0 \quad &\text{or}\quad d &\leq 20 \\ 35 + d &\geq 0 \quad &\text{or}\quad d &\geq -35 \\ 115 - 3d &\geq 0 \quad &\text{or}\quad d &\leq \tfrac{115}{3} \\ -95 - 3d &\geq 0 \quad &\text{or}\quad d &\leq -\tfrac{95}{3} \end{aligned}$$

and the range of d that satisfies all four of these constraints is

$$-35 \leq d \leq -\tfrac{95}{3}$$

Over this range, the optimal program is $\mathbf{x}^* = (0, 35 + d)$, the same as for the previous range investigated.

Our next step is to set $d = -35$ and generate a new tableau. When $d = -35$, the tableau in Table 19-12 has a zero right-hand side element for x_2 and we may use the multidimensional dual simplex to pivot on x_2 and ρ_2. *However*, if d is equal to or less than -35, the right-hand side of G_2 will become negative; thus, we are not interested in this range. Table 19-13 summarizes the results.

Table 19-13. SUMMARY OF RESULTS FOR EXAMPLE 19-8

Range of d	*Optimal Program*	*Optimal Achievement Vector*
$-\frac{95}{3} \leq d \leq 20$ (see Table 19-11)	$\mathbf{x}^* = (0, 35 + d)$	$\mathbf{a}^* = (0, 305 + 3d, 20 - d)$
$-35 \leq d \leq -\frac{95}{3}$ (see Table 19-12)	$\mathbf{x}^* = (0, 35 + d)$	$\mathbf{a}^* = (0, 115 - 3d, 20 - d)$

Our conclusion is then that for any values of *d of interest* (i.e., from -35 to $+20$), x_1 is always zero while x_2 is equal to $35 + d$. In most problems the ranges that must be investigated (and the computational burden) will be far greater than those in the example. However, the procedure remains the same.

Summary

In this chapter we have indicated, through example, the manner in which sensitivity analysis is performed for the lexicographic linear goal programming model. Such analysis is not only useful from the conventional point of view, but also allows us to examine, to a degree, the impact of the preemptive-priority structure and weights that were initially assigned. In Chapter 20 we carry this relaxation even further and, in fact, may totally discard the preemptive-priority concept.

REFERENCE

1. IGNIZIO, J. P. *Goal Programming and Extensions*. Lexington, Mass.: Heath (Lexington Books), 1976.

PROBLEMS

19.1. Compare sensitivity analysis in single-objective linear programming (Chapter 10) with that in lexicographic linear goal programming. What are the differences, if any? Is the sensitivity analysis in goal programming in any way more limited than that available in conventional linear programming?

19.2. List the inverse of the basis matrices (i.e., $\mathbf{B}^{-1}$) for Tables 17-13 and 17-14.

19.3. What is the inverse of the basis matrix for Table 17-21?

19.4. *All* methods of optimization, be they directed toward the single- or multiple-objective model, have certain inherent limitations, restricting assumptions, and disadvantages. When solving a model of any real problem, we realize that its coefficients, weights, and right-hand side values are virtually always estimates only. This holds true for goal programming as well as for all alternative methods of optimization. However, some critics of goal programming state that the requirement for specifying weights on the goal deviation variables (i.e., formulate the achievement function) is a "serious limitation" of goal programming. Develop an argument that logically and systematically counters such criticism.

19.5. In Problem 17.9, determine the effect of a discrete change in the weighting factor, at priority level 4, for ρ_1 from its present value of 1 to a value of 3.

19.6. In Problem 17.9, investigate the effect, if any, if the weighting factor at priority level 2, for η_3, is changed from a value of 1 to a value of 100.

19.7. In Problem 17.9, what is the effect of a change in $c_{1,2}$ from 1 to 5? A change in $c_{1,1}$ from 2 to -1?

19.8. In Problem 17.10, how would a change in both b_1 and b_2 from 100 and 90 to 75 and 80 affect the original solution? A change to 150 and 75?

19.9. Given the following problem, determine the set of optimal policies over all reasonable ranges in the value of the parameter d.

$$\text{Lexicographically minimize} \quad \mathbf{a} = \{(\eta_1 + \rho_1), (2\rho_2 + \rho_3)\}$$

subject to

$$\begin{aligned} x_1 - 10x_2 + \eta_1 - \rho_1 &= 50 - d \\ 3x_1 + 5x_2 + \eta_2 - \rho_2 &= 20 - 2d \\ 8x_1 + 6x_2 + \eta_3 - \rho_3 &= 100 \\ \mathbf{x}, \boldsymbol{\eta}, \boldsymbol{\rho} &\geq \mathbf{0} \end{aligned}$$

19.10. Determine the set of optimal policies for a range in the weighting factor of η_3 from zero up to 100 for the following problem:

$$\text{lexicographically minimize} \quad \mathbf{a} = \{(\eta_1), (\rho_2), [(1 + d)\eta_3 + 5\eta_4], (\rho_1)\}$$

subject to

$$\begin{aligned} x_1 + x_2 + \eta_1 - \rho_1 &= 100 \\ x_1 + x_2 + \eta_2 - \rho_2 &= 90 \\ x_1 \qquad + \eta_3 - \rho_3 &= 80 \\ x_2 + \eta_4 - \rho_4 &= 55 \\ \mathbf{x}, \boldsymbol{\eta}, \boldsymbol{\rho} &\geq \mathbf{0} \end{aligned}$$

19.11. Solve the model in Problem 19.10 with the following achievement function:

$$\mathbf{a} = \{(\eta_1), (\rho_2), (d_1\eta_3 + d_2\eta_4), \rho_1\}$$

19.12. Consider the example problem of Chapter 19, as defined by functions (19.1) through (19.6) subject to the following changes:

(a) The addition of a new goal, G_5, at priority level 2 (wherein the new goal is assumed to be commensurable), where

$$G_5: \quad x_1 + x_2 + \eta_5 - \rho_5 = 50$$

and ρ_5 is to be minimized with its minimization considered to be twice as important as that of η_4.

(b) In part (a), consider the following form of the achievement function:

$$\text{lexicographically minimize} \quad \mathbf{a} = \{(\rho_1 + \rho_2), (\eta_3 + 2_1\eta_4 + d\rho_5), (\eta_1)\}$$

(c) Consider the addition of a new variable to the problem, variable x_3, wherein the goals are now:

$$\begin{aligned} x_1 \qquad + x_3 + \eta_1 - \rho_1 &= 20 \\ x_2 \qquad + \eta_2 - \rho_2 &= 35 \\ -5x_1 + 3x_2 - x_3 + \eta_3 - \rho_3 &= 220 \\ x_1 - x_2 \qquad + \eta_4 - \rho_4 &= 60 \\ \mathbf{x}, \boldsymbol{\eta}, \boldsymbol{\rho} &\geq \mathbf{0} \end{aligned}$$

(d) Given the original formulation for the example problem, determine the impact of now considering negative as well as positive values for x_2 (i.e., x_2 is now unrestricted).

19.13. Consider Problem 2.16. First formulate this problem as a linear goal programming model wherein (1) the market limits are considered absolute, and (2) profit maximization is considered to be twice as important as minimizing mpg rates below 27. Then:

(a) Solve the problem as stated.

(b) Evaluate the uncertainty about the respect weights given to profit and mpg rates below 27.

(c) Consider profit and the mpg levels to be at *separate* priority levels and determine the effects on the permutation of these priorities.

(d) For the original model, in (a), what is the impact of an error in regard to the estimated mpg of the compact cars? That is, rather than a rate of 36 mpg, their actual rate is 32 mpg.

19.14. Consider Problem 17.15. Reverse priorities 2 and 3 and solve for the resultant solution. Is there any effect on the previous optimal program and solution?

EXTENSIONS AND RELATED TOPICS

Introduction

The multiple-objective model and methodology with which we have dealt in the preceding four chapters is restricted to linear multiple-objective models with continuous-valued variables and for which a preemptive-priority structure is appropriate. Extensions of the goal programming methodology that would be desirable would then include:

1. The consideration of integer (or discrete)-valued variables.
2. The consideration of nonlinear forms (in both the goals and the achievement function).
3. A relaxation of the preemptive-priority concept, as encompassed via the lexicographic minimum of an ordered vector.
4. The consideration of "ranges" in establishing aspiration levels.

Extensions and/or refinements to goal programming to encompass models with discrete variables and/or nonlinear forms are relatively straightforward. However, since this text is dedicated to the investigation of linear models, we shall not address such topics. Those who desire further details on these aspects are directed to an earlier book by the author [6].

20

The Main Thrust of the Chapter

The two aspects of the multiple-objective methodology to be considered herein are:

1. The relaxation of the preemptive-priority concept via a combination of lexicographic linear goal programming with the efficient solution (or nondominated or Pareto optimal) technique.
2. The consideration of "ranges of aspiration-level values" as achieved through extensions of either (a) the methodology of fuzzy sets or (b) goal programming.

By means of these approaches, we develop the "augmented goal programming," "fuzzy programming," and "interval goal programming" methods, respectively.

The Efficient Solution Technique

Prior to the discussion of augmented goal programming, let us describe briefly the method of efficient solutions [2, 3] as referred to in Chapter 16. The efficient solution technique deals with a special multiple-objective model, the vector-maximum (or vector-minimum) model. The vector-maximum model can be developed from the baseline model of relationships (4.1) through (4.4) by means of the following assumptions:

1. All of the goals within (4.3) (i.e., those goals that are to be "satisfied") are assumed to be rigid constraints.
2. The remaining set of objectives [i.e., in (4.1) and (4.2)] are all converted to the maximization form.

One example of a vector-maximum formulation as given in (20.1) through (20.4) is as follows [11]:

Find x_1 and x_2 so as to

$$\text{maximize} \quad z_1 = 2x_1 + x_2 \tag{20.1}$$

$$\text{maximize} \quad z_2 = -x_1 + 2x_2 \tag{20.2}$$

subject to

$$\left.\begin{aligned} -x_1 + 3x_2 &\leq 21 \\ x_1 + 3x_2 &\leq 27 \\ 4x_1 + 3x_2 &\leq 45 \\ 3x_1 + x_2 &\leq 30 \end{aligned}\right\} \tag{20.3}$$

$$\mathbf{x} \geq \mathbf{0} \tag{20.4}$$

An efficient solution to the foregoing models is any value for **x**, satisfying (20.3) and (20.4), such that the resultant values for z_1 and z_2 are not dominated by any other **x**.

Specifically, given K objectives to be maximized within the vector-maximum model, and given that the value of each of these objectives is denoted as z_k (for the kth objective), consider two solutions to the model:

$$\mathbf{Z}^{(1)} = (z_1^{(1)}, \ldots, z_K^{(1)})$$

$$\mathbf{Z}^{(2)} = (z_1^{(2)}, \ldots, z_K^{(2)})$$

Note that $z_k^{(s)}$ represents the objective function value for objective k in solution s. Further, the *order* of the values within $\mathbf{Z}^{(s)}$ is unimportant as long as they are all consistent.

If all objectives are to be maximized, $\mathbf{Z}^{(1)}$ is said to dominate $\mathbf{Z}^{(2)}$ if:

$$z_k^{(1)} \geq z_k^{(2)} \qquad \text{for all } k$$

$$z_k^{(1)} > z_k^{(2)} \qquad \text{for at least one } k$$

For example, $\mathbf{Z}^{(A)} = (8, 7, 6, 12)$ dominates $\mathbf{Z}^{(B)} = (2, 7, 5, 4)$. Any solution, s, for which $\mathbf{Z}^{(s)}$ is not dominated by any other solution is then designated as an efficient (or nondominated or Pareto-optimal) solution.

The premise fundamental to the approach is, then, that a rational decision maker would (or at least should) always prefer an efficient solution to one that is dominated. Knowing nothing else about the preferences of the decision maker, the "optimal" solution to the vector-maximum problem must be one of the efficient solutions. The advocates of such an approach cite, as the main advantage, the fact that one does not need to either weight or rank the objectives. From a practical point of view, the primary disadvantage is fairly obvious (i.e., the number of efficient extreme points alone are enormous for even modest-size problems). As a result, the most computationally efficient and practical exploitations of this approach have been arrived at by means of heuristic adjuncts [7, 9, 10]. We describe just one of these in the following section.

Augmented Goal Programming [7][a]

Since we are already familiar with lexicographic linear goal programming, the combination of the efficient solution set technique with this tool should be easy to follow. However, there is also a more objective reason to consider this particular combination. Extensive empirical evidence indicates that although the lexicographic minimum may not always be optimal (with respect to the decision maker's actual preferences and values), it is often within the "neighborhood" of the optimal solution. Consequently, the lexicographic minimum, as achieved through the solution of the goal programming model, is generally an extremely good starting point in the search for efficient solutions.

The General Procedure

The augmented goal programming method follows the steps given below:

Step 1. Convert the vector-maximum model into a (lexicographic) linear goal programming model. Rank the goals as appropriate, with *each* goal at a *separate* priority level.

Step 2. Solve the goal programming model for the initial solution (this should also be a nondominated or efficient solution).

Step 3. Determine the "leeway" or allowable degradations in the achievement of the goals (from the solution in step 2) that would be acceptable to the decision maker.

Step 4. Generate the remaining set of efficient solutions within this allowable degradation region.

We discuss these steps further below and illustrate them via implementation of the example given in (20.1) through (20.4).

[a]Based on the paper of reference [7]. Permission granted by Pergamon Press Ltd.

The Goal Programming Equivalent of the Vector-Maximum Model

Starting with the vector-maximum problem

$$\text{maximize} \quad \mathbf{Z} = \mathbf{Cx} \tag{20.5}$$

subject to

$$\mathbf{Ax} \leq \mathbf{b} \tag{20.6}$$

$$\mathbf{x} \geq \mathbf{0} \tag{20.7}$$

we form the equivalent (lexicographic) linear goal programming model as:

Find the lexicographic minimum of

$$\mathbf{a} = \{g_1(\boldsymbol{\eta}, \boldsymbol{\rho}), \ldots, g_K(\boldsymbol{\eta}, \boldsymbol{\rho})\} \tag{20.8}$$

subject to

$$\mathbf{Ax} + \boldsymbol{\eta}^{(r)} - \boldsymbol{\rho}^{(r)} = \mathbf{b} \tag{20.9}$$

$$\mathbf{Cx} + \boldsymbol{\eta}^{(s)} - \boldsymbol{\rho}^{(s)} = \mathbf{z}^0 \tag{20.10}$$

$$\mathbf{x}, \boldsymbol{\eta}, \boldsymbol{\rho} \geq \mathbf{0} \tag{20.11}$$

where

S = number of original objectives

K = number of priority levels $(1 + S)$

$\boldsymbol{\eta}^{(r)}, \boldsymbol{\rho}^{(r)}$ = deviation variables associated with the original set of rigid constraints

$\boldsymbol{\eta}^{(s)}, \boldsymbol{\rho}^{(s)}$ = deviation variables associated with the original set of objectives

$\mathbf{z}^0$ = aspiration levels associated with the original set of objectives

Using the example given in (20.1) through (20.4), we form the following linear goal programming equivalent:

1. Set an aspiration level for z_1 of 40 units and an aspiration level for z_2 of 20 units. (Other values might have been selected, but this serves to demonstrate the process.)
2. We assume that z_1 is more important than z_2.

The result is then as follows:

$$\text{lexicographically minimize} \quad \mathbf{a} = \{(\rho_1 + \rho_2 + \rho_3 + \rho_4), (\eta_5), (\eta_6)\} \tag{20.12}$$

subject to

$$\text{original constraints}\begin{pmatrix} -x_1 + 3x_2 + \eta_1 - \rho_1 = 21 \\ x_1 + 3x_2 + \eta_2 - \rho_2 = 27 \\ 4x_1 + 3x_2 + \eta_3 - \rho_3 = 45 \\ 3x_1 + x_2 + \eta_4 - \rho_4 = 30 \end{pmatrix} \tag{20.13}$$

$$z_1(\mathbf{x}):\quad 2x_1 + x_2 + \eta_5 - \rho_5 = 40 \quad \text{original} \tag{20.14}$$

$$z_2(\mathbf{x}):\quad -x_1 + 2x_2 + \eta_6 - \rho_6 = 20 \quad \text{objectives} \tag{20.15}$$

$$\mathbf{x}, \boldsymbol{\eta}, \boldsymbol{\rho} \geq \mathbf{0} \tag{20.16}$$

Solving the foregoing model, we develop the final goal programming tableau as shown in Table 20-1. As may be seen, the solution (the initial solution in the augmented goal programming procedure) is

$$\mathbf{x}^* = (9, 3)$$

$$\mathbf{a}^* = (0, 19, 23)$$

Table 20-1. FINAL GOAL PROGRAMMING TABLEAU FOR INITIAL EFFICIENT SOLUTION

			P_3									
			P_2									
			P_1			1	1	1	1			
P_3	P_2	P_1	V	η_4	η_3	ρ_1	ρ_2	ρ_3	ρ_4	ρ_5	ρ_6	$\mathbf{x}_B$
			η_1	3	-2	-1	0	2	-3	0	0	21
			η_2	$\frac{9}{5}$	$-\frac{8}{5}$	0	-1	$\frac{8}{5}$	$-\frac{9}{5}$	0	0	9
			x_2	$-\frac{4}{5}$	$\frac{3}{5}$	0	0	$-\frac{3}{5}$	$\frac{4}{5}$	0	0	3
			x_1	$\frac{3}{5}$	$-\frac{1}{5}$	0	0	$\frac{1}{5}$	$-\frac{3}{5}$	0	0	9
	1		η_5	$-\frac{2}{5}$	$-\frac{1}{5}$	0	0	$\frac{1}{5}$	$\frac{3}{5}$	-1	0	19
1			η_6	$\frac{11}{5}$	$-\frac{7}{5}$	0	0	$\frac{7}{5}$	$-\frac{11}{5}$	0	-1	23
			P_1	0	0	-1	-1	-1	-1	0	0	0
			P_2	$-\frac{2}{5}$	$-\frac{1}{5}$	0	0	$\frac{1}{5}$	$\frac{3}{5}$	-1	0	19
			P_3	$\frac{11}{5}$	$-\frac{7}{5}$	0	0	$\frac{7}{5}$	$-\frac{11}{5}$	0	-1	23

It may be proven [7] that the solution to the lexicographic linear goal programming model is an efficient solution, *given that the aspiration level values established for the objectives are set high enough.* Thus, this particular solution, depicted as $\mathbf{X}^{(1)}$ in Figure 20-1, is an efficient solution to the original vector-maximum model. Actually, in Figure 20-1, points $\mathbf{X}^{(1)}$, $\mathbf{X}^{(2)}$, $\mathbf{X}^{(3)}$, and $\mathbf{X}^{(4)}$ are all efficient, whereas $\mathbf{X}^{(5)}$ and $\mathbf{X}^{(6)}$ are dominated.

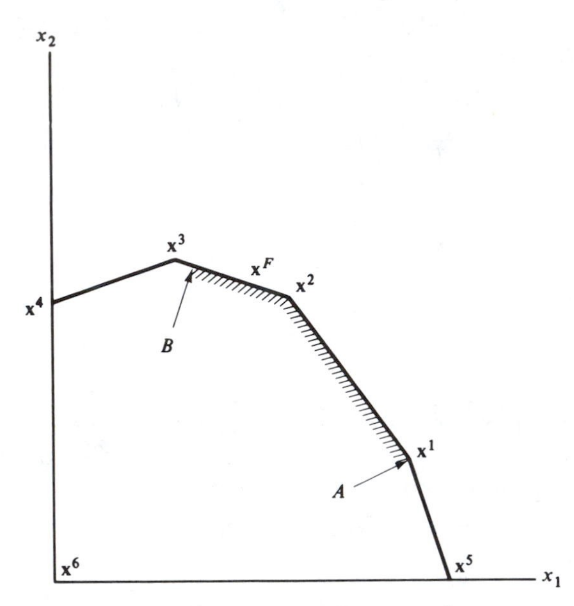

Figure 20-1. Graph for example.

Deriving a Subset of Efficient Solutions

If two adjacent extreme points are efficient, any point on the edge joining these two points is also efficient. As a result there can be an infinite number of efficient solutions. We thus restrict our attention to the finite number of efficient *extreme points*. However, as mentioned earlier, even this number will, in general, be excessive for computational purposes. Consequently, our derivation of efficient extreme points will be further restricted to those within the "region of allowable degradation." *One* approach to the development of the information needed to determine this region is given below. Although not an elegant, strictly analytical process, it has still worked well in actual practice.

The approach we discuss is to interface with the decision maker via the following procedure.

1. The efficient solution arrived at via step 2, above, is presented to the decision maker. (Since this solution was developed using the decision maker's initial input with regard to the ordering of the objective set, this solution is usually either optimal or, at least, within the neighborhood of the true optimal solution.)
2. The decision maker is then asked to indicate the amount of degradation, Δ_k, that he or she would allow for a_k if this would provide a "significant" improvement in some a_t, $t \neq k$.[b]

[b]This phase of the procedure has been accomplished in several ways, ranging from a simple question and answer period with the decision maker up to fairly lengthy interactive efforts employing parametric goal programming.

Using this information, the search for efficient solutions is restricted as follows:

Given the (initial) linear goal programming efficient extreme point solution, designate the associated achievement vector as

$$\mathbf{a}^0 = (a_1^0, \ldots, a_K^0)$$

Let

$\boldsymbol{\Delta} = (\Delta_1, \ldots, \Delta_K)$ be the vector of allowable degradations (i.e., *increases* in $\mathbf{a}$)

Given some other solution, $\mathbf{a}^t$, this solution is considered only if

$$a_k^t \leq a_k^0 + \Delta_k \qquad \text{for all } k$$

$$\mathbf{a}^t \text{ is efficient}$$

As a result of the employment of these rules, the efficient solution set generated is restricted to that subset of efficient extreme points which satisfies the allowable degradations on the achievement of the goal set. Typically, this reduces the number of efficient extreme points generated by a substantial degree.

We are now ready to discuss the basic features of the augmented goal programming algorithm. These steps involve, primarily, the search for the subset of efficient extreme points (within the allowable degradation region) as achieved by movement from a given efficient extreme point to an adjacent extreme point, as arrived at via the conventional simplex pivoting rules.

Step 1. Given: An efficient extreme point within the allowable degradation region.

Step 2. Generate an adjacent extreme point that satisfies the following:

(a) *Rule One.* An exchange or pivot (i.e., of a basic variable for a nonbasic variable) should not be made if it would cause a_1 to become positive (i.e., violate the constraints).

(b) *Rule Two.* An exchange should not be made if it will lead to a previously investigated basis.

(c) *Rule Three.* If *none* of the shadow prices ($R_{k,s}$ values) under a nonbasic variable are positive, then an exchange involving that variable should not be made, as it will lead to a dominated solution.

Step 3. Continue steps 1 and 2 until the efficient solution subset is generated (i.e., there are no more efficient extreme points to be found within the allowable degradation region).

Unfortunately, the foregoing algorithm is not totally general, as, for example, it does not deal with the problem of a degenerate basis. For a more detailed discussion of the development of the set of efficient extreme points, the reader may consult the references [2, 3].

Example 20-1

We are now ready to solve the vector-maximum model of (20.1) through (20.4), as represented by the equivalent lexicographic linear goal programming model of (20.12) through (20.16). First, let us specify the allowable degradation region. Examining Table 20-1, we see that our initial efficient extreme point is

$$\mathbf{X}^{(1)} = (9, 3)$$

with

$$\mathbf{a}^{(1)} = (0, 19, 23)$$

We assume that our hypothetical decision maker will allow a degradation on a_2 of 4 units (i.e., from 19 up to 23) and a degradation on a_3 of 5 units (i.e., from 23 units up to 28). Thus,

$$\mathbf{\Delta} = (0, +4, +5)$$

We now examine Table 20-1 for any efficient extreme points within this region. From rule one, we see that no exchanges involving ρ_1, ρ_2, ρ_3, or ρ_4 should be made.[c] From rule three, it should be apparent that no exchange involving η_3, ρ_5, or ρ_6 should be made. (For example, if an exchange involving η_3 were made, we obtain the following *dominated* solution: $\mathbf{X}^{(5)} = \{10, 0\}$, $\mathbf{a}^{(5)} = \{0, 20, 30\}$.)

Consequently, in Table 20-1, the only nonbasic variable presently of interest is η_4. If η_4 is brought into the basis, η_2 must depart, leading to the solution:

$$\mathbf{X}^{(2)} = (6, 7)$$
$$\mathbf{a}^{(2)} = (0, 21, 12)$$

This is *both* an efficient solution *and* within the degradation allowances.

Our next step, which we shall discuss only, is to construct the new tableau as associated with $\mathbf{X}^{(2)}$. We then repeat our process of searching for the efficient extreme points *within the degradation allowance*. The result of such search shows that $\mathbf{X}^{(3)}$ (see Figure 20-1) is efficient *but not within the degradation allowances*. Thus, we interpolate between $\mathbf{X}^{(2)}$ and $\mathbf{X}^{(3)}$ to determine the allowable efficient region. The total set of efficient solutions within the allowable region is the crosshatched "face" shown in Figure 20-1.

Some Difficulties

All variations of the efficient solution technique, including those limited to an investigation of just a subset of efficient extreme points (as with augmented goal programming), have difficulty in solving problems of other than small to modest size. The augmented goal programming method has, in addition, a few other problems. One, in particular, is that associated with finding the *initial* efficient extreme point. Unless the aspiration levels set for the objectives within

---

[c] And, in fact, from the column drop rule of Chapter 17, the columns for ρ_1, ρ_2, ρ_3, and ρ_4 may be dropped.

the vector-maximum model are high enough, we could obtain an inferior (i.e., dominated) solution. In practice this has not occurred with any frequency. Generally speaking, if the achievement level for an original objective is satisfied (i.e., some zero a_k values, $k > 1$), this indicates the possibility of an initial dominated solution, and the aspiration levels should be raised.

Weighted Linear Goal Programming

Although alluding to the original weighted linear goal programming method, we have yet to discuss it at length. In essence, the weighted model is the same as the lexicographic, weighted model presented in the earlier chapters except that the goals are weighted *but not ranked.* For example, consider the following vector-maximum model:

$$\begin{aligned}
&\text{optimize} \quad z_1 \text{ (i.e., maximize or minimize according to the problem)}\\
&\quad\vdots\\
&\text{optimize} \quad z_S\\
&\qquad\qquad \text{subject to}\\
&\qquad\qquad\qquad \mathbf{Ax} \leq \mathbf{b}\\
&\qquad\qquad\qquad \mathbf{x} \geq \mathbf{0}
\end{aligned}$$

Adding aspiration levels and deviation variables to each objective and weighting each resultant goal, we obtain:

$$\text{minimize} \quad a = \sum_{s=1}^{S} (u_s \eta_s + w_s \rho_s)$$

$$\text{subject to}$$

$$\begin{aligned}
z_1(\mathbf{x}) + \eta_1 - \rho_1 &= z_1^0\\
&\vdots\\
z_S(\mathbf{x}) + \eta_S - \rho_S &= z_S^0\\
\mathbf{Ax} &\leq \mathbf{b}\\
\mathbf{x}, \boldsymbol{\eta}, \boldsymbol{\rho} &\geq \mathbf{0}
\end{aligned}$$

where u_s = weighting factor for the negative deviation of goal s

w_s = weighting factor for the positive deviation of goal s

z_s^0 = aspiration level for objective s

Since the resultant model is in a conventional (i.e., single-objective) form, conventional methods of solution may be applied. The obvious criticism of the strictly weighted model is in the selection of *valid* weighting factors.

There is an alternative way in which the weighted model may be formed. Rather than multiplying each deviation variable by a constant weight, we may, instead, raise each deviation variable in the achievement function to some power. This results in a polynomial form for the achievement function, such as:

$$\text{minimize} \quad a = \rho_1^3 + \eta_2^3 + \rho_3^2$$

The intent is essentially the same in both cases, and the questions raised with regard to the weighting factors are now just as applicable in regard to the choice of exponents.

Fuzzy Linear Programming

A fairly recent attempt at modeling and solving the multiple-objective problem is that known as fuzzy programming [8, 11]. The approach is similar, in many respects, to the weighted linear goal programming method previously discussed, differing primarily in the manner in which the importance of the goals are considered. Weighted linear goal programming depends on the development of weights, whereas fuzzy programming utilizes a concept known as the fuzzy membership function. The fuzzy programming model and solution method, upon examination, bears a striking resemblance to the approach discussed in Chapter 11 for fitting a surface to a set of data wherein we wish to minimize the maximum deviation of any data point to that surface.

The first step in modeling via the fuzzy linear programming method is to transform the baseline model of Chapter 2 into the vector-maximum (or -minimum) model as we did with the efficient solution set technique. Our next step is to assign, for each objective, two values, which we shall term U_k and L_k, where:

$$U_k = \text{aspired level of achievement for objective } k$$

$$L_k = \text{lowest acceptable level of achievement for objective } k$$

$$d_k = U_k - L_k$$

or

$$d_k = \text{degradation allowance, or leeway, for objective } k$$

For example, if our desired level of profit per week was \$1,000 and we would not wish it to be any lower than \$800, then

$$U = \$1{,}000$$

$$L = \$800$$

$$d = \$1{,}000 - 800 = \$200$$

Once the aspiration levels and degradation or leeway for each objective have been specified, we have formed what is termed the fuzzy model. Our next step is to transform the fuzzy model into a "crisp" model (i.e., a conventional mathematical programming model such as a linear program). The easiest way to explain this procedure, without consuming a major portion of this chapter, is to demonstrate the process on our vector-maximum model of (20.1) through (20.4). In this illustration, we find the values of U and L as follows:

1. Solve the vector-maximum problem as a linear programming problem using, each time, only one of the objectives (ignore all others).
2. From the results in step 1, determine the corresponding values for every objective at each solution derived.
3. From step 2 we may find, for each objective, the best (U) and worst (L) values corresponding to the set of solutions.

Examining the problem in (20.1) through (20.4) and considering Figure 20-1, we see that for

$$z_1: \quad U_1 = 21, L_1 = 7, d_1 = 14$$

$$z_2: \quad U_2 = 14, L_2 = -3, d_2 = 17$$

Our *initial* fuzzy model is then given by associating the aspiration levels with each objective as follows:

$$z_1 \geq 21 \qquad \text{or} \qquad 2x_1 + x_2 \geq 21 \tag{20.17}$$

$$z_2 \geq 14 \qquad \text{or} \qquad -x_1 + 2x_2 \geq 14 \tag{20.18}$$

$$\left.\begin{aligned} -x_1 + 3x_2 &\leq 21 \\ x_1 + 3x_2 &\leq 27 \\ 4x_1 + 3x_2 &\leq 45 \\ 3x_1 + x_2 &\leq 30 \end{aligned}\right\} \tag{20.19}$$

$$\mathbf{x} \geq \mathbf{0} \tag{20.20}$$

Relationships (20.17) and (20.18) correspond, of course, to objectives 1 and 2 with their associated aspiration levels (U_k), while (20.19) is the original set of rigid constraints.

Associated with each goal is a fuzzy membership function [8, 11], which is denoted as $\mu_k(\mathbf{x})$, or

$$\mu_k(\mathbf{x}) = \begin{cases} 0 & \text{if } z_k \leq L_k \\ 0 < \mu_k(\mathbf{x}) < 1 & \text{if } L_k < z_k < U_k \\ 1 & \text{if } z_k \geq U_k \end{cases}$$

That is, $\mu_k(\mathbf{x})$ reflects the degree of achievement (values of 1 for perfect achievement) or nonachievement (values of zero for strong nonachievement) of a given goal. The actual form of the fuzzy membership function may be either linear or nonlinear. In this case, we presume a linear form so as to transform our fuzzy model into a linear programming model. That is, we let

$$\mu_k(\mathbf{x}) = 1 - \left(\frac{U_k - z_k}{d_k}\right) \tag{20.21}$$

and note that $\mu_k(\mathbf{x})$ will take on the value of 1 if z_k is achieved (i.e., its value is U_k), a value of zero if not (if the value of $z_k = L_k$), and some intermediate values otherwise.

Thus, given any solution, **x**, to the fuzzy model, we wish to maximize the minimum value of $\mu_k(\mathbf{x})$. That is, we desire to minimize the worst underachievement of any goal. This can be accomplished by using the dummy variable λ, where we

$$\text{minimize} \quad \lambda$$

$$\text{subject to}$$

$$\lambda \geq \left(\frac{U_k - z_k}{d_k}\right) \qquad \text{for all } k$$

Thus, the equivalent linear programming formulation for (20.17) through (20.20) is:

Find **x** so as to

$$\text{minimize} \quad \lambda \tag{20.22}$$

$$\text{subject to}$$

$$\lambda \geq \frac{21 - (2x_1 + x_2)}{14} \tag{20.23}$$

$$\lambda \geq \frac{14 - (-x_1 + 2x_2)}{17} \tag{20.24}$$

$$\left.\begin{aligned} -x_1 + 3x_2 &\leq 21 \\ x_1 + 3x_2 &\leq 27 \\ 4x_1 + 3x_2 &\leq 45 \\ 3x_1 + x_2 &\leq 30 \end{aligned}\right\} \tag{20.25}$$

$$\mathbf{x} \geq \mathbf{0}, \ \lambda \geq 0 \tag{20.26}$$

It may also be proven [11] that the solution to the model above is an efficient solution. For this case, the solution is given by $x_1 = 5.03$, $x_2 = 7.32$, as may be found via any linear programming algorithm. (This solution is shown as $\mathbf{X}^F$ in Figure 20-1.)

Advantages and Disadvantages

Again, a major advantage of fuzzy linear programming is that (given the linear forms for the fuzzy membership functions as in the example) it may be transformed into a conventional linear programming model. Another advantage is that the treatment of objectives (once they are transformed into goals via the aspiration level) is symmetric with the treatment of the constraints. We could, if we wished, relax the rigidity of our conception of constraints and allow some leeway about constraint satisfaction via an associated fuzzy membership function. Finally, the concept of an aspiration level subject to some allowable degradation seems consistent with actual decision making.

Some of the disadvantages associated with the approach include:

1. The underachievement of just one goal can have a major impact on the solution, since we attempt to minimize the maximum underachievement.
2. The form of the membership function, $\mu_k(\mathbf{x})$, is open to question. In fact, one can perceive the membership function as an alternative to weighting and thus subject to similar criticisms.
3. Obtaining good information on the aspiration levels and allowable degradation is vital to the credibility of any results obtained (as is true with most alternative approaches).

Interval Goal Programming

At roughly the same time that fuzzy programming was being developed, investigators [5] were engaged in the formulation of a straightforward technique, based on goal programming, that is very much similar in basic intent. This method, denoted as interval goal programming, also addresses the desire to consider the satisfaction over a *range* of aspiration levels rather than a single value. However, unlike fuzzy programming wherein we attempt to *minimize the maximum deviation* from a set of ranges, we instead try to *minimize the weighted sum* (or, alternatively, a prioritized set) of deviations from the set of ranges. This can be explained best by means of an example.

Consider a simple problem in which we have but two objectives, z_1 and z_2, as follows:

$$z_1: \quad x_1 + x_2 \geq t_1$$

$$z_2: \quad 3x_1 + x_2 \leq t_2$$

where

$$x_1, x_2 \geq 0$$

Now assume that:

x_1 and x_2 are the control variables

t_1 = target (aspiration level) value for objective z_1

t_2 = target (aspiration level) value for objective z_2

Given a specific value for both t_1 and t_2, objectives z_1 and z_2 are transformed into goals. However, let us assume that the target values are not precise and the easiest way in which to specify them is by means of a satisfactory *range*. That is,

$$L_1 \leq t_1 \leq U_1$$
$$L_2 \leq t_2 \leq U_2$$

where the definitions of U_i and L_i are similar to those encountered in fuzzy programming. That is,

$U_i =$ upper limit on the target value (i.e., the upper end of the range)

$L_i =$ lower limit on the target value (i.e., the lower end of the range)

What we would like then is to find a solution, **x**, that lies in the *interval* specified by the intersection of these two ranges. Assume, for example, that

$$10 \leq t_1 \leq 12$$
$$16 \leq t_2 \leq 20$$

We may then construct these two intervals as shown in Figure 20-2. The cross-hatched region is the intersection and thus the interval of satisfying solutions.

Depending upon the particular problem, goal programming provides us with a flexible tool for various formulations. We demonstrate with a few of these below.

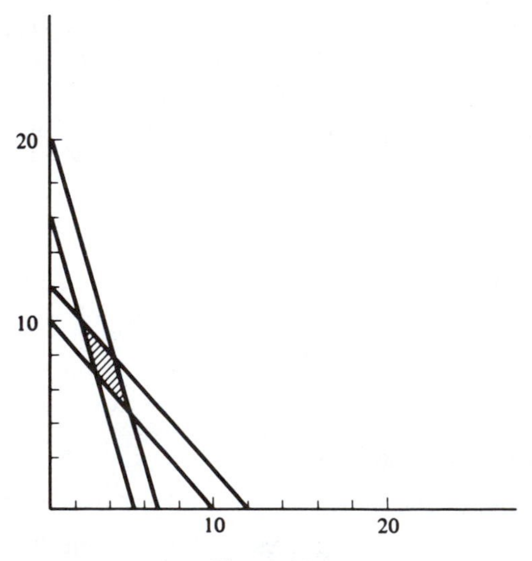

Figure 20-2

Interval Goal Programming: The Weighted Model

Consider first the following, weighted formulation:
Find **x** so as to

$$\text{minimize} \quad a = \sum_{k=1}^{K} (u_{k,1}\eta_{k,1} + w_{k,2}\rho_{k,2}) \tag{20.27}$$

subject to

$$z_{k,1} + \eta_{k,1} - \rho_{k,1} = L_k \qquad \text{all } k \tag{20.28}$$

$$z_{k,2} + \eta_{k,2} - \rho_{k,2} = U_k \qquad \text{all } k \tag{20.29}$$

$$\mathbf{x}, \boldsymbol{\eta}, \boldsymbol{\rho} \geq \mathbf{0} \tag{20.30}$$

where $z_{k,1} = z_{k,2}$ = expression for objective k

$\eta_{k,1}, \eta_{k,2}$ = negative deviations

$\rho_{k,1}, \rho_{k,2}$ = positive deviations

$u_{k,1}$ = weighting factor for the negative deviation for goal $z_{k,1}$

$w_{k,2}$ = weighting factor for the positive deviation for goal $z_{k,2}$

By specifying the weights above, we will tend to drive the final solution within the intervals as specified.

Example 20-2

Returning to our previous problem, wherein:

$$z_1: \quad x_1 + x_2 \geq t_1; \quad 10 \leq t_1 \leq 12$$

$$z_2: \quad 3x_1 + x_2 \leq t_2'; \quad 16 \leq t_2' \leq 20$$

We first rewrite this as

$$z_1: \quad x_1 + x_2 \geq t_1; \qquad 10 \leq t_1 \leq 12$$

$$z_2: \quad 3x_1 - x_2 \geq t_2; \quad -20 \leq t_2 \leq -16$$

and then formulate it as follows:
Find **x** so as to

$$\text{minimize} \quad a = u_{1,1}\eta_{1,1} + u_{2,1}\eta_{2,1} + w_{1,2}\rho_{1,2} + w_{2,2}\rho_{2,2}$$

subject to

$$\begin{aligned} x_1 + x_2 + \eta_{1,1} - \rho_{1,1} &= 10 \\ -3x_1 - x_2 + \eta_{2,1} - \rho_{2,1} &= -20 \\ x_1 + x_2 + \eta_{1,2} - \rho_{1,2} &= 12 \\ -3x_1 - x_2 + \eta_{2,2} - \rho_{2,2} &= -16 \\ \mathbf{x}, \boldsymbol{\eta}, \boldsymbol{\rho} &\geq \mathbf{0} \end{aligned}$$

Obviously, prior to solving we should take care of the negative right-hand sides. Any solution found will then reflect an attempt to find a compromise within the specified intervals and will be highly dependent on the actual selection of the weighting factors in the achievement function.

Interval Goal Programming: The Lexicographic Model

It should be obvious that one transparent extension to the weighted interval model of (20.27) through (20.30) may be arrived at by means of prioritizing the various goals. This would be useful, for example, when it is determined that *any* deviation outside the specified target interval for certain goals cannot be tolerated. The model should be self-evident.

Simplifications

The form of the interval goal programming models discussed is at an elementary level and intended purely for discussion. Any actual model could be refined in a number of ways by an assortment of conventional techniques.

Goal Range Programming

The models described herein indicate just a few of many possible extensions to goal programming for the general purpose of considering intervals for aspiration levels. One recent extension in this spirit has been denoted, by its developers [4], as "goal range programming," and it may be of interest, for some readers, to examine the references wherein applications of this approach are provided.

Interactive Multiple-Objective Methods

One approach to the multiple-objective problem that has generated a fair amount of interest is termed, by its advocates, "interactive (multiple-objective) programming" [1]. Although the technique may vary, the central theme is to actively involve the decision maker in the decision-making process. That is, various data and results are presented to the decision maker at different stages of the process. The process is then continued in, hopefully, the general "direction" indicated by the decision maker.

It becomes obvious that, based on this very general description, virtually all of the methods previously presented would qualify as "interactive methods," since there is always *some* interaction with the decision maker. However, the advocates of such an approach appear to distinguish between approaches as based on the degree of active interaction.

Although the author finds such approaches interesting, he is dubious as to the practicality of such methods when they require considerable time and effort on the part of a busy decision maker. His own experience indicates that, although such approaches may be feasible in certain "pilot studies" or in research efforts (wherein the role of the decision maker is taken by a graduate student or colleague), very few real-world decision makers have been willing to take the time to participate in these approaches.

A Classification and Evaluation of Multiple-Objective Methods

The various multiple-objective methods that we have discussed or alluded to in this and the preceding four chapters may be classified by noting certain philosophies that serve to distinguish their basic approach (a somewhat different classification was presented in Chapter 16). We list these below.

1. GEOMETRIC MEASURES OF GOAL DEVIATION
   a. MINSUM: In this approach we attempt to minimize the *sum* of undesirable deviations from the goal set. For example, in weighted goal programming, we try to minimize the weighted sum of absolute deviations from the goals. In lexicographic goal programming we try to minimize a prioritized set of absolute deviations.
   b. MINMAX: Our purpose here is to minimize the *maximum deviation* from any one goal within our total set of goals. Such an approach was used for curve fitting in Chapter 11 and underlies the fuzzy programming method of this chapter.
   c. INTERVALS: Here, rather than minimizing some geometric measure of deviation from a single value (i.e., the aspiration level), we consider an interval or range of acceptable values. Either MINSUM (as in interval goal programming) or MINMAX (as in fuzzy programming) may be used.
2. EFFICIENT SETS: Here we are unwilling to either rank or weight objectives or, in fact, to establish any measure of achievement. Instead, we seek to generate all efficient or nondominated solutions.
3. INTERACTIVE METHODS: Any of the multiple-objective methods may be modified to utilize a high degree of input from the actual decision maker in an attempt to refine or "point" the search in a, hopefully, better direction.
4. A COMBINATION OF APPROACHES: Any of the individual methods discussed may be combined to form, in some cases, a method perhaps better fitted for the actual problem at hand. For example, augmented goal programming is simply a combination of the minsum concept with the efficient solution set technique.

Evaluation of any method for multiple-objective optimization cannot, and should not, be made without regard to the total decision structure and environment. Each and every approach has its own drawbacks as well as its own strengths, and any rational choice as to which should be used is almost always dependent on at least two vital considerations:

1. The type and size of the problem.
2. The characteristics of the ultimate decision maker(s).

The efficient solution approach avoids entirely the problems and criticisms associated with the development of weighting factors and/or rankings. Unfortunately, these advantages are usually offset by both computational inefficiency and the imposition on the decision maker(s). In any real-life problem, the vast number of efficient extreme points alone makes the generation of the efficient solution set impractical. Further, even if one could generate this set, the result would be a severe case of "data overload" for the decision maker. Most decision makers have difficulty in comparing more than five or six alternatives much less the hundreds, thousands, or millions that would be generated in even a modest-size problem via the efficient set approach.

The minsum concepts can solve far larger problems than the efficient set approaches but are criticized for their reliance on weights and/or rankings. These factors are certainly drawbacks to such approaches but not nearly to the degree often cited by some critics. The author's two decades of experience in a variety of multiple-objective approaches have indicated that, in the majority of cases, it has been easier and required less time of the decision maker to develop estimates of weights and/or rank goals than it has taken to evaluate a large set of efficient solutions. Further, with the use of sensitivity analysis and priority relaxation, it is not necessary for weights and rankings to be precise or fixed.

The minmax approach and fuzzy programming in particular would seem to overcome both the limitations of the efficient solution method (i.e., computational burden and data overload) and that of lexicographic goal programming (i.e., reliance on weights and rankings). That is, fuzzy programming requires no weightings or rankings and generates only a single (efficient) solution. Unfortunately, it too has disadvantages (as cited earlier) in regard to the proper form of the membership function, the establishment of good aspiration levels and leeways, and the validity of the minmax concept itself.

Those techniques that attempt to deal with aspiration intervals rather than single values (such as fuzzy programming and interval goal programming) have their own attraction. Often, the concept of an interval is more consistent than that of a single target value. Unfortunately, such approaches tend to require larger formulations, and this may limit their computational efficiency.

SUGGESTIONS

Based on these considerations, as well as the author's personal experience and biases, it is believed that, at least at present, the most robust and attractive candidates for real-world implementation are

—Lexicographic goal programming
—Fuzzy programming
—Interval goal programming

The choice between the three is made on the basis of the decision maker and the actual problem under consideration. If the problem is relatively small, one might consider a combination of approaches, such as augmented goal programming or Steuer's [9, 10] heuristic adjunct to efficient sets.

REFERENCES

1. DYER, J. S. "Interactive Goal Programming," *Management Science*, Vol. 19, No. 1, 1972, pp. 62–70.
2. ECKER, J. G., AND KOUDA, I. A. "Finding Efficient Extreme Points for Linear Multiple Objective Programs," *Mathematical Programming*, Vol. 8, No. 3, 1975, pp. 375–377.
3. GAL, T. "A General Method for Determining the Set of All Efficient Solutions to a Linear Vectormaximum Problem," *European Journal of Operational Research*, Vol. 1, No. 5, 1977, pp. 307–322.
4. GROSS, J., AND TALAVAGE, J. "A Multiple-Objective Planning Methodology for Information Service Managers," *Information Processing and Management*, Vol. 15, 1979, pp. 155–167.
5. IGNIZIO, J. P. "Interval Goal Programming and Applications," Working Paper, Pennsylvania State University, December 1974.
6. IGNIZIO, J. P. *Goal Programming and Extensions*. Lexington, Mass.: Heath (Lexington Books), 1976.
7. IGNIZIO, J. P. "The Determination of a Subset of Efficient Solutions via Goal Programming," *International Journal of Computers and Operations Research*, Vol. 3, No. 1, 1981, pp. 9–16.
8. KICKERT, W. J. M. *Fuzzy Theories on Decision Making*. The Hague: Martinus Nijhoff, 1978.
9. STEUER, R. E. "Multiple Objective Linear Programming with Interval Criterion Weights," *Management Science*, Vol. 23, No. 3, 1976, pp. 305–316.
10. STEUER, R. E. "Goal Programming Sensitivity Analysis Using Interval Penalty Weights," Research Report 78-EES-3, Princeton University, 1978.

11. ZIMMERMAN, H. J. "Fuzzy Programming and Linear Programming with Several Objective Functions," *Fuzzy Sets and Systems*, Vol. 1, 1978, pp. 45–55.

PROBLEMS

20.1. Given a linear vector-maximum problem with 5 variables, 3 objectives, and 6 constraints, what is the maximum number of efficient solutions that might exist? What is that number if there are 50 variables, 5 objectives, and 30 constraints?

20.2. Can a solution *not* on the boundary (i.e., within the interior) of the constraint set ever be efficient in a linear vector-maximum problem? Why or why not? Can an optimal solution to a linear goal programming problem ever be dominated (i.e., not efficient according to the efficient solution concept)?

20.3. Can an efficient solution to a vector-maximum problem having *nonlinear* objectives ever exist within the interior of the constraint set? Construct a graphical illustration of such a situation if it can happen.

20.4. If the deterministic model of a linear vector-maximum problem is in error (e.g., some error in the coefficients or right-hand side values in the constraints), it is possible that an "efficient" vertex may actually be dominated. Construct a graphical illustration of such a condition.

20.5. Construct the graphical representation of the vector-maximum problem given below and determine the set of efficient vertices as well as the efficient "face" (i.e., the boundary of the constraint set that is efficient):

$$\begin{aligned}
\text{maximize } z_1 &= x_1 + 80x_2 \\
\text{maximize } z_2 &= -x_1 - 6x_2 \\
&\text{subject to} \\
2x_1 + x_2 &\geq 4 \\
3x_1 + 4x_2 &\leq 24 \\
-x_1 + x_2 &\leq 10 \\
x_1 + 2x_2 &\leq 40 \\
5x_1 + x_2 &\leq 60 \\
\mathbf{x} &\geq \mathbf{0}
\end{aligned}$$

20.6. Use the results of Problem 20.5 to explain why it is important to avoid an attempt to optimize a single objective without regard to the resultant impact on other objectives.

20.7. Construct the goal programming equivalent of Problem 20.5.

20.8. Construct the fuzzy linear programming equivalent of Problem 20.5.

20.9. In Problem 20.7, assume that:

1. Objective z_1 is ranked before z_2.
2. The respective aspiration levels for z_1 and z_2 are set at 1,000 units and 0 units.

Then solve the resultant model via the multiphase simplex algorithm.

20.10. Solve the fuzzy linear programming model of Problem 20.8.

20.11. Given that the rigid constraints must be satisfied but that we would tolerate a reduction of 100 units in z_1 and 20 units in z_2, use the augmented goal programming algorithm to solve for the resultant subset of efficient vertices for Problem 20.9.

20.12. Read reference [4] and compare the technique described with that of interval goal programming.

INDEX

M

N

O

P

R

V

W